A Question of Rigs, of Rules, or of Rigging the Rules?

A Question of Rigs, of Rules, or of Rigging the Rules?

Upstream Profits and Taxes in US Gulf Offshore Oil and Gas

JUAN CARLOS BOUÉ

With EDGAR JONES

Published by the Oxford University Press
For the Oxford Institute for Energy Studies
2006

OXFORD
UNIVERSITY PRESS
Great Clarendon Street, Oxford OX2 6DP

Oxford University Press is a department of the University of Oxford.
It furthers the University's objective of excellence in research, scholarship
and education by publishing worldwide in

Oxford New York

Auckland Cape Town Dar es Salaam Delhi Hong Kong Karachi
Kuala Lumpur Madrid Melbourne Mexico City Nairobi
New Delhi Shanghai Taipei Toronto

With offices in

Argentina Austria Brazil Chile Czech Republic France Greece
Guatemala Hungary Italy Japan Poland Portugal Singapore
South Korea Switzerland Thailand Turkey Ukraine Vietnam

Published in the United States
by Oxford University Press Inc., New York

First published 2006

British Library Cataloguing in Publication Data
Data available

Library of Congress Cataloguing in Publication Data
Data available

Cover designed by Clare Hofmann
Typeset by Philip Armstrong, Sheffield
Printed by The Alden Group, Oxford

ISBN 0-19-920575-2 978-0-19-920575-2

1 3 5 7 9 10 8 6 4 2

CONTENTS

LIST OF FIGURES

LIST OF TABLES

ABBREVIATIONS

API	American Petroleum Institute
ASOGB	Alabama State Oil and Gas Board
AVO	Amplitude vs. Offset
AWL	Areawide Leasing
BD	Barrels per day
BCF	Billion cubic feet
BCFD	Billion cubic feet per day
BCFGE	Billion cubic feet of gas equivalent
BIT	Bilateral Investment Treaty
BLM	Bureau of Land Management
BNOC	British National Oil Corporation
BOED	Barrels of Oil Equivalent per day
CBO	Congressional Budget Office
CIS	Commonwealth of Independent States
CNG	Compressed Natural Gas
COWPT	Crude Oil Windfall Profits Tax
CRRT	Commonwealth Resource Rent Tax
CZMA	Coastal Zone Management Act
DOE	Department of Energy
DOI	Department of the Interior
DSDP	Deep Sea Drilling Programme
DTI	Department of Trade and Industry
DWRRA	Deepwater Royalty Relief Act
DWT	Deadweight tons
EEZ	Exclusive Economic Zone
EIA	Energy Information Agency
E&P	Exploration and production
EMPCO	ExxonMobil Pipeline Company
EPAA	Emergency Petroleum Allocation Act
F&D	Finding and development
FERC	Federal Energy Regulatory Commission
FLNG	Floating Liquid Natural Gas
FOGRMA	Federal Oil and Gas Royalty Management Act
FOIA	Freedom of Information Act
FONDESPA	Fondo Especial de Desarrollo (Venezuela)
FPS	Floating production system
FPSO	Floating production, storage and off–loading

FSO	Floating Storage and off-loading
GAO	Government Accountability Office (formerly General Accounting Office)
GOM	Gulf of Mexico
HOOPS	Hoover Offshore Pipeline System
IBP	Initial Boiling Point
ICA	Interstate Commerce Act
IEA	International Energy Agency
IMF	International Monetary Fund
INRF	International Natural Resources Fund
IRR	Internal Rate of Return
JOIDES	Joint Oceanographic Institution for Deep Earth Sampling
LDC	Less Developed Country
LDNR	Louisiana Department of Natural Resources
LMOGA	Louisiana Mid-Continent Oil and Gas Association
LOOP	Louisiana Offshore Oil Port
LSU	Louisiana State University
LNG	Liquid Natural Gas
MB	Thousand barrels
MBD	Thousand barrels per day
MBOED	Thousand barrels of oil equivalent per day
MEM	Ministerio de Energía y Minas (Venezuela)
MLOT	Money left on the table
MMB	Million barrels
MMBD	Million barrels per day
MMBOE	Million barrels of oil equivalent
MMBTU	Millions of British Thermal Units
MMCFD	Million Cubic Feet per Day
MMS	Minerals Management Service
MMUSD	Millions of dollars
MOIP	Mandatory Oil Import Programme
MRM	Minerals Revenue Management
NAFTA	North American Free Trade Agreement
NGL	Natural Gas Liquid
NGPA	Natural Gas Policy Act
NPC	National Petroleum Council
NPD	National Petroleum Directorate
NPV	Net Present Value
NYMEX	New York Mercantile Exchange
OBPA	Outer Banks Protection Act
OCS	Outer Continental Shelf

OCSLA	Outer Continental Shelf Lands Act
OPEC	Organisation of the Petroleum Exporting Countries
OS&T	Offshore Storage and Treatment
OTA	Office of Technology Assessment
OTC	Offshore Technology Conference
PADD	Petroleum Administration for Defense District
PdVSA	Petróleos de Venezuela S.A.
PIFUA	Powerplant and Industrial Fuel Use Act
Ppm	Parts per million
PRT	Petroleum Revenue Tax
PSA	Production Sharing Agreement
psia	Pounds per square inch, absolute
QMA	Qualified marketing agent
RCT	Railroad Commission of Texas
R&D	Research and Development
RIK	Royalty in Kind
RRT	Resource Rent Tax
RSVP	Royalty Suspension Viability Programme
SCF	Standard Cubic Foot
SLA	Submerged Lands Act
SOCAL	Standard Oil of California (later Chevron)
SONJ	Standard Oil of New Jersey (later Exxon)
SOO	Suspension of Operations
SPR	Strategic Petroleum Reserve
STB	Stock Tank Barrel
TCF	Trillion Cubic Feet
TLP	Tension Leg Platform
TN	Tract Nomination
USD	US Dollars
USD/B	US Dollars per barrel
USGS	US Geological Survey

ACKNOWLEDGEMENTS

There are a number of individuals without whose collaboration and suggestions this study could not have been completed in its present form. All of their contributions were valuable, some were decisive. In no particular order, these individuals are Iván Sandrea (OPEC, formerly Merrill Lynch), Paul Horsnell (Barclays Capital), Daniel Johnston (Daniel Johnston and Co.), Bernard Minkow (formerly of McKinsey and Company), Willi Olsen (formerly Statoil), Pablo Galante (formerly BP Trading), James Hostetler (Poseidon Pipeline Co.), David Pursell (Pickering Energy Partners), Max Cummings (Allocation Specialists Inc.), John Higgins (LLOG), George Kahale (Curtis, Mallet-Prevost, Colt & Mosle LLP) and, as ever, Adrián Lajous, John Mitchell, Bernard Mommer and Charles Henderson. Among these names are a few who were very generous in providing access to their time, data and their expertise, but the changing scope of the project in the end meant that their contributions could not be utilised in full, or even in part. Edgar Jones was responsible for putting together, handling and then coaxing usable and useful information from the very large databases that underlie this study (many of which were originally put together by Paul Horsnell and later expanded by Gerardo Luyando). The original secondment of Edgar Jones to the Oxford Institute of Energy Studies, and its extension on a couple of occasions, was due entirely to the generosity of Eduardo Martínez del Río, Raúl Cardoso, Bernardo de la Garza and Rosendo Zambrano, who went out of their way to reaffirm the longstanding support for the work of the Institute by Petróleos Mexicanos (in this case through its trading arm, PMI Comercio Internacional S.A. de C.V.). Bob Skinner did yeoman work in revising numerous drafts, and the finished product has been much improved by his suggestions and patience. To all of them go heartfelt thanks and the exemption of any responsibility as to the contents or analytical thrust of the study, as well as any errors of fact or appreciation. All of these should be put down to the author's many shortcomings and entrenched prejudices.

Sonia, Bidu and Pia gave me the most important contribution of all: a sense of perspective.

CHAPTER 1

INTRODUCTION

With around 56 thousand exploration and production wells sunk and a cumulative production of 16 billion barrels of oil and 167 trillion cubic feet (TCF) of gas (to 2004), the Gulf of Mexico (GOM) is – by some distance – the most intensively explored, drilled and developed offshore petroleum province in the world. At the end of 2005, the 8220 producing leases and 3909 active production platforms within the area of GOM under the jurisdiction of the US Federal government (called the Outer Continental Shelf, henceforth OCS) produce over 90 percent of US offshore oil and virtually all of its offshore natural gas, an output equivalent to around a quarter of the total US production of these two primary energy sources. In turn, the 662 active fields in the GOM Federal OCS hold around 13 percent of the US total proven reserves, and they constitute the single most important combined source of oil and gas for the USA, larger by some distance than any other US state or foreign supplier.

From a fiscal standpoint, upwards of 90 percent of all OCS mineral lease payments are generated in GOM, making petroleum activities in the region the second most important individual source of revenue for the US Federal government after general income taxation (admittedly, it is a distant second place).[1] Even in years of low oil and gas prices, the revenues that the US Minerals Management Service (MMS) receives from oil and gas activities in GOM would place the agency squarely among the first 100 firms in *Industry Week*'s survey of the 500 largest US manufacturing companies. Furthermore, the OCS offshore leasing programme constitutes by far the largest non-financial auction market in the world, in constant dollar terms.[2]

During the 1990s, GOM arguably became one of the premier frontier exploration and production plays for the world oil industry, with large finds in deepwater zones that have more than compensated for declines in production in its mature shallow zones. Thus, while in 1992 there were only six active deepwater projects, by 1997 this figure had grown to 17 and by 2003 it had again multiplied fivefold, to 86 producing projects (as of March 2006, the figure is 118 projects). Out of the 8109 currently active GOM leases, 42 percent lie at depths of 1000 feet or greater and, in 2003, 29 drilling rigs were active in the

GOM deepwaters (in 1992, the equivalent figures were 27 percent out of 5600 active leases, and three drilling rigs).

Thanks to recent quantum advances in upstream technology, the prolific GOM deepwater fields have been making a major contribution to US oil and gas reserves and, more importantly, to US production. Between 1993 and 2000, for instance, overall US oil production declined by nearly 1.1 MMBD, but deepwater GOM production increased by 640 MBD over the same period (in 1999, deepwater GOM output surpassed shallow water output for the first time in history, accounting in the process for about 12 percent of total US production). Indeed, deepwater production rates have risen by well over 100 MBD of oil and 400 MMCFD of gas every year since 1997 and up to 2002. Overall, from 1992 to 2002, deepwater oil and gas production increased by 840 percent and 1600 percent, respectively (deepwater production rates have remained flat since 2002 though). Furthermore, the vast majority of the leases issued during the record-setting sales of the mid-1990s will soon see the expiry of their primary terms without a single well having tested their potential (out of the approximately 3200 deepwater leases issued from 1996 through 2000, for instance, only 6.5 percent had been drilled to the end of 2004). This means that, from 2006 onwards, large extensions of attractive deepwater acreage will once again become available for leasing. Therefore, even though the majority of GOM deepwater finds have yet to enter production, it is clear that MMS was not exaggerating back in 2000 when it claimed that 'the deepwater of the Gulf of Mexico can rightly claim to be America's new frontier and has truly emerged as a world class hydrocarbon province'.[3]

From its very inception as an oil and gas province, the GOM offshore has caught the attention of the oil industry actors and its observers – not to mention policymakers – all over the world. In part, this can be ascribed to its standing as the oldest offshore petroleum province: GOM has offered a template for the development of other provinces, not least because it has served as a crucible and testing ground for new technologies that are now employed worldwide. Its prominence also reflects the fact that 'because of its strength, whatever the United States does ... is scrutinised everywhere'[4] (as former US Secretary of State Warren Christopher put it in an early discussion on the institutional framework of the Federal OCS). More recently, the remarkable revival of GOM production has once again put the region firmly back in the spotlight.

Given all of the above, it is hardly surprising that the literature on oil and gas activities in GOM is vast. While a large proportion of it is to be found in trade journals and specialised industry literature, the academic literature on the subject is also voluminous. As one would expect, if one is interested in geology, then not even a lifetime would suffice to read all the published material available. Social science studies are also reasonably abundant, but their coverage is uneven. By far the best represented discipline in the bibliography is economics, with a very significant majority of articles and papers concentrating on the outcomes of the offshore acreage lease sales held regularly by the US Federal government since the late 1950s, and the behaviour of firms participating in such sales. This reflects the fact that, since the inception of the offshore leasing programme, the US Department of the Interior (DOI) has maintained extraordinarily detailed and publicly available records not only on the acreage auctions themselves but also on the multiple activities undertaken on the assigned leases (exploratory wells drilled, development wells drilled, oil, gas, lease condensate, natural gas liquids, and water, abandonment and so on). In terms of their scope, breadth and depth, these extensive databases (most of which are now downloadable from the internet) have no peer in any other country or industry[5] and, in the words of Hendricks, Porter and Boudreau, they constitute 'an excellent source of [field] data ... [to model] the strategic behaviour of firms in situations of imperfect and asymmetric information'.[6] This mother lode has been very profitably exploited by researchers throughout the years, by no means all of them economists by profession (for instance, the key concept of the 'winner's curse' was coined by three petroleum engineers,[7] and papers on OCS leasing have also figured prominently in advanced operations research literature). Paradoxically, the very bountifulness of this treasure trove has led much of the recent economic literature on OCS leasing down an empirical cul-de-sac. The format of the data for lease sales held up to 1983 inclusive is better suited for sophisticated modelling of bidding behaviour, so researchers have increasingly tended to focus their efforts on these older data even though they are not very germane to understanding the here and now of the offshore upstream sector, let alone to divining its future prospects (see Chapter 10).

The literature dealing with the institutional framework underpinning GOM oil and gas activities is surprisingly sparse and scattered, and much of it tends to concentrate on arcane jurisdictional issues. At first glance, these issues appear to be of interest primarily to a small specialist audience of lawyers[8] but in fact they have a much wider relevance, especially to political scientists and political economists looking at the principles and rules governing the access to natural resources.[9] Inevitably, a historical approach characterises this politico-legal strand to GOM academic literature. The same, however, cannot be said for the literature dealing with the industrial organisation aspects of the GOM upstream. MMS has sponsored a sizable number of studies dealing with these topics on a chronological basis (the reader will come across these studies throughout this book), but the scope of their enquiries has always been very specific. Aside from this, offshore operations are conspicuous by their absence from the abundant corpus on the history of the industrial economics of oil. This was true even during the heyday of this type of literature in the USA (which saw the publication of the landmark studies by Adelman, De Chazeau and Kahn and McLean and Haigh,[10] for instance, as well as numerous official and semi-official oil company histories), with the main exception to the rule being an official history of Kerr-McGee.[11] In this case, the prominence of offshore operations in the book was a reflection of the pioneering role that this company played in opening up the GOM region to exploration and production (E&P) activities. Nevertheless, as late as 1997, a doctoral thesis could still make a reasonable claim to be the first 'in-depth study ... of the history of the offshore petroleum industry'.[12]

This claim proved ephemeral, because 1997 also saw the publication – under the aegis of the Foundation for Offshore Studies – of an excellent historical survey commemorating fifty years of offshore petroleum technology. This effort by Hans Veldman and George Lagers was prompted by the perception that the historical fraternity had done scant justice to the eventful history of offshore petroleum activities, as projects, developments and milestones had been written about 'in a fragmented manner, generally lacking adequate overview and the context in which ... developments took place'.[13] Veldman and Lagers succeeded in remedying this deficiency, producing a vivid history 'painted in broad strokes of the brush with selective colouring of a number of details' but which, by their own admission, in no way constituted 'a definitive account of the first fifty years of offshore history'.[14] E&P activities in the GOM offshore region also figured prominently in a contemporary official history of the offshore oil services company Brown & Root,[15] which amounted to a succession of interesting vignettes more than a definitive history. At the time of writing, and within the framework of a cooperative agreement with the Coastal Marine Institute at Louisiana State University (LSU), MMS is underwriting a major research effort whose aim is to produce an exhaustive study[16] that will tell 'the story of offshore Gulf of Mexico ... from the perspective of the managers, geoscientists, and surveyors who pioneered path-breaking exploration technologies, took the risks, found the oil, and made the play'.[17] However, if the terms of reference of this study are anything to go by, it still does not address one of the major gaps in the OCS academic literature; namely, a study that brings together and combines a historical and institutional approach to the development of GOM oil and gas activities with the wealth of statistical and analytical material put in the public domain by various agencies of the US Federal or state governments: the Alabama State Oil and Gas Board (ASOGB), the Louisiana Department of Natural Resources (LDNR), the Railroad Commission of Texas (RCT), the Congressional Budget Office (CBO), the Government Accountability Office (GAO, until very recently known as the General Accounting Office), the Bureau of Land Management (BLM), the Federal Energy Regulatory Commission (FERC), the US Geological Survey (USGS), the defunct Office of Technology Assessment (OTA) of the US Congress, the US Department of Energy (DOE) and, first and foremost, DOI and MMS.

The present study is an exercise in applied economics whose main objective is to fill this gap in the literature. Originally, the study was meant to encompass both the upstream and the downstream dimensions of the oil industry in the GOM region, as well as the structure and behaviour of the markets for imported and domestic crude oils and petroleum products (and their linkages with both oil financial markets and the international oil market at large). But in the same way that no plan of battle can survive contact with the enemy, no research plan can emerge intact after having come into contact with statistical sources, not to mention unfolding events. In this particular case, for instance, no provision had been made in the original study plan to say anything about natural gas in the GOM region. However, during the long process of drafting some of the sections dealing with oil, a looming natural gas supply problem suddenly assumed critical proportions. Given the implications associated with this development, it did not seem reasonable to ignore it merely to adhere to the original plan.

The decision to alter the course of the study in midstream was made easier by the fact that, as far as energy data of any description go, the USA presents the researcher with a veritable embarrassment of riches: the centrality of oil and natural gas to US economic activity means that there is a wealth of data, much superior in quality and accuracy to that available for any other industry, anywhere in the world. Again, most of this information is put in the public domain by either Federal or state agencies and is effectively available for free. Moreover, much of it is micro-analytic in character, disaggregated down to the level of individual firms. All of this makes it possible for researchers on US oil and gas topics to decide what they might want to say about a certain issue and then find the statistics to support their case (as opposed to the situation that is far more commonly encountered, in which the availability of statistics largely dictates the topics that a researcher is in a position to tackle).

Of course, the change to the original study plan meant that the drafting process was more protracted than expected (and a test to the sponsors' patience). In addition, it contributed to the length of the study in its final form. By way of an *apologia pro obra sua*, however, the authors would like to suggest that the length of the study is chiefly a function of its revisionist character. As a general principle, we are as ready as anybody to accept that what is good and brief is twice as good. However, brevity is not necessarily well suited to revisionism, for the simple reason that questioning in a convincing fashion the conventional wisdom on any subject ideally requires that this conventional wisdom be expounded rigorously and then deconstructed piecemeal (otherwise, revisionist statements tend to sound either too sibylline or else steeped with conspiratorial overtones). This, by the way, is also the explanation for the inordinate size of the study's critical apparatus. After all, to quote John Kenneth Galbraith on the subject, footnotes 'provide an exceedingly good index of the care with which a subject has been researched'. Admittedly, Galbraith also said that there 'is a line to be drawn between adequacy and pedantry', and perhaps there is a tangible risk that some readers will feel that this study ended on the wrong side of that demarcation. In the study's discharge, we can offer a paraphrase of Galbraith's pithy response to the suggestion that 'readers might be offended by footnotes': we 'have no desire to offend or even in the slightest way discourage any solvent customer'.[18]

The order of the study is as follows. It is divided into eleven chapters, of which this introduction is the first. Chapter 2 gives background information on the political and administrative organisation of the GOM Federal OCS. It also gives an overview of the key producing areas in the region, but not on a play-by-play basis (at the time of writing, 65 individual oil and gas plays have been identified in GOM, and these harbour around 25,000 reservoirs in over 13,500 sand units). Rather, we have divided the GOM region into four sub-provinces on the basis of the factors that have the greatest incidence on the economics of specific upstream projects, and which can be expressed in a mixture of bathymetric (water depth), geological (sub or suprasalt) and even fiscal (presence or absence of royalty relief) criteria. These four subdivisions are as follows: a suprasalt province in very shallow and shallow waters (depths of less than 100 feet, and depths between 100 feet and 1000 feet, respectively); a subsalt province that extends from the GOM shallows up to the deepwater boundary (1000 feet); a shallow water deep gas province (up to a water depth of 330 feet but with reservoirs found at least 15,000 feet below sea level) and, finally, a sub and suprasalt deepwater province (1000 feet of water or more).

Chapter 3 gives a chronological overview of the long-running and fruitful relationship between cutting-edge technology and oil entrepreneurship in GOM, where it has been the key driver for oil activities to a far greater extent than has been the case in other offshore provinces (where oil companies have by and large been content to apply the innovations and lessons originally pioneered and learned in GOM). This chapter concludes that, in view of the rates at which reserves are being added and existing fields are declining, the traditional shallow water sub-province will be all but played out in a relatively short space of time. This poses a considerable problem in the context of the aforementioned natural gas supply crisis, because the shallow water province accounts for close to 20 percent of total US natural gas production.

Chapter 4 deals with the shallow subsalt province. Significant discoveries in this province (above all the Mahogany field) during 1993 were responsible in part for the resurrection of GOM as one of the hottest exploration provinces in the world. At the time, it was thought that the contribution of the shallow subsalt province to total GOM output could approximate that of the deepwater province. However, the shallow subsalt has turned out to be a disappointment: early successes were followed by a spate of costly failures, whose negative effects on the companies most active in this sub-province were magnified by the 1998 price crisis. Although drilling activity in the shallow subsalt resumed after the enforced hiatus brought about by this event, it never really recovered and the province is no longer a priority for oil companies. The larger among these reached the conclusion not only that more reserves could be found in the deepwater province, but also that they could be more economically brought on stream. Smaller companies, for their part, have turned their sights back to the shallow waters, to look for deep gas fields, whose peculiarities and productive potential are the subject of Chapter 5.

The rise to prominence of the deep gas sub-province was a consequence of the growing realisation that, without dramatic change in exploration, development and production patterns in the shallow water GOM, production from the region would not be able to meet its quota of future supply, as required by very bullish demand scenarios that had largely determined investment patterns in electricity generation capacity started in the late 1990s. Indeed, the limits of the deep gas sub-province were fixed by administrative fiat, with a prime objective in mind: using royalty relief to accelerate the exploration and production of non-deepwater gas in GOM. However, the deep gas royalty relief initiative is unlikely to make an appreciable difference in terms of the US supply/demand balance for natural gas during the critical 2005–2009 window: the 1.2 BCFD figure posted in 2003 probably represents a peak that will not be exceeded for some time without the discovery of some totally new geologic play. According to MMS expectations, the oil industry should drill around 130 deep gas wells per year over the 2003–2009 timeframe. At an average of 53 wells per year, observed drilling data for the period 2000–2003 – even with higher prices – fall well short of this target. Therefore, it is most unlikely that, as MMS hopes, incremental deep gas production will compensate for 60 percent of the decline in the traditional shallow water sub-province over the 2004–2013 period. Thus, one has to conclude that the USA will face a natural gas supply/demand imbalance throughout the remainder of the present decade, whose seriousness will be exacerbated by the fact that natural gas-fired generating units will have to meet virtually all of the incremental electricity needs of the American economy up to the year 2015.

The next five chapters, which form the core of the study, revolve around E&P activities in the deepwater sub-province. Chapter 6 consists of an historical overview of deepwater activities. Although the GOM deepwater province burst seemingly out of nowhere to become, in a short space of time, the hottest worldwide exploration play at the turn of the twentieth century, deepwater E&P activities in GOM in fact had a very long gestation period whose origins can be traced back to a couple of scientific initiatives of the late 1950s to mid-1960s. The first genuine deepwater development project in GOM (and, indeed, anywhere in the world) came on stream in 1979. However, up until the mid-1990s, deepwater production in the region expanded very slowly, although the projects responsible for this expansion made crucial contributions in terms of the advancement of deepwater exploration and production technology. Once protracted teething problems were overcome in the province, however, incremental deepwater production began to dominate the path of GOM's overall oil output profile. The initial phase in the development of the GOM deepwater province came to an end with Shell's announcement of its development plans for the Mars field (1993). From then on, thanks to relentless technological progress, the industry has been able not only to discover and unlock new reserves but also to enhance recovery at existing fields. Nevertheless, although deepwater production and the number of discoveries have increased substantially, some key measures of deepwater activity have been slowing down perceptibly (there have been decreases in the average bid amount per block, in the average number of rigs operating, in the number of wells drilled, and in the number of deepwater plans submitted). Thus, the chapter concludes by positing that MMS's bullishness as to the future prospects of the deepwater (i.e. the as yet undiscovered deepwater resource base) not only appears too optimistic but does not seem to be shared by the majority of either oil companies or, perhaps even more importantly, oil service companies.

Chapter 7 addresses the issues of the GOM deepwater resource endowment, the production profile of GOM deepwater fields, the evolution through time of deepwater oil and gas output, and the impact on future deepwater production and reserve addition trends derived from the foreseeable allocation of exploration capital between oil and gas activities in North America. A couple of problematic issues are highlighted and discussed. First, whereas production at larger deepwater developments has tended on the whole to exceed initial expectations, the exact opposite appears to be the rule for smaller projects (which are responsible for a fair share of incremental production). Second, despite the significant increases in deepwater gas production achieved in recent years, the lack of giant non-associated gas discoveries in this province will probably never allow production to reach the levels necessary to counteract increasing declines in non-associated gas production, while simultaneously meeting a rising US demand.

Chapter 8 is a discussion on the province's long-term productive potential. This issue is analysed from four distinct – albeit closely related – angles. First, the prospect that the output contribution of ultradeepwater frontier areas to GOM production may increase significantly in the short to middle term, thereby delaying the peak point in the province by a few years. Second, the role that Floating Production Storage and Offloading (FPSO) vessels may or may not play in opening up these frontier areas to development activities. Third, the production impact associated with significant improvements in resource recovery rates from fields exploited by means of subsea facilities. Last is the role that aggressive fiscal incentives could play in ensuring that, even if industry perceptions about the attractiveness in geological terms of GOM deepwater deteriorate, investment capital would continue to flow into the sub-province, thereby sustaining – and possibly even increasing – production. The key findings of the chapter are as follows. First, costs in ultradeepwater areas are still so high that, even in a high oil price environment, prospects (even largish ones) may be unable to generate enough revenue to pay their way unless they are developed by means of an FPSO. Second, the contribution that FPSO development projects in these areas can make will, at best, involve smoothing the post-peak production decline, because the introduction of this production method in GOM will be frustrated by conservation strictures pertaining to associated deepwater natural gas. Third, improved subsea recovery will greatly prolong the longevity of some fields but, by the same token, it will not allow them to increase their pre-decline production rates to any great extent. Finally, no matter how much the GOM fiscal regime is adjusted in order to make it worthwhile for oil companies to continue to pour investment funds into the deepwater sub-province, historical precedent shows decisively that it is not possible to 'buy' greatly increased output purely by means of tax breaks like deepwater royalty relief.

Chapter 9 turns to the question of deepwater economics. We examine the cost structure in the province, and its evolution through time. In terms of finding and development (F&D) costs, the main thrust of the analysis involves coming to grips with three key questions: namely, just how far these costs have fallen, why they have fallen and, finally, what their behaviour is likely to be in the future in light of recent discovery trends. As far as lifting costs are concerned, the characteristics that set GOM apart from other deepwater provinces are highlighted, and an assessment offered as to whether they are advantageous or otherwise. The revenue generation capabilities of GOM deepwater fields are then assessed in the light of this cost structure. Finally, these various strands will be woven together into an account that explains why the GOM deepwater has been the most profitable petroleum province in the world for the past decade or so. The key conclusion of the chapter is that, even if it is true that GOM deepwater geology has been very favourable, the fact that rates of return in the province are so far above the industry's cost of capital is *fundamentally* attributable to the benevolent fiscal environment created by a fall in acreage acquisition costs, which itself is a product of the radical changes that the Reagan administration grafted onto the GOM fiscal and institutional framework during the early 1980s, with the introduction of the policy known as areawide leasing (AWL).

Chapter 10 deals with the effects that AWL has had on the structure of the offshore oil industry. In particular, we focus on the influence that the programme has had in terms of the way that oil companies compete among themselves for offshore oil and gas leases. Our aim is to show that AWL has sapped the vigour of competition in the offshore sector. Whereas the method used for selling acreage until 1982 inclusive was highly effective in inducing advantaged players into revealing their ideas about the prospects of different areas and then transmitting this information to other players, the Reagan reforms destroyed the conditions under which such transmission of information could take place, thereby exacerbating the marked informational asymmetries prevalent in the market for offshore leases. Paradoxically, in seeking to reduce a certain type of entry barrier through the easing of restrictions on access to offshore acreage (namely, very large up-front signature bonus bids), these reforms *raised* entry barriers overall, for the benefit of a few large oil companies (instead of the small and medium-sized firms that AWL was supposed to help). Furthermore, the unforeseen and acrimonious political conflicts that the reforms unleashed led to leasing and drilling moratoria that have effectively closed off the majority of the OCS (outside GOM) to oil and gas activities.

Finally, in Chapter 11, we bring the study to a close by reflecting on the lessons that governments and policymakers in other petroleum provinces can draw from the GOM experience, concentrating in particular on whether policies similar to AWL might be of interest to those governments that wish to boost their oil revenues by achieving a higher oil output. Our main conclusion in this regard is that such governments would find the fiscal sacrifices that this approach entails to be crippling, and totally out of proportion to the results that a policy like AWL can bring about. That does not mean that the GOM experience is totally bereft of elements that might be worthy of imitation. However, if governments in producing countries want to open up access to their upstream sectors in a manner that is fiscally rewarding but at the same time promotes a competitive industry and efficient market outcomes, then they should focus on trying to devise licensing policies modelled after the tried and tested methods that the US Federal government so unwisely discarded in 1983.

As far as oil companies are concerned, the most important lesson to be derived from the AWL saga is that their pretension to give the governments of less developed countries (LDCs) as little patrimonial retribution as possible in exchange for access to these countries' petroleum resources, is a short-sighted and counterproductive policy that will (and indeed, already has in some places) led to a drying up of investment opportunities, and all that this entails: declining organic reserve replacement rates, increases in finding and development costs, and excessive reliance on acquisitions for growth. The enduring legacy of AWL, even more so than the deepwater bonanza in the GOM region, is the closing off to exploration and production activities of a vast area of the OCS, potentially quite rich in hydrocarbons. This outcome is a pointer of what is likely to happen in those producer countries which, having invoked their sovereign powers to grant access to their natural resources, now find themselves stripped of other meaningful attributions of sovereignty and eminent domain (notably the power to tax extractive industries located within their territories), through a combination of contractual provisions and legal fetters incorporated into bilateral investment treaties (BITs) and multinational investment agreements. Access to resources for the oil industry will tend to be compromised if the populations of the territories from which they are extracted feel that they are not getting their fair share of the bounty generated by the liquidation of their mineral assets.

NOTES

1 In 2000, for instance, GOM natural gas royalties currently accounted for approximately 45 percent of all petroleum and non-petroleum royalties paid to the Federal government (GAO 2000: 11).
2 Quoted by Priest 2004: 30.
3 Baud, Doyle, Peterson and Richardson 2000: ix.
4 Christopher 1953: 32.
5 The GOM Federal OCS monthly production data since 1947 exist at the level of individual wells, for instance.
6 See the extensive annotated bibliography in Saidibaghgandomi 1987, as well as the bibliography in this study.
7 Capen, Clapp and Campbell 1971.
8 See for instance Bartley 1953, Fitzgerald 2001.
9 See Gramling 1995, Sollen 1998, Boué 2002.
10 Adelman 1972, de Chazeau and Kahn 1959, McLean and Haigh 1954.
11 Ezell 1979.
12 Kreidler 1997: 13.
13 Veldman and Lagers 1997: 5.
14 *Ibid.*: 6.
15 Pratt, Priest and Castaneda 1997.
16 See MMS 2004b.
17 Priest 2004: 54.
18 Galbraith 1979: xxix.

CHAPTER 2

THE US GULF OF MEXICO: A GEOGRAPHICAL PRIMER

The Gulf of Mexico is a partially landlocked sea resembling an inverted hat with a broad and shallow rim, measuring approximately 1600 kilometres from east to west and 900 kilometres from north to south. With a total surface area of 1.5 million square kilometres, it is the largest such body of water in the world. It is bordered by the US states of Florida, Alabama, Mississippi, Louisiana, and Texas to the north, by the Mexican states of Tamaulipas, Veracruz, Tabasco, Campeche and Yucatán to the west, and by Cuba to the southeast. Around twenty major river systems drain into GOM, and annual freshwater inflow amounts to approximately 280 trillion gallons. The bulk of this flow (around 85 percent) comes from US rivers with the Mississippi River system alone contributing 64 percent of the total. In addition, each year, the Mississippi deposits about 500 million tons of new sediment in its enormous delta, and this deposition advances the Louisiana shoreline seaward by about 6 miles every hundred years or so. Over millions of years, this deposition has given rise to a vast wedge of sediment that has weighed down the crust beneath GOM. Gravels deposited around 45 million years ago on the surface have been encountered at depths of over 17,000 feet, as well as in a number of cores from the Deep Sea Drilling Programme (DSDP), taken more than 200 miles away from shore. GOM, of course, is also the birthplace of the Gulf Stream, a marine current that moves 100 times as much water as all the rivers in the world put together and from which, on any given year, Western Europe will obtain a third as much warmth as it has received from the Sun.

GOM's main marine shoreline from Cape Sable, Florida to Cabo Catoche in Yucatán extends for over 5700 kilometres, but if bays and other inland waters are included in this calculation, GOM's total shoreline increases to over 27,000 kilometres in the US portion alone. Nearly 40 percent of the surface of the American sector is classed as shallow inter-tidal areas. The continental shelf, continental slope and abyssal areas each account for roughly 20 percent of the total surface. Mean water depth in these other areas is around 5300 feet, with the greatest depths being found in the southwestern part of the Sigsbee Escarpment (17,000 feet). With a total surface of 19.3 million acres, the

GOM Federal OCS accounts for less than 10 percent of the 1.9 billion acres covered by the US Exclusive Economic Zone (EEZ). However, GOM's significance to the US economy is out of all proportion to its extension. Its waters provide 40 percent of the entire US commercial fisheries harvest. It is home to the top ten US ports for value of landings (with Houston in first place by a long distance), and four out of the top five ports for volume landed. It also supports an extensive tourist industry (encompassing thousands of businesses and tens of thousands of jobs) worth over USD 20 billion annually.

And then, of course, there is oil and gas...

2.1 Political and Administrative Division

Offshore petroleum production within the GOM region takes place in waters under the jurisdiction of the US Federal government, as well as waters belonging to three GOM coastal states (Alabama, Louisiana and Texas). From the early 1950s to the early 1980s, the output recorded for leases in Louisiana and Texas state waters accounted for 39 and 19 percent of the cumulative US offshore oil and gas production, respectively.[1] Since then, however, the overall significance of output from state waters in the US supply picture has dwindled and nowadays, production in Federal waters accounts for the bulk of offshore oil and gas production.[2] Thus, this study will only focus on oil activities within the GOM Federal OCS.

For mineral (i.e. sulphur, sand and gravel as well as petroleum) leasing and administrative purposes, the GOM portion of the EEZ is divided into three planning areas: Eastern Gulf of Mexico (covering 75.6 million acres), Central Gulf of Mexico (covering 47.8 million acres) and Western Gulf of Mexico (covering 35.9 million acres). The latter two areas, lying offshore Louisiana and Texas, respectively, account for the bulk of GOM petroleum resources (the locations and names of the subdivisions within these planning areas are shown in Figure 2.1). Development in the Eastern planning area (which lies mostly off the coasts of Florida) has lagged far behind that of the other two regions, in part because of its lack of prospectivity[3] but mainly due to the effects of various drilling moratoria.

The developmental stalemate in the Eastern Planning Area was supposed to have changed with Eastern Gulf sale 181, held at the end of 2001, ostensibly to offer up for auction a large area contiguous to producing areas in the Central Planning Area. As things turned out, though, the surface actually offered in this sale was but a fraction of

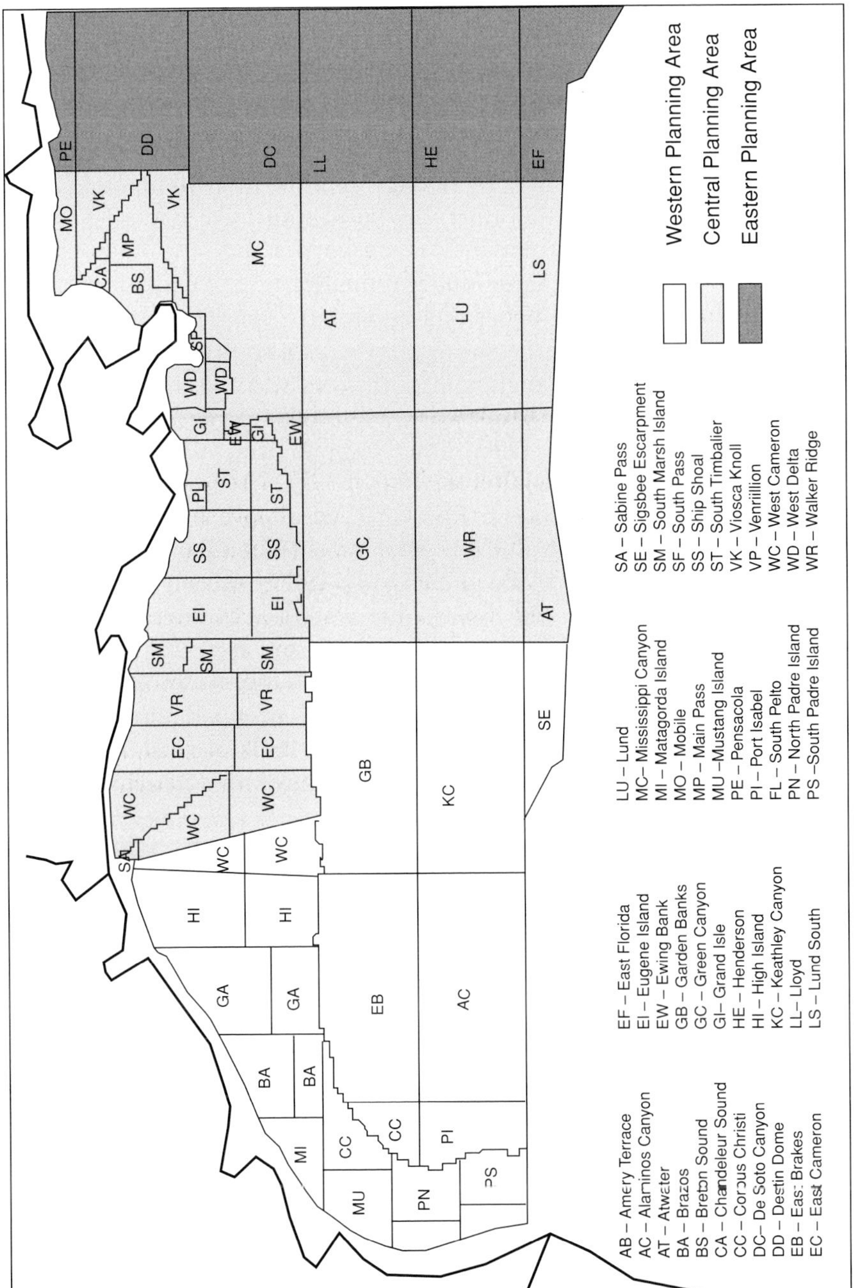

Figure 2.1: Gulf of Mexico Federal OCS Planning Areas and Administrative Subdivisions

the acreage originally intended.[4] This was because, in spite of his unabashedly pro-oil stance, President George W. Bush understandably decided that he was better off not alienating the constituents from the state that effectively put him in the Oval Office in the contentious and divisive 2000 presidential race (in addition, the incumbent Republican governor of the state – the President's brother, as it happened – was to seek re-election less than one year away from the date of the lease sale).

MMS held another lease sale in the Eastern Planning Area in December 2003, and a further one in March 2005. However, the only blocks involved in these sales were ones not taken up in Sale 181 and, unsurprisingly, industry interest in both sales was minimal. Moreover, since this sliver of territory was reopened to oil and gas leasing in 2001, drilling results there have been rather disappointing, with a succession of dry holes and non-commercial discoveries.[5] For the foreseeable future, therefore, the Eastern Planning Area will continue to be of little importance both to the US petroleum industry and the oil and gas markets at large and, because of this, our study will ignore it and instead concentrate on the petroleum activities in the Central and Western planning areas.

The GOM continental shelf area slopes seaward at an angle of less than 1 degree, forming a broad plain of relatively shallow water (which ranges in breadth from 12 miles off the alluvial fan of the Mississippi River to as much as 140 miles off the mouth of the Crystal River in Florida). In the area close to the Texas–Louisiana marine border, for instance, water depths can be below 500 feet more than 120 miles out from shore.

As a result of the above, even at a very early stage of development, a vast area in GOM was open to exploration and development, without requiring any mastery of deepwater operations on the part of oil companies. Beyond 500 feet of water, however, the seafloor dips sharply down an escarpment. Thus, a relatively small surface area is to be found at water depths between 500 and 2000 feet (Figures 2.2 and 2.3). In consequence, the technological advances necessary to exploit resources located at water depths greater than these had to take place in other offshore petroleum provinces of the world, and only then were they brought back to GOM to be applied to the abyssal plain that lies beyond the base of the continental slope, and whose depths go down to 17,000 feet at the outer edge of the EEZ.

The depth distribution of active leases within the Central and Western planning areas at the time of writing is shown in Figures 2.4 and 2.5. As can be appreciated, a significant proportion of active

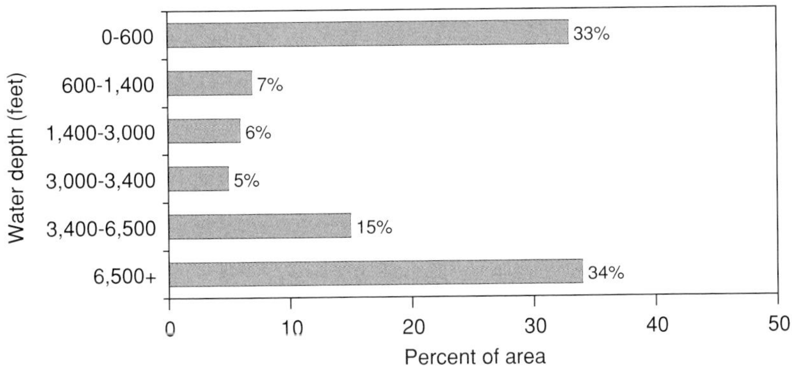

Figure 2.2: Central Gulf of Mexico Planning Area Distribution by Depth

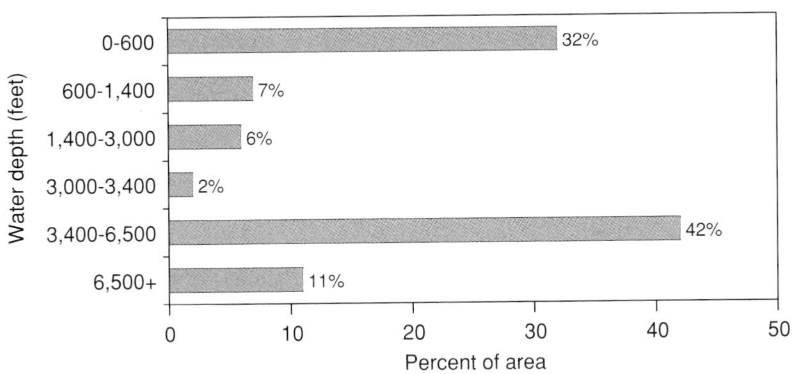

Figure 2.3: Western Gulf of Mexico Planning Area Distribution by Depth

leases lie in deepwaters (1000 feet or more). This is a relatively new development. In 1970 the average production weighted depth was just 100 feet, and it was still below 200 feet in 1980. As late as 1990, it had barely reached 250 feet. However, the trend towards a greater production depth accelerated significantly during the early 1990s, with the weighted average reaching the 1000 foot milestone in 1998 (at which point deepwater production became the norm, rather than the exception, in GOM). At the time of writing, production routinely takes place in 5000 feet of water, and drilling in 9000 feet and beyond (the current record for deepwater drilling in GOM stands at 10,011 feet of water, which ChevronTexaco achieved in an exploration well drilled in Alaminos Canyon block AC951 during 2003).[6]

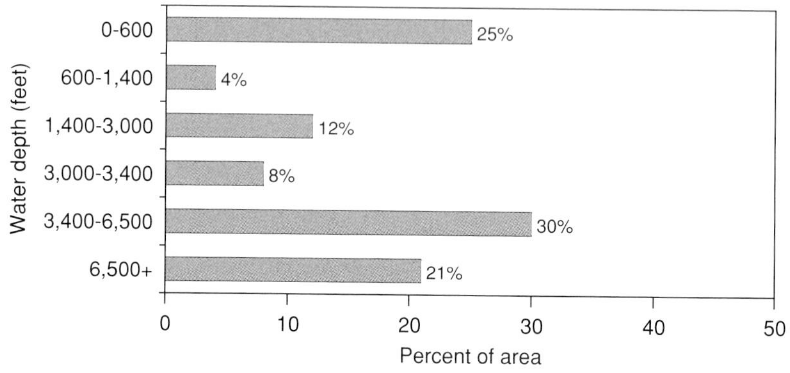

Figure 2.4: Central Gulf of Mexico Planning Area Extension Under Lease by Depth

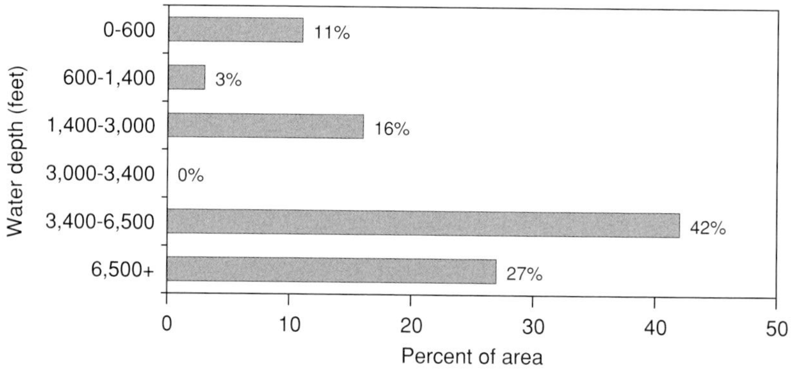

Figure 2.5: Western Gulf of Mexico Planning Area Extension Under Lease by Depth

The Central and Western planning areas comprise 26 and 15 administrative subdivisions, respectively.[7] Offshore Texas – the Western Gulf – can be characterised as being predominantly gas prone. Offshore Louisiana – the Central Gulf – has mainly oil with associated gas in shallower waters, and large oil fields with relatively little gas in deeper waters (in absolute terms, this region has also proven far more bountiful, even as far as natural gas is concerned). The central region has a crude oil to gas production ratio approximately three times higher than the western region.

At the southernmost tip of the Central and Western planning areas can be found three smaller subdivisions, the Sigsbee Escarpment, Amery Terrace and Lund South (the first of these has been assigned to the Western planning area, while the latter two belong to the Central planning area).[8] These three ultradeepwater zones (10,000 to 12,000 feet) constitute the northern part of the so-called 'Western Doughnut Hole',[9] which came into being with the signing in 1978 of a USA–Mexico treaty establishing the maritime borders between both countries. As the Western Doughnut Hole lay beyond the 200 nautical mile limit of both countries' EEZs, it had to be divided somehow between both of them (Figure 2.6). However, because of the US Senate's refusal to ratify the aforementioned treaty, this partition was delayed for two decades.

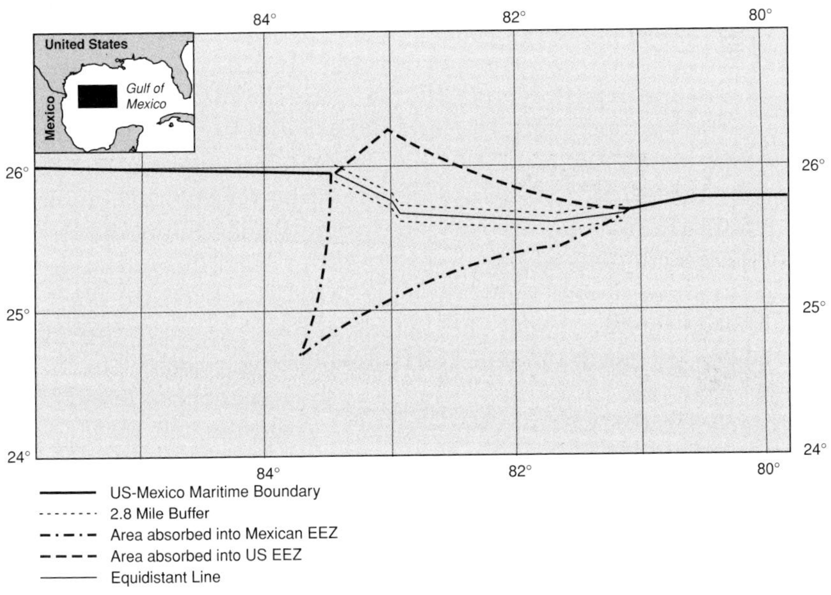

Figure 2.6: Gulf of Mexico Western Gap

As Applegate explains, 'when the treaty was negotiated, the US State Department was primarily concerned about finding rights and obtaining the best possible boundary in the Pacific, where rich fishing banks were at stake'. The treaty naturally also covered subsea resources, but these 'were considered of secondary importance',[10] not least by major oil companies (who thought – rather shortsightedly, as it turned out – that 'since drilling technology did not allow production in the water depths at stake in negotiations', the whole thing had to be seen merely as 'a fishing rights treaty'). In Senate hearings held in

1980 to examine the provisions of the treaty, this was still very much the received wisdom concerning the matter. However, the eminent American geologist Hollis Hedberg raised a dissenting voice, arguing that the treaty was fatally flawed because 'it gave away areas possibly containing important mineral resources for the sake of expediency in resolving boundary negotiations'.[11]

With 20/20 foresight, Hedberg pointed out that, in his expert opinion, the sediments located in deepwaters beyond the continental shelf slope would prove to be 'abundantly petroliferous'.[12] He recommended therefore that the USA negotiate 'a separate mineral resource boundary' with Mexico (rather than lose the area through a treaty that did not 'specifically consider mineral resources'[13]) and, for good measure, he drew up a number of alternative boundaries that were much more advantageous, in territorial terms, for the USA. Hedberg insisted that the border as drawn up was unfair because Mexico had staked its claim using a small group of islands off the coast of Yucatán (the Alacrán Atoll) as the baseline point. However, he neglected to mention that this was merely a negotiated quid pro quo for the US request that the Pacific border be measured from several islands offshore California, an action that led to its obtaining the rights to 'prime Pacific fishing banks'.[14]

Hedberg went to great lengths to establish a legal-theoretical distinction between territorial rights deriving from islands on the shelf (i.e. Alacrán), on the one hand, and islands in the deep ocean (like those off California), on the other. As Hedberg saw it, 'in the case of the former …the base of the slope should be the boundary, whereas in the latter each island has its own boundaries'.[15] Unfortunately for Hedberg, this fine display of casuistry 'fell on deaf ears in the committee which unanimously approved the treaty and sent it to the Senate floor for ratification',[16] and it found no resonance in international legal circles either.[17] Crucially, though, during the ratification hearings several key senators did side with Hedberg and they succeeded in delaying any signing until the USGS prepared an assessment on the matter.[18] This study was eventually completed in 1982, by which time all sense of urgency for ratification was gone, and the treaty was more or less shelved and forgotten. Paradoxically, this meant that, by the time the US oil industry developed a keen interest in these deepwater areas (i.e. the early 1990s), the 1978 provisional boundaries had effectively assumed a de facto character, thereby making it impossible to fix a baseline using 'the base of the [Continental] slope'[19] in a manner that would extend the reach of US resource jurisdiction.

By the end of the 1980s, quite a few people had begun to 'wonder whether the tuna fishing off the Pacific coast, so important to US negotiators in 1978, was worth the seafloor given up in [GOM]'.[20] They reached the conclusion that Hedberg had had the right idea about this trade-off, but by then it was too late to do much about it. Then, during April 1997 (three years after the deepwater boom in GOM got underway in earnest), MMS attempted to lease tracts within the area of the doughnut hole expected to fall under US jurisdiction. The tracts failed to attract any bids, making it clear to MMS that companies would steer clear of the area until they were sure that the US government was fully entitled to hold lease sales there in the first place.[21] Thus, on 23 October 1997, at the instigation of oil lobbyists and senators from GOM states who were eager to see the initiation of E&P activities in the US portion of the doughnut hole, the Senate at long last ratified the *Treaty about Maritime Boundaries between the United States of America and the United Mexican States*, whereupon the road was clear for the division of the doughnut hole.[22] The partition officially took place on 9 June 2000.

The treaty that divided the 5092 square nautical miles of the Western Doughnut Hole gave the USA 1913 square nautical miles (about 38 percent of the total), while Mexico received 3179 square nautical miles. This treaty also established a small buffer zone (with a 1.4 nautical mile extension) on each side of the new boundary, in recognition of the possible existence of trans-boundary oil and gas reservoirs. Both countries agreed to a 10-year moratorium on oil and gas exploration and production in the buffer area. Upon expiration of this agreement, each country may permit drilling in its respective buffer zone, but must notify the other when any buffer area is made available for drilling. As of the time of writing, no wells have been drilled anywhere within the US sector of the erstwhile doughnut hole, and only around twenty blocks have been leased in the area (all of them in Amery Terrace).

2.2 Key Producing Areas

The Gulf of Mexico formed approximately 300 million years ago, in the wake of the separation of the North American tectonic plate from the African and South American plates. Theories regarding the details of the formation and evolution of GOM's petroleum systems abound; indeed, they have proliferated as data and knowledge about the region have expanded. As McBride, Weimer and Rowan observe,

> although the northern Gulf of Mexico Basin is one of the most extensively studied and explored sedimentary basins in the world, no consensus exists regarding the mechanics of its petroleum systems. This lack of consensus is

> primarily the result of the basin's large size, thickness of strata, and complex structural evolution. In the offshore regions, much of the data needed for a petroleum systems analysis are difficult to compile because of the relatively shallow well penetrations (source rocks occur at much greater depths), the ambiguity of deep structure, difficult stratigraphic correlations, and the generally poor understanding of the evolution of specific allochthonous salt systems.[23]

Clearly, a study like the present one is not the place to address these various theories, or to go into any great detail apropos key geological features of the Gulf. Suffice it to say that the oil industry's long experience in both the American and the Mexican sectors of GOM confirms that

> small oceanic basins are among the most promising areas in the world for petroleum accumulation. Proximity to land and large rivers has ensured thick sedimentary sections with accumulations of both terrestrial and marine organic matter even in their central parts. Their restricted nature favours limited circulation and the preservation of organic matter under bottom reducing conditions or as a result of rapid burial of sediments ... They are generally situated in tectonically mobile environments ... [and] their restricted character also has been favourable to the formation of sealing evaporite deposits.[24]

Evaporite deposits occupy a prominent place in the GOM geological framework, to an even greater extent than in other offshore hydrocarbon provinces. The region's petroleum systems bear a distinctive imprimatur, which is the product of the deposition, evolution and migration of halite (salt) deposited in Jurassic times.[25] On the petroleum generation front, for instance, 'the high thermal conductivity of salt retard[ed] the thermal maturation of subsalt petroleum source rocks and cause[d] late generation and migration from them'. Salt also determined the places where hydrocarbon reservoirs formed, as its 'impermeability ... prevent[ed] vertical petroleum migration and cause[d] migration pathways to be deflected laterally up the dip of base salt', while in those places where 'salt welds form[ed], petroleum migration [was] unimpeded and continue[d] vertically'.[26] The location of salt bodies explains many of the endearing features of GOM oilfields (their large net pay and reservoir pressure values) as well as their limitations (the relatively small mean field size in the GOM deepwater when compared to values registered in similar provinces elsewhere, as shown in Figure 2.7, is because GOM traps tend to be confined between salt bodies).[27]

The interplay in geological time of salt deposition and migration, on the one hand, and faulting and tectonic movements, on the other, gave rise to a highly irregular bathymetry and an even more complex

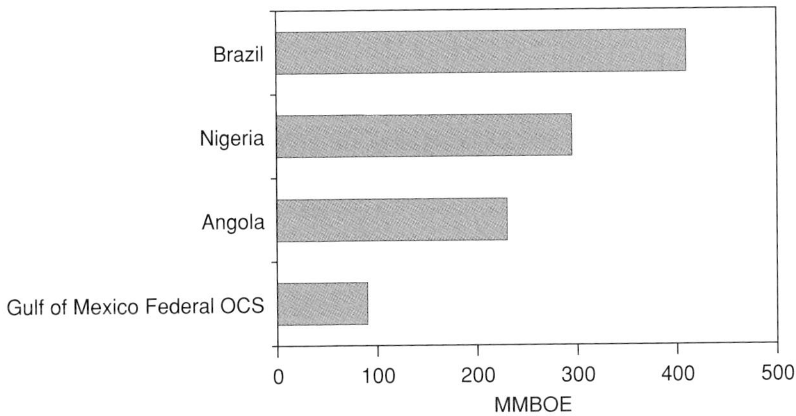

Figure 2.7: Average Size of Deepwater Fields in Selected Oil Provinces

sub-surface stratigraphy. As Anderson and Boulanger note, 'horizontal velocities of salt movement to the south are in the several centimetre per year range, making this supposedly passive margin as tectonically active as most plate boundaries'.[28] Within the GOM region, therefore, there is to be found a bewildering variety of tectonic basins, plays, trends, productive horizons, regions and so on. To cite but one concrete example: the EI330 field is a discrete sub-basin encompassing more than 25 productive Pleistocene sandstones at depths ranging from 4300 to 12,000 feet, and separated by faults and seals into more than 100 oil and gas reservoirs.[29] Across the Federal GOM region, there are over 25,000 reservoirs in over 13,500 sands, belonging to 59 different chronozones, and distributed in more than 65 different plays at the latest count(see Table 2.1).

Table 2.1: GOM Federal OCS. Distribution of Reserves and Production by Geologic Age (as of 2000)

	Proved Reserves		*Cumulative Reserves*		*Remaining Proved Reserves*	
	% Oil	*% Gas*	*% Oil*	*% Gas*	*% Oil*	*% Gas*
Pleistocene	36	40	36	40	34	35
Pliocene	30	16	31	16	29	17
Miocene	34	43	33	43	37	44
Pre-Miocene, Cretaceous and Jurassic	0	1	0	1	0	4

Source: MMS

Given the sheer number of productive intervals from which petroleum is extracted in GOM, it should come as no surprise that the physical characteristics of the crude oils and condensate streams available in the region exhibit a considerable degree of variation. A chemical analysis of a very broad sample of GOM oils (a total of 355), carried out in the late 1980s, indicates that it was possible to distinguish between four genetic families of petroleums in the region, on the basis of their carbon and sulphur isotope ratios as well as nickel and vanadium content.[30] Three of these families comprised a small number of rather unusual oils: nine super-mature condensates from offshore Texas (Brazos Blocks BA578 and BAA47) with relatively light carbon isotope ratios, four condensates from offshore Texas (from blocks in Galveston Bay, Matagorda Island and Mustang Island) with unusual heavy carbon isotope ratios and, finally, a single crude from offshore Florida (state lease 826Y in the Marquesas Key area) with a light sulphur isotope ratio and a high sulphur content (5.25 percent), displaying genetic affinities to crudes from the (onshore) Smackover formation.

The great majority of the GOM oils analysed in this study (341) turned out to be indistinguishable between each other in terms of their carbon and sulphur isotope ratios. They could therefore be said to belong to a single broad superfamily.[31] Within this family, two sub-groups were clearly distinguishable on the basis of their respective vanadium content (this variable was clearly bimodal, divided at a value for V/(V+Ni) of 0.5), as well as their sulphur content. The first of these groups included the greater number of oils (286), all of which had a relatively low vanadium concentration (0.27 ppm $\pm$ 0.10 ppm) and sulphur content (0.25% $\pm$ 0.15%). The second group consisted of oils characterised by significantly higher vanadium and sulphur concentrations (0.67 ppm $\pm$ 0.10 ppm and 0.73 % $\pm$ 0.43%, respectively). Furthermore, this second group showed clear chemical affinities to crudes from some wells in the Sound of Campeche (Abkatún, Akal and Ixtoc), as well as with crude shows that the DSDP encountered in the Challenger Knoll area in 1969.

Crudes coming from deepwater fields discovered since the aforementioned study was undertaken have also turned out to belong to the high vanadium and sulphur sub-group. Their inclusion in the group has broadened the range of its characteristic vanadium and sulphur signature, thereby highlighting to an even greater extent the affinities between crudes from the Louisiana continental shelf break and slope, on the one hand, and crudes from the Sound of Campeche, on the other. Indeed, chemical compositions of oils from both of these groups are well within the range of variability of a single family, which suggests

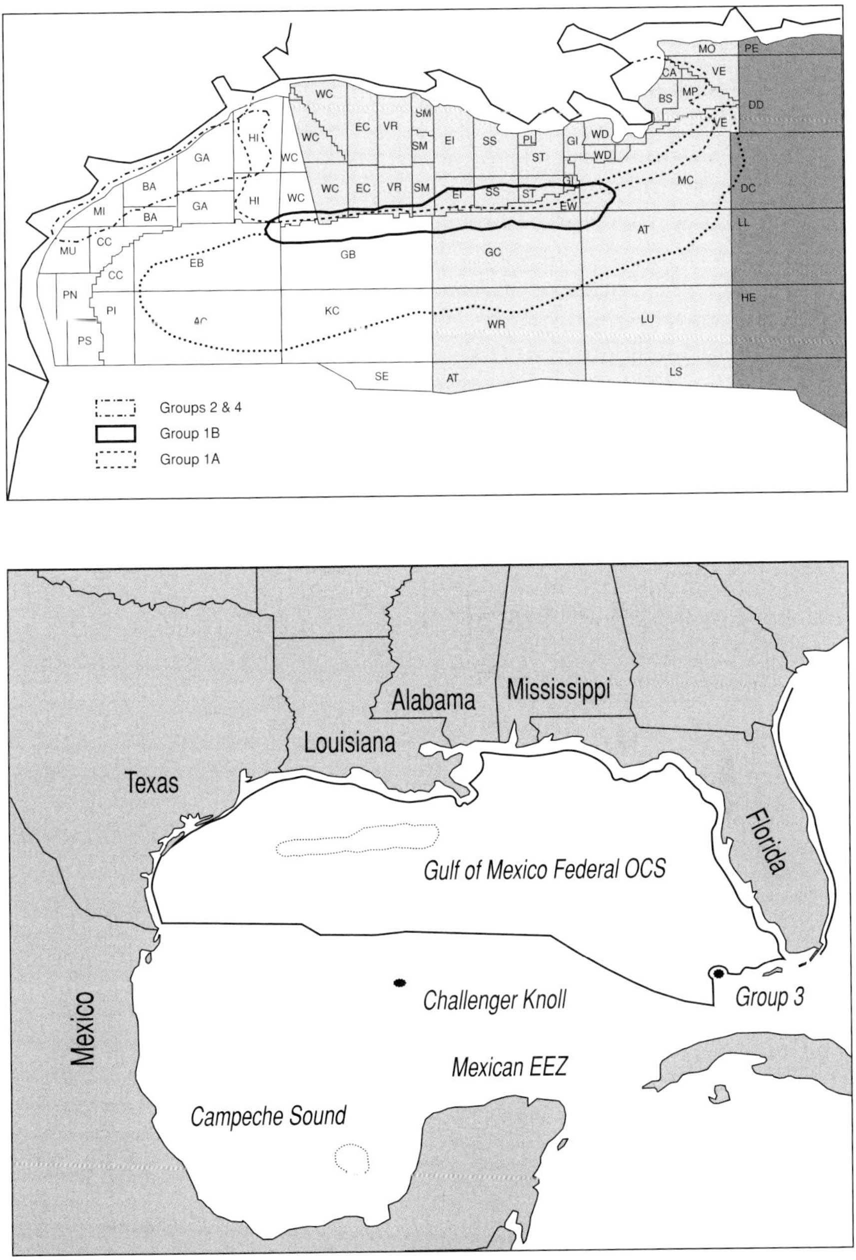

Source: Brooks Kennicutt and Thompson, 1990

Figure 2.8: Distribution of Crude Oil Families in GOM

that all of them might be derived from the same or similar source facies[32] (this, of course, has interesting implications for future deepwater E&P activities in the Mexican sector of the Gulf).

Figure 2.8 shows the geographical distribution of the various families of oils in the northern GOM. As a very broad rule, the lightest and sweetest oils tend to be found relatively close to shore, broadly distributed across the shallow Louisiana–Texas continental shelf. As one moves further away from shore, towards the Louisiana shelf edge and continental slope regions, one encounters oils that are still quite light (with gravities ranging from the low to mid-thirties), but also moderately sour and metal-laden. Oils from the deep and ultradeep sectors have more sulphur and metals still, and their typical gravity ranges from the upper twenties to the lower thirties. The tendency towards progressively more adverse quality parameters reaches its peak at the other end of GOM, in the crudes found in the Sound of Campeche.

This study does not analyse upstream oil and gas activities according to the geological or geophysical criteria mentioned in the preceding paragraphs, however.[33] Instead, it concentrates on the factors that have the greatest influence on the economics of actual projects: these are a mixture of bathymetric (water depth), geological (sub- or suprasalt) and even fiscal (presence or absence of royalty relief) criteria. On the basis of these criteria, the northern GOM can be divided into four major sub-provinces (see Figure 2.9).

These sub-provinces are not mutually exclusive in a physical sense. In bathymetric terms, for instance, the first three occupy the same space: the shallow portion of GOM. Even in geophysical terms, the degree of overlap between these sub-provinces is considerable: salt formations, for instance, overlie around 60 percent of the deepwater tracts currently under lease in GOM, and also a significant proportion of the deep structural closures that might harbour deep gas reservoirs in shallower waters.

The distinctions between the four categories are not entirely arbitrary, though. Rather, the dividing lines have been drawn in a way that reflects the weight of certain factors that are more or less exclusive to each of the areas, and which have a great bearing on the costs of projects located within them. In the shallow water province, for instance, exploration and drilling are both relatively straightforward, while taking oil to market is cheap (because of the proximity of shore and the density of offshore pipeline infrastructure); however, this province has been under intensive exploitation for around fifty years, and so both reserve additions and production rates tend to be modest. In the shallow subsalt province, proximity to shore is still an advantage,

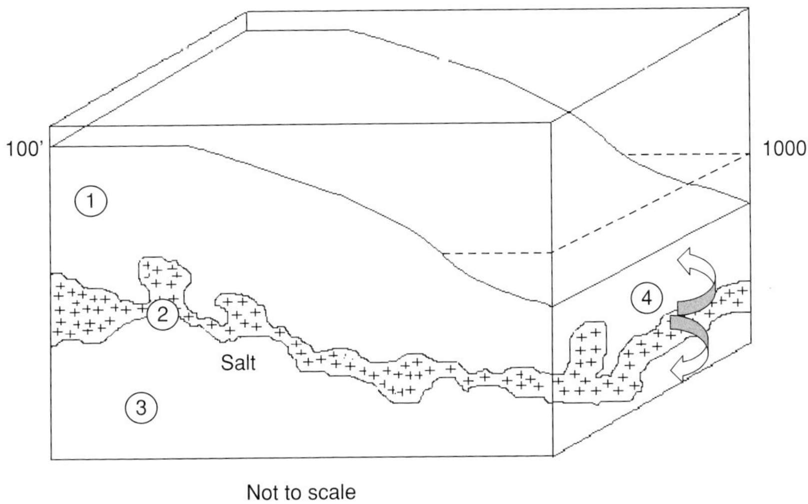

1) a suprasalt province in very shallow and shallow waters (depths of less than 100 feet, and depths between 100 feet and 1000 feet, respectively)
2) a subsalt province that extends from the GOM shallows down to the deepwater boundary (1000 feet)
3) a deep gas province in the GOM shallows (water depths down to 330 feet, but with reservoirs found at least 15,000 feet below sea level)
4) a sub- and suprasalt deepwater province (1000 feet of water or more)

Figure 2.9: Schematic Representation of GOM Sub-Provinces

as are larger field sizes and more attractive production rates; however, both seismic interpretation and drilling pose major problems whose solution adds greatly to E&P costs. In the deepwater province, the main economic challenges come from a given project's distance from shore and extant infrastructure, and the water depth at which it lies (in the deepwater, drilling costs might very well be 30 percent higher for a subsalt well compared to a suprasalt one, but even though this is a large relative difference, it is overshadowed in magnitude by the incremental costs that a project faces the further it lies from shore and the deeper is its drilling target). Finally, in the deep gas province, exploration and production activities can be just as complicated as in the subsalt province (indeed, many deep gas wells will also be subsalt wells), but wells tend to be prolific and producers face more favourable cash flow conditions as a result of royalty relief. The following chapters will be dedicated to analysing, in greater detail, petroleum activities in each sub-province.

NOTES

1 Louisiana has always accounted for most of the wells drilled and total hydrocarbons produced in state waters in the USA. Texas leased its first offshore tract in 1922, but had to wait until 1940 to see the first commercial production from its territorial waters.
2 Currently, wells in Louisiana waters produce about 35 MBD of oil and 0.4 BCFD of natural gas. Output in Texas state waters is both modest and highly skewed towards gas (0.1 BCFD versus 1.6 MBD of oil).
3 A concise summary of oil industry activities and resource prospectivity offshore Florida can be found in Gohrbrandt 2001.
4 The 233 blocks offered (covering approximately 1.3 million acres) were no closer than 100 miles from Alabama's shores and no more than 24 miles east of the eastern boundary of the Central Planning Area. MMS received 190 bids on about 547,000 acres, and eventually awarded 95 leases (about 41 percent of the tracts offered).
5 See *PON* 24 September 2003: 3.
6 The drilling of this well marks the first time that a drilling rig has explored for petroleum in more than 10,000 feet of water.
7 Each one of these subdivisions has a designated abbreviation for identification purposes and for use on maps and in databases. In addition, there are a number of smaller subdivisions/fields that have received individual identifiers because of their respective locations relative to state/Federal borders. In the Central and Western Gulf Planning areas, these fields are Coon Point (CP), Lighthouse Point (LP) and Tiger Shoal (TS).
8 These areas are still commonly referred to as NG 15-8, NG 15-9 and NG 16-7, respectively.
9 There is also an Eastern Doughnut Hole, bordered by the EEZs of Mexico, Cuba and the USA. For obvious reasons, it does not appear that this particular region will be partitioned any time soon, so no one in the USA is seriously considering its exploitation for mineral purposes at the moment.
10 Applegate 1997: 70.
11 *Ibid.*: 71.
12 *Ibid.* In support of this view, Hedberg cited the fact that one of the holes drilled under the auspices of DSDP in the Challenger Knoll area had encountered oil-saturated caprock (Hedberg, Moody and Hedberg 1979: 295).
13 Applegate 1997: 72.
14 *Ibid.*: 70.
15 *Ibid.*: 71.
16 *Ibid.*
17 As Applegate (*ibid.*: 72) points out, 'although many of Hedberg's ideas were incorporated in the Law of the Sea, his island theories were not, and international law [now] precludes the sort of boundaries ...[he] sought'.

18 Powers 1981.
19 Applegate 1997: 72.
20 *Ibid.*
21 There are unsubstantiated reports that blocks within the zone did receive bids in the April 1997 lease sale, but that MMS returned these to their submitters unopened.
22 The Mexican government had stated that there could be no negotiation on the doughnut holes until the 1978 treaty was ratified, for the simple reason that 'one cannot change boundaries that have never been fully approved' (Applegate 1997: 70).
23 McBride, Weimer and Rowan 1998: 1083.
24 Hedberg, Moody and Hedberg 1979: 295
25 See Wu 1993.
26 McBride, Weimer and Rowan 1998: 1083
27 The mean field size of the largest 19 GOM deepwater finds is less than that of all the fields discovered thus far in Brazil's Campos basin, for instance.
28 Anderson and Boulanger 2002: 9.
29 See Holland, Leedy and Lammlein 1999.
30 Brooks, Kennicutt and Thompson 1990.
31 *Ibid.*: 189.
32 *Ibid.*: 194.
33 For such an analysis, consult Bascle, Nixon and Ross 2001.

CHAPTER 3

THE SHALLOW WATER SUPRASALT

The development of shallow water GOM petroleum resources spans all of the post-Second World War history of the world offshore upstream sector. However, its origins can be traced back to the 1930s, when various oil companies started drilling in Galveston Bay and off the Texas and Louisiana coasts, in response to surveys that had identified possible traps and salt domes under the shallow waters of the Gulf. These structures held the promise of harbouring pools of the sort that had become progressively more and more difficult to find onshore.[1]

The GOM shallow waters were the logical and in many ways the ideal place for the world offshore oil sector to take off, and for offshore technology to mature. For one thing, they were located in what was the undisputed epicentre of the world petroleum industry until the late 1950s (a location that allowed many specialised construction and service firms to emerge from existing companies). For another, the marine environment in the region was benign, characterised as it was by a gentle slope with shallow water far out from shore and a relatively flat bottom (the first GOM well drilled out of sight of land was a full 10 miles away from shore, but sat in just 18 feet of water). GOM waters also had the advantage of being calm most of the time (in fact, relative to subsequent offshore environments, GOM has been characterised as a veritable millpond!).

The move offshore was no Sunday outing, though. Oil companies had to contend with unstable sediments on the sea floor, active faults, and a treacherous and complex underlying karst topography of caves and sinkholes. Above all, they had to find ways for their operations to emerge unscathed from those brief but highly eventful periods when GOM waters cease to be calm and are whipped up by hurricanes (the obvious problems that hurricanes cause above the sea surface are mirrored below, as their passage over the outermost reaches of the Mississippi delta frequently unleashes colossal mudslides and turbidity flows down the slope of the delta that can easily uproot production facilities).[2] Dealing with these problems in effect made it impossible for American oil companies to apply the undemanding approach to oil and gas production in a seashore or lacustrine environment. As a result, there began a long-running and fruitful relationship between cutting-edge technology and oil entrepreneurship, which saw the former becoming the key driver for oil activities in GOM, to a far greater extent than has been the case in other offshore provinces (where oil companies have by and large been content to apply the innovations and lessons originally pioneered and learned in GOM).

3.1 From Tidelands to 'Dead Sea'

The oil industry's move offshore in GOM initially followed the pattern that had been established in California, Baku, Lake Maracaibo and, closer to home, the Louisiana bayous. In these places, oil companies had sought to recreate land-like drilling sites in water, mainly through the use of piers and trestles. This method worked reasonably well in the sheltered waters of Galveston Bay which, unfortunately, proved to be largely barren. Thus, the industry was forced to move towards more exposed waters. There, companies encountered more demanding conditions and the traditional approach to offshore operations began to reveal serious shortcomings.

First of all, building trestles, piers and platforms for exploratory drilling activities proved too costly, not least because a dry hole meant having to write off the value of infrastructure that, as a rule, was not salvageable. Secondly, the amount of building material required in GOM was far greater than that necessary in the other provinces mentioned above, partly because of distance to shore considerations, but mainly because the very thick silty organic mud bottom of the shallow GOM provided insufficient purchase and support to counter the vibrations associated with drilling machinery. Since friction between pilings and the silty seabed was the only thing that could hold a platform in place, the use of at least a couple of hundred pilings was required in order to obtain an acceptably firm foundation for drilling in shallow waters (moreover, drilling in deeper waters was not possible using this method). Finally, trestles and piers were not structurally sound: aside from being very vulnerable to wood boring worms, they were unable to withstand the full force of a hurricane.

These drawbacks notwithstanding, approximately 25 wells were drilled from conventional pile foundations in shallow water off the Gulf Coast from 1937 to 1942.[3] Eleven of these were located in the giant wooden platform that the Pure Oil Company built to tap the Creole field. This particular project represented the culmination of the 'onshore operations offshore'[4] approach but, at the same time, it was a technological dead-end[5] that made it quite clear to the industry that technology had to be devised to permit 'the installation and servicing of structures built at sea and not connected to land'.[6] GOM was to become the first offshore province where this dream was realised, partly thanks to the unique advantages mentioned above but also because, to quote an early offshore pioneer, 'there had never been a time when anyone was crazy enough to try to build a platform in the open ocean and place men and equipment on it'.[7] However, the intervention of the Second World War meant that the US oil industry was only able to leverage GOM's locational and topographic advantages more than ten years after the startup of the Creole field.

The end of the War brought about the resumption of offshore leasing in Louisiana and Texas. Industry response to the availability of acreage was very favourable, not least because advances in offshore technology had by then started off anew. The greatest breakthrough involved the successful adaptation of the submersible drilling barge – first used in 1933 in a Louisiana estuary – to a marine environment (the first successful refloating of such a barge took place in 1949). As Gramling observes, 'all the previous technological innovations (mobility, transparency to wave action, offshore scheduling of work, and the inclusion of living quarters) came together ... to produce the first self-contained, reusable, offshore drilling machine'.[8] However, drilling from fixed platforms with floating tenders continued to be the prevalent mode of operation even in the early 1950s, mainly because of the restricted depth at which early submersible barges could operate.

The days of such platforms were numbered, in particular because the use of tenders gave rise to major complications whenever circumstances were less than optimal (as the tenders were not self-propelled, they could easily turn into the proverbial loose cannons in rough seas).[9] Nevertheless, the availability of large numbers of surplus and inexpensive military landing craft meant that the technological development of this production method continued apace, notably through the introduction of modularisation (which involved the construction of separate units that could be loaded and bolted onto complete platforms, with considerable time savings) and also through the use of large tender barges that could hold all the things necessary for drilling (this not only meant that platforms could be reduced dramatically in size but also that they could be at least partly salvaged, moved and re-erected at a different location if the well turned out dry). Indeed, it was with one of these large barge/small platform combinations that Kerr-McGee achieved the historic milestone of producing oil out of sight of land for the first time, in November 1947.

According to the history of offshore oil services company Brown&Root,

17 companies invested more than USD 260 million in 5 million acres' worth of GOM leases[10] and exploration and development work between 1947 and 1951 (an imposing sum equivalent to more than USD 1.4 billion in money of 2000).[11] Major oil companies were responsible for a significant part of this expenditure, and their enterprise was rewarded with large finds located around and above relatively large salt domes lying beneath 30 feet or less of water: Shell's South Pass Blocks SP24 and SP27, and Eugene Island Block EI18, SOCAL's Bay Marchand Block BM2 and Main Pass Block MP69, and Humble's Grand Isle Block GI18. Overall, major oil companies drilled over 90 percent of the wildcat wells in the GOM Federal OCS over the period 1951–60, and accounted for nearly 100 percent of the discoveries.[12]

The boom in offshore activities ground to a halt in the wake of a US Supreme Court decision that upheld a challenge of the Federal government to the title of Texas and Louisiana to submerged lands beyond the low water mark. This dispute marked one of the most serious breaches in relations between state and federal governments (indeed, it went so far as to prompt some talk of secession among hotheads in Texas!). The details surrounding the development and resolution of the Tidelands controversy are not germane to the present study.[13] Suffice it to say that the Federal government thought that there was nothing less at stake than the viability of the nation-wide control of petroleum production that had so painstakingly been put in place throughout the 1930s, after the discovery of the East Texas field.

The US Federal government was aware that some very prolific fields might be discovered offshore the Gulf Coast (between 1947 and 1952, leases in the territorial waters of Louisiana and Texas had already produced 20 MMB and 0.6 MMB of oil, respectively). Thus, the Tidelands dispute was all about preventing the multi-layered national scheme for the control of production (centred on the Connally Hot Oil Act, the Interstate Oil Compact and the Railroad Commission of Texas) from being undermined by new flows whose magnitude could be expected to be much greater if the leasing process were in the hands of revenue-hungry coastal oil states, as opposed to the steadier hands of the Federal government. Everette DeGolyer, founder of Amerada and assistant deputy petroleum administrator during the Second World War, succinctly summed up the fears of the Federal government (fears that the larger oil companies for the most part shared) in this regard when he stated that 'he preferred federal development of the tidelands if that meant a more gradual development'.[14] There were sound national security reasons to opt for a slower pace of development in Federal hands as well, of course, and there have even been allegations that President

Eisenhower's preferences in this regard might have been motivated by rather base political considerations![15] Moreover, it cannot be questioned that the devil-may-care mode of development and exploitation of onshore pools in the oil patch (prompted by the economic imperatives of the Rule of Capture) would have been particularly catastrophic had it ever been let loose in a marine environment.

With the passage by the US Congress of the Submerged Lands Act (SLA) and the Outer Continental Shelf Lands Act (OCSLA) in 1953, the US Federal bureaucracy saw the whole Tidelands matter as closed, and so both BLM and USGS got round to the serious business of offering OCS acreage for sale. The first sale was held at the BLM offices in Washington during October 1954, and it saw 23 companies submitting USD 129.5 million in high bids for 417,221 of the 748,000 acres offshore Louisiana that were on offer.[16] The second lease sale, for acreage offshore Texas, was held one month later, and even though only four companies participated, the BLM still managed to end up with USD 23.4 million in high bids for 107,730 of the 215,460 acres on offer. The third sale, held in New Orleans during July 1955, saw the collection of USD 100 million for 532,980 acres offshore Louisiana and USD 8.4 million for 153,090 acres offshore Texas.

Notwithstanding the success of the first three OCS lease sales, all was not plain sailing for the offshore leasing programme. Almost simultaneously with the 1955 offering, Louisiana had held a lease sale in which nine of the 22 tracts offered lay over the 3 miles seaward line as defined by the Federal government (in 1954, Louisiana had chosen to define its coastline in a way that extended its jurisdiction 37 miles seaward in some places).[17] Before long, bitter protests and recriminations were flying between Washington D.C. and Baton Rouge, and the escalation of the conflict led Louisiana to obtain an injunction which forced the cancellation of the Federal lease sale planned for June 1956. The aggrieved parties reached an interim agreement to break the impasse reasonably quickly, but just at that point 'economic recession, an oversupply of crude, a series of hurricanes, and declining oil finds in deeper waters ... forced a slowdown in offshore exploration'[18] which saw the percentage utilisation of the drilling rig fleet dropping from 100 percent in 1957 to 37 percent in 1958.[19] In an uncanny parallel to what was to transpire during the early 1990s, this slump led some observers to write off the GOM as a play,[20] when in fact a boom beyond anybody's wildest reckonings lay just around the corner.

Despite the uncertainty surrounding the issue of state versus Federal jurisdiction on GOM waters throughout the mid- to late-1950s, the rate of technological progress in GOM upstream activities over this period was remarkable, no doubt thanks to the attractive economics of offshore production, encapsulated in an encouraging success rate for wildcat exploratory wells,[21] quite reasonable field development costs (although both dry hole and capital costs were markedly higher for any prospect located in water depths beyond 60 feet) and generous production allowables.[22] These favourable conditions prompted major innovations across the whole upstream operational spectrum,[23] which translated into spectacular exploration success (during the years 1954, 1955 and 1956, respectively, American oil companies discovered 34, 57 and 72 new GOM oilfields). On the exploration drilling front, there was the introduction of jackup rigs, drillships and submersible and semi-submersible drilling rigs. In terms of data acquisition and processing, one can count the introduction of continuous velocity well logs, the use of magnetic tape for recording seismic sound waves, the invention of techniques to enhance seismic reflections and filter out noise, the substitution of dynamite with less extreme and dangerous methods of sound generation, and the introduction of analogue computers for processing seismic data.

On the development front, there was the substitution of wood by steel as the main building material, and the relocation of the major construction activities to onshore facilities, where the major support structures (jackets) would henceforth be built on the basis of templates. The great expansion in the size and weight of production platforms that this entailed led to advances in the ancillary processes used in the launching, flotation and installation of offshore structures. By the same token, the improved sturdiness of platforms opened the possibility not only of having numerous wells in each one, but also of attaching living quarters to them. This, in turn, prompted the appearance of the peculiar pattern of shiftwork characteristic of the offshore oil industry, and it also did wonders for the companies that produced those ubiquitous offshore workhorses: helicopters. As far as transportation activities were concerned, there was a rapid move towards pipelines, which supplanted the previous scheme centred on storing output in tanks and then taking it to shore on barges. Naturally, the expansion in the number, extension and size of pipelines was also accompanied by advances in related areas, notably materials and underwater pipe-laying.

All of these advances did not come cheaply: the first all-steel production platform, Humble's Grand Isle Block GI18, cost USD 1.23 million in 1948, an unprecedented sum for the time. In view of such hefty price tags, it is not surprising that many smaller companies clung to the much cheaper small platform and tender combination until well into the 1960s (for instance, Kerr-McGee's trailblazing platform cost a total of USD 230,000[24]) or that the 'wildcat spirit' that had traditionally driven such companies in onshore basins was much attenuated in the offshore.[25] But even though offshore costs were higher by an order of magnitude than in those places where independent producers had their traditional strongholds (i.e. East Texas), it was actually companies like Kerr-McGee, Phillips and Tenneco (then called Tennessee Gas) who 'set the initial pace to establish the offshore frontier'[26] – especially before the landmark 1960 and 1962 lease sales – with the vital collaboration of drilling contractors (who quickly 'became data clearinghouses for various disciplines and fields of expertise, and took on a technological sophistication that often surpassed their land-based cousins'[27]). By the late 1960s, independents were drilling approximately 30 percent of wildcat wells in the GOM Federal OCS. However, these companies could not have blazed this trail without 'the willingness of drilling contractors to invest in new equipment' because, as Kreidler notes, 'operators shied away from experimenting with technology and handling their own drilling operations ... [while] the majors had avoided capital investment in drilling equipment and personnel'.[28]

Although activity in the GOM area (in both Federal and state waters) only took off in a major way after the BLM conducted the first post-settlement general lease in 1960, firm indications of the bright future ahead became apparent in May 1959, when BLM held a small sale (39,000 acres), and obtained USD 88 million in bonuses for only 19 blocks (Shell paid 30 percent of the total for a single, half-block tract adjoining some of its producing leases in the South Pass area). The 1960 sale saw the re-entry of major oil companies in the GOM offshore sector, and their renewed interest immediately manifested itself in the form of bonus payments that independents found both 'inordinately high' and 'puzzling' (i.e. objectionable).[29] The sale, involving 1.17 million acres offshore Louisiana and 437,000 acres offshore Texas, generated USD 285 million in high bids (of which USD 249 million were spent on tracts located offshore Louisiana), easily doubling the next highest amount spent in a Federal lease sale up to that point.

The success of the 1960 sale owed a great deal to the participation of major oil companies, but not only in their capacity as bidders. Originally, the BLM had intended to offer a very limited amount of acreage, all of it lying in less than 100 feet of water, as the USGS had expressed an opinion that offering leases located at greater depths would be tantamount to promoting speculation. It was the USGS' view that no company had the wherewithal to conduct E&P activities at such depths, but unbeknownst to the agency, Shell Oil had already secretly designed and commissioned a semi-submersible drilling vessel

– *Bluewater 1* – capable of coping with such depths. After the initial call for block nominations had gone out, Shell managed to convince DOI to withdraw it, and instead issue a new set of leasing maps (drawn with Shell's assistance) that appended 'south additions' to the original leasing subdivisions. The majority of the area covered by these additions, which attracted intense bidding, lay in 300 feet of water or more.[30]

The next sale, held in March 1962, was an even greater success: it elicited a response so enthusiastic that it – wrongly – went down in industry lore as having required two days for all the bids to be read.[31] A total of 420 blocks (covering nearly 2 million acres) were assigned after the sale, making it unnecessary for DOI to hold another lease sale during the subsequent five years. Bonus payments came to an astonishing USD 445 million (the equivalent of around USD 2.1 billion – and more than USD 1000/acre – in money of 2000).[32] Many of these blocks, in newly opened areas like Eugene Island, South Marsh Island and Ship Shoal, were found at hitherto unprecedented water depths (the average water depth of leases in the 1962 sale was 125 feet, compared to 67 feet for the 1954–1955 sales and 89 feet for the 1960 sale).

Overall, the end result of the first decade (1953–1962) of oil activities in the Federal GOM OCS was,

> a growing demand for improved technology ... [for] greater numbers of exploratory rigs, more efficient ways to bring development platforms on line, pipelines to transport offshore production, and a massive support sector to support these other demands. The economic forces at work brought about integrated changes in the technology, infrastructure base, and physical environment and concomitant changes in the social and economic environments of the coastal Gulf of Mexico ... to produce a massive offshore-onshore system in a remarkably short period of time.[33]

Indeed, by the mid-1960s, more than a thousand production platforms had been installed in the GOM shallow waters, and the industry's progress into ever deeper waters advanced relentlessly (whereas in 1957, the maximum depth had been 100 feet of water, by 1965 it had more than doubled to 225 feet and it reached 300 in 1969). In the aftermath of the 1962 sale, the oil industry is estimated to have spent USD 1 million *per day* on drilling alone. Admittedly, the attraction of the GOM Federal OCS as a wildcat province faded somewhat in comparison to the late 1950s (in 1956, for instance, the success rate for wildcat exploratory wells offshore Louisiana was an exceptional 34 percent). Nevertheless, the success rate for exploratory drilling offshore Louisiana during the early 1960s was still comparable to the US average, but with a great advantage in terms of the reserve finding rate per exploratory well: during the early 1960s, it only took the drilling of 155 wells offshore Louisiana to discover a 100 million barrel field, whereas in the USA as a whole it took 3773 wells.[34]

Throughout the 1960s, the massive production and transportation system in GOM was put through its paces, as hydrocarbons output raced ahead to reach a figure of 2.6 MMBOED by 1969. In this year, oil production reached 916 MBD (up from 350 MBD in 1962), with the lion's share of the increase coming from blocks leased in 1962.[35] Crude oil output would peak only two years later, at 1.02 MMBD, but natural gas output would continue to expand unabated, on the back of reserve addition rates that averaged 1 billion barrels of oil equivalent per year up until the late 1970s.

And if things looked rosy for GOM producers on the output front throughout the 1960s and 1970s, on the price front they were even better. For much of this period, US oil prices exceeded world oil prices by a handsome margin (thanks to the Mandatory Oil Import Programme or MOIP), and the eventual correction of this anomaly did not entail any hardship for US producers (as world oil prices first levelled with US prices, and then continued to climb relentlessly). On top of this, GOM gas output had a readily accessible and stable (regulated) market outlet, at a time when finding gas was still seen as an expensive nuisance almost anywhere else in the world.

In the early 1980s, as gas output peaked, reserve addition rates in the shallow water GOM began to decline. From 1984 onwards, the industry never managed to replace produced reserves. On the positive side, oil production posted small year-on-year increases between 1982 and 1986 (averaging about 50 MBD). However, the impact of the 1986 world oil price crash brought about the return of hundreds of leases to the US government and the drastic curtailment of E&P expenditure, thereby putting an end to this temporary recovery. In 1987, only seventy platforms were installed throughout GOM, whereas in prior years platform installations had averaged between 150 and 160 units. In addition, a record number of rigs were removed from the fleet during 1987. As a result of this contradiction, many of the larger offshore drilling contractors were forced into merging and restructuring, while smaller companies simply vanished from the scene, as their small rig fleets were repossessed one by one (in early 1987, for instance, 56 offshore rigs were solely in the hands of financial institutions). Similarly, only a handful of major fabrication yards remained open in the Gulf Coast, as many facilities were forced to shut down due to lack of new orders. In all, as an inside observer wryly put it, 'ten-dollar oil made the [offshore] industry one time, and the second time it damned near broke it'.[36]

The oil price collapse did not derail the march of technological progress (always a key driver for GOM oil activities). The year 1987, for instance, saw the installation of the first floating production system (FPS) in GOM. Emblematically, the company that achieved this milestone (the Hunt brothers' Placid Oil) was bankrupt at the time, and the project (which involved the conversion of a semisubmersible drilling rig into a floating production unit, and its emplacement at a record water depth of 1540 feet in Green canyon block GC29) could only go ahead after Placid defeated a challenge by its creditors in the Federal bankruptcy court. Ironically enough, the project was plagued by problems and had to be abandoned less than two years after coming on stream.

As the oil price stabilised around 18 USD/B, offshore drilling activity in GOM started to show signs of recovery. Independent oil and gas companies were able to farm into shallow acreage held by the majors on leases due to expire during 1988 and 1989. However, the shift in the centre of gravity for GOM oil activities towards progressively deeper waters (foreshadowed by the Placid project mentioned above) gathered pace during these same years. Over the decade of the 1980s, the average depth of oil discoveries grew to 548 feet, more than double what it had been during the previous decade.[37] In contrast, the average water depth of gas discoveries *decreased* from the 1970s to the 1980s, going from 179 feet to 136 feet. These contrasting trends are a reflection of 'the emphasis of major oil companies during the 1980s of exploring the high-potential, oil prone prospects of the deeper water areas', while the independents increasingly shifted their exploration activity to 'low-risk, low-cost, relatively modest potential, shallow water gas prospects easily targeted by new seismic techniques'[38]

By the early 1990s, the long-running and fruitful relationship between cutting-edge technology and oil entrepreneurship in the GOM shallow waters appeared to be running out of gas (both literally and figuratively). The year 1992, for instance, marked the first time that the number of platforms removed in the GOM region exceeded that of platforms installed (Figure 3.1). In that year, bonus payments for acreage reached their nadir (USD 84 million, as compared to their 1981 peak of USD 4.9 billion), a situation that led industry participants to start referring to GOM as the 'Dead Sea'. Ironically, the beginning of the deepwater boom was merely one year away when this sobriquet was coined, and this event gave a new lease of life even to the shallow water province, as majors divested more and more of their mature properties in order to devote their full attention to the deepwater.

Another boost came in the form of the natural gas market, which left

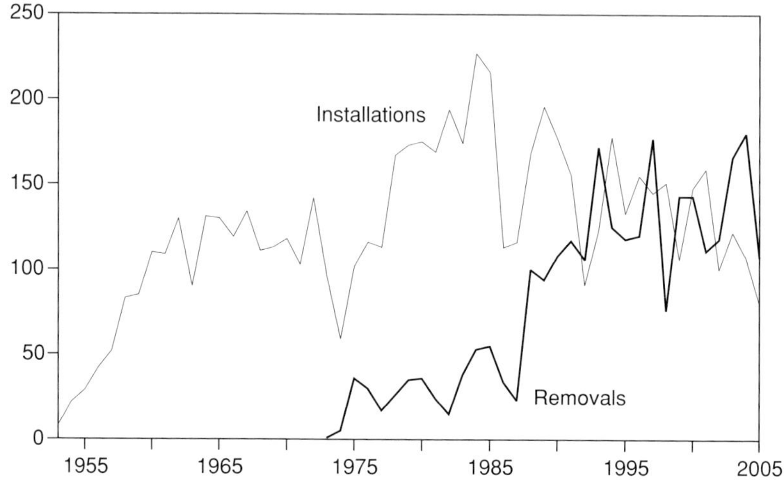

Source: MMS

Figure 3.1: Annual Platform Installations and Removals in the GOM Federal OCS, 1953–2005

behind its 'bubble' phase and prices began to be driven by nation-wide depletion rates. In addition to this, some of the technological advances that the industry applied in the novel subsalt and deepwater provinces also proved very fruitful in the shallows. This was particularly the case of three-dimensional seismic techniques (which allowed better identification of both bypassed and/or deeper pay zones in compartmentalised reservoirs with multipay horizons)[39] and the AVO (amplitude vs. offset) method of comparing the amplitudes of near, middle and far seismic traces to identify the presence of hydrocarbons.

Since the beginning of the 1990s, independent producers in the shallow water province have posted some impressive success rates using these technologies, whose economic attraction is further leveraged by the abundance of infrastructure in the province (wellbores can be used again and again to drill deviated wells to tap small hydrocarbons pockets located with 3-D seismic techniques, for instance). Even though strike rates have been quite high, the pickings have tended to be slim (in absolute terms), and the economics of this type of mopping-up operation have proven to be exceptionally vulnerable to low oil (and gas) prices. Nevertheless, thanks to the intensive application of new technology, oil companies have managed, apparently against all odds, to keep on discovering reserves (and not in negligible magnitudes either).

Having said all that, shallow water reserves added in recent years have been significantly below even those added between 1983 and

1990. And, of course, they have come nowhere near making a dent in the production decline rate (Figure 3.2). This goes to show that, even in an economic environment characterised by attractive gas prices, the intensive application of new seismic technologies has, at best, delayed the day of reckoning for the shallow water province. Indeed, a persuasive case has been made that these technological advances *accelerated* decline rates overall: the pervasive impression that known fields were being exploited more effectively was valid in the case of a few easily identifiable opportunities but, once these had been grasped, the new technologies merely drove exploration towards untapped marginal reservoirs.[40] In other words, the new technology has not enhanced the ability of the industry to add reserves as such, and has only allowed it to drain discovered fields faster (thus, far from being reversed, production decline curves after peak have actually accelerated).

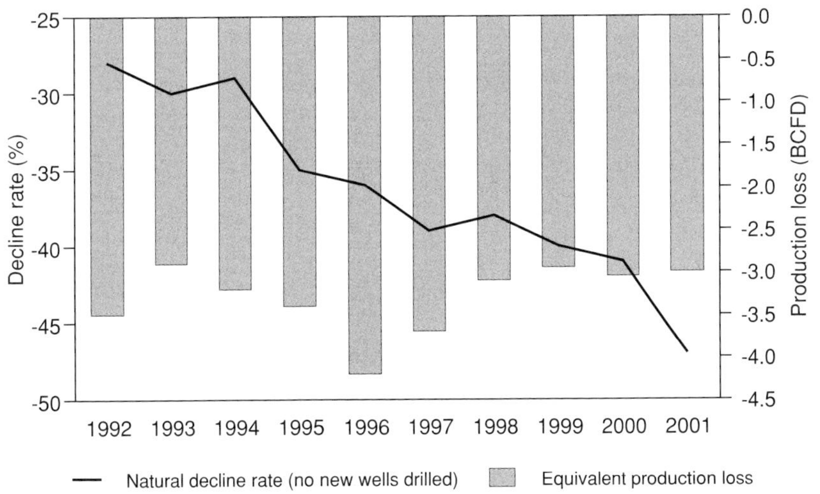

Source: NPC 2003

Figure 3.2: Natural Decline Rate in the GOM Federal OCS Shallow Water Gas Fields and Equivalent Potential Production Loss absent New Drilling, 1992–2001

The tail-end of shallow water oil production in GOM will definitely not be as long as it has been in onshore provinces like Texas or Oklahoma: offshore cost structures cannot support genuine stripper well operations, so the decline function may never assume a genuinely asymptotic profile. Granted, up until the early 1950s, the population of wells producing 15 BD or less still represented a significant percentage of the total population, but that was only because at that point most wells were still located very close to shore. As distances to shore grew, costs did as well, and as a result the percentage share of this production bracket in the total well population fell below 10 percent by the mid 1950s, a level that it has never exceeded since (when a well falls in this production bracket, this is a signal of the imminent end of its useful life).

According to the latest well productivity data compiled and put in the public domain by the energy consulting firm IHS Energy on behalf of DOE, oil wells producing 15 BD or less account for around 8 percent of GOM population but less than one-tenth of 1 percent of output, whereas the comparable figures for onshore Texas are 86 and 39 percent, respectively. These statistics also show that in 2002, 72 percent of all GOM oil wells produced 200 BD or less. This group of wells (all of which are in shallow waters), accounted for slightly over 11 percent of total output in 2002, whereas in 1988 they represented a similar proportion of the well population but were responsible for 24 percent of output (Figure 3.3).

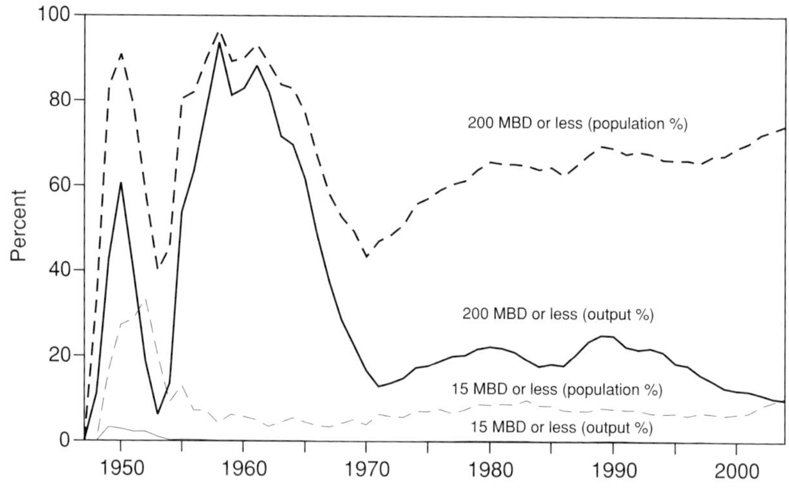

Source: DOE/IHS

Figure 3.3: Oil Well Productivity Indicators for GOM Federal OCS by Production Rate Bracket, 1947–2004

In light of the rate at which shallow water oil reserves are being added, the overall attrition rate for wells, the rate at which productivity in this well category is declining, and finally, the rate at which productivity of deepwater wells (which account for almost all the wells with an output of 500 barrels of oil equivalent per day (BOED) or greater) is increasing, it seems reasonable to conjecture that, as far as oil is concerned, the shallow water sub-province in its 'traditional' incarnation will be fully played out sometime around 2015 (Figure 3.4). Granted, past forecasts of GOM output have consistently understated future production and their inaccuracy was magnified with the passage of time. For instance, DOE estimates of 1995 GOM production drawn up in 1990 undershot the mark by 12 percent, while 1985 estimates were off by 15 percent (notwithstanding the fact that the latter were prepared before the 1986 oil price collapse). This time around, though, it does appear as if the shallow water oil province is very much in the terminal stages of depletion.

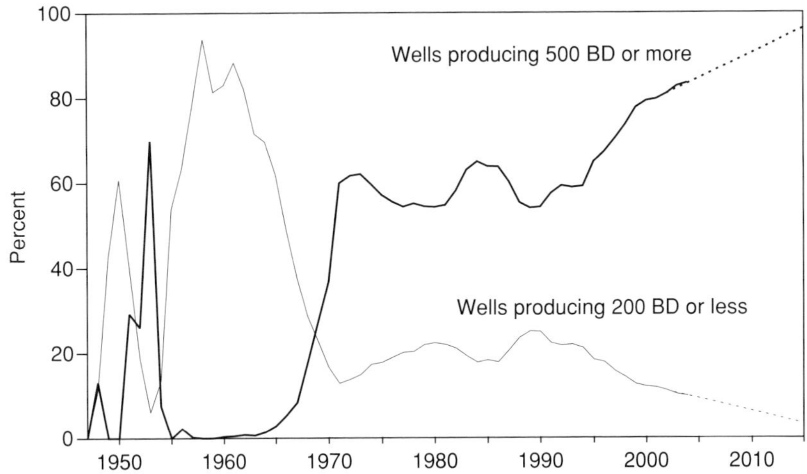

Source: DOE/IHS

Figure 3.4: Percentage of Output for GOM Federal OCS Oil Wells by Production Rate Bracket, 1947–2015

At first glance, the situation for shallow water natural gas appears less fraught, which is just as well given the reliance of the US economy on shallow water gas produced in the GOM Federal OCS. The IHS data show that non-associated wells producing 15 BOED or less of gas currently account for around 12 percent of the GOM well population and contribute less than 1 percent of the total output. This production bracket merely represents the number of gas wells that are being put out of commission in any given year. Far more important is the fact that, at around 30 percent, the share of wells producing respectable volumes of natural gas (400 BOED or more) is significantly higher than the proportion of oil wells producing equivalent volumes of crude

(14.4 percent). Moreover, output from these high-productivity gas wells has accounted for a relatively stable share of total GOM production for a very long time, and the simple extrapolation of the data for the most recent ten years results in a trendline whose profile is not all that alarming (Figure 3.5).

Apart from the above, gas production activities in the shallow water

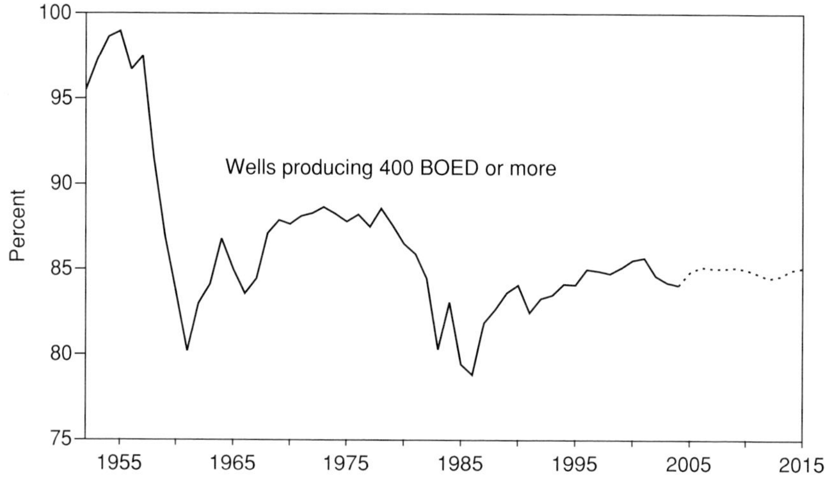

Source: DOE/IHS

Figure 3.5: Percentage of Output for GOM Federal OCS Natural Gas Wells by Production Rate Bracket, 1952–2015

sub-province display other outward signs of health. For instance, a significant proportion of the scores of GOM wells that produce 1.6 MBOED or more of gas are located in shallow water, and the share of GOM output accounted for by such wells stands at 51 percent of the total production in the province (up from 32 percent in 1992). Similarly, the GOM-wide share of wells producing 6.4 MBOED or more of gas has gone from 1.1 to 19 percent and, again, such wells have by no means been restricted to the deepwaters. Output figures for the critical first years of well operation have remained quite steady over time (indeed, the cumulative average production for the first three years after on-stream date was 10 percent higher for wells drilled in 1996 than that for wells drilled in 1972). Likewise, a study published by DOE in 2000 showed that peak output in the shallow water province as a whole had been on a rising trend over the 1990s (Table 3.1), going from an average of 4.9 MMCFD per well in 1988 to 6.1 MMCFD per well by 1996.

Table 3.1: Average Production Indicators for Natural Gas Wells in the GOM Federal OCS by Vintage, 1976–1996

Vintage	*Peak output (MMCFD)*	*Production decline after peak (%)*	
		1 year	*2 years*
1976	5.6	24	38
1980	5.5	35	51
1988	4.9	35	51
1996	6.1	46	70

Source: EIA 2000

All of these figures may look superficially reassuring, but they conceal a most unwelcome sting in the tail; namely that 'the Gulf is a treadmill ... characterised by fast gas', with most of the output coming from wells less than three years old, tapping very small fields.[41] Peak production capacity may indeed have increased, but the other side to this coin has been an extremely rapid acceleration in production decline: two years after peak production, output from wells drilled in 1996 averaged only 30 percent of peak, whereas the equivalent indicator for wells drilled in 1972 was 63 percent of peak.[42] Figure 3.6 shows the production paths for 1996 and 2000 as being similar in their time to peak, their

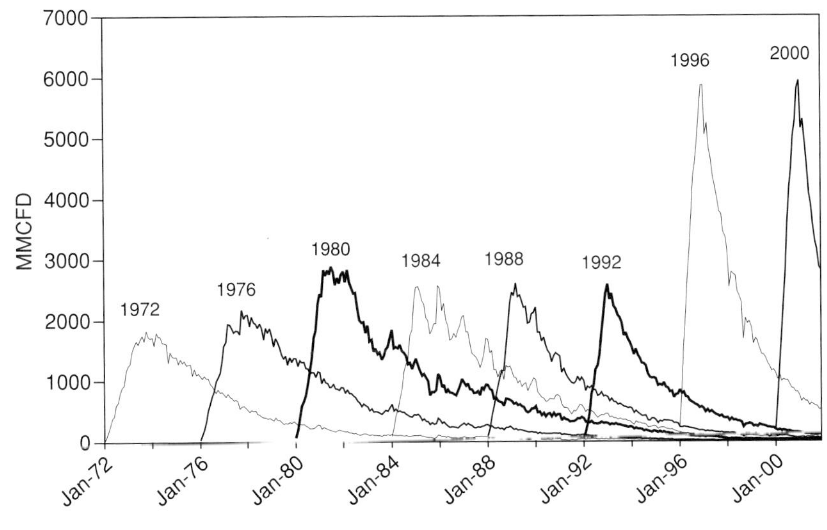

Source: MMS

Figure 3.6: Combined Monthly Natural Gas Output for all Wells in the GOM Federal OCS by Drilling Year Vintage, 1972–2000

peak production rate and their rate of decline. However, this similarity is slightly misleading, because of the far greater weight of deepwater wells in 2000. In actual fact, in the shallow water province, both peak rates and plateau times reached their apogee during 1996–7. Since then, peak rates have fallen (albeit marginally) while plateaus have shortened to an even greater extent than the graph implies.[43]

As a result of the highly accelerated decline profile of GOM shallow water gas wells, the hydrocarbons reserves to production ratio from GOM shallow water reservoirs (less than 700 feet) has been dropping steadily for the past twenty years, as progressively smaller gas fields have been produced at progressively faster rates (Figure 3.7). The shallow water decline rates also mean that even maintaining shallow water output constant would presuppose unattainable exploration and production targets: i.e. the completion of around 1200 additional successful wells per year, with each one of these wells producing 6 MCFD of gas at peak (during the 1990s, the average number of successful well completions in GOM was only 940).[44] This means that, for all of the outward signs of health that it gave off up to the late 1990s, natural gas production in the traditional shallow water sub-province could very well cease to be a viable economic proposition even earlier than the production of crude oil, and even though no substitute for this crucial component in the USA's energy mix is in sight.

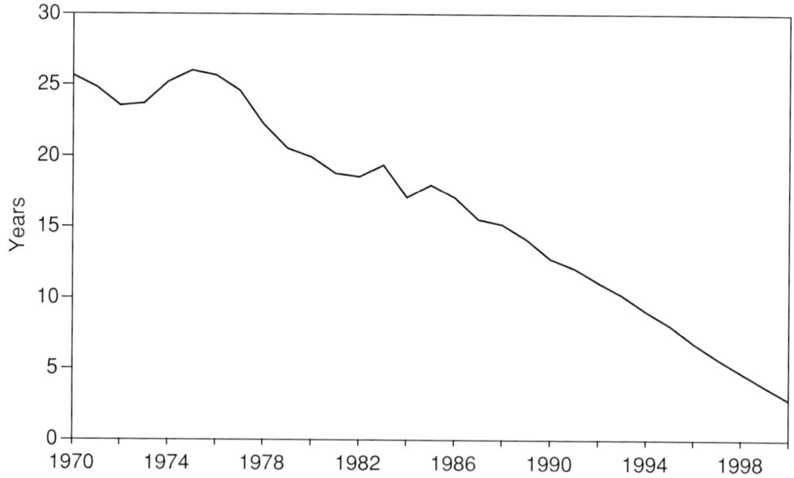

Source: MMS

Figure 3.7: Total Hydrocarbons Reserves to Production Ratio in the GOM Federal OCS Shallow Water Sub-province, 1970–2000

3.2 Trends in Hydrocarbons Production and Major Producing Fields

The evolution over time of shallow water GOM output is shown in Figure 3.8. As can be appreciated, production of crude oil and condensate on the one hand, and natural gas, on the other, were comparable up to the mid-1960s. From this point onwards, the output of natural gas began to leave that of crude oil far behind. Shallow water oil production peaked in 1971, at slightly over 1 MMBD, whereas gas production continued to increase, reaching its apogee of 13.2 MMCFD (2.4 MMBOED) only in 1981. Natural gas production in this province only began to decline in earnest from 1990 onwards.

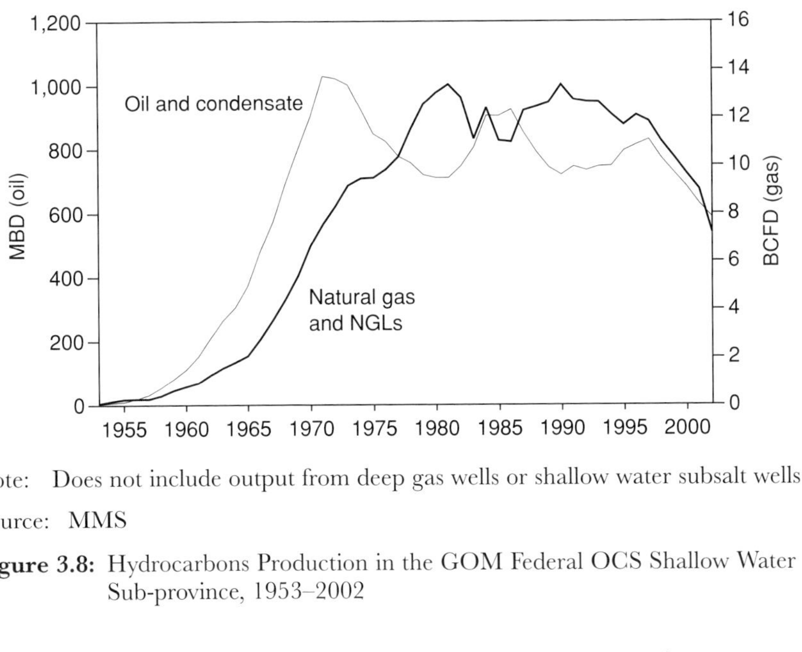

Note: Does not include output from deep gas wells or shallow water subsalt wells

Source: MMS

Figure 3.8: Hydrocarbons Production in the GOM Federal OCS Shallow Water Sub-province, 1953–2002

The shallow water province was always very gas-prone (gas accounts for around 65 percent of cumulative production), and has become increasingly so over time.[45] For the GOM region as a whole, this is true even in fiscal terms: the annual royalties from sales of natural gas have been larger in absolute terms than those from crude oil sales since 1977. As if this were not enough, when one considers the different contributions that GOM gas and oil production increasingly make to the US energy balance, it is clear that the former is vested with an even greater economic significance than the latter. GOM natural gas output has accounted for 25 percent of total US domestic output for the last twenty years, whereas GOM crude output as a proportion of total US output used to be around 10 percent, and has only recently surpassed the 20 percent mark, in the wake of the deepwater boom and the ongoing decline in production elsewhere (particularly Alaska). Furthermore, the share of GOM gas production in the total US domestic demand for this fuel is much greater than the equivalent share for GOM crude production. Finally, GOM's gas output is far more difficult to replace with imports, on grounds of logistics. Having said all that, the largest fields found in the GOM shallow water have tended to be oilfields, as the ranking in Table 3.2 shows.

In 1970, oil production was dominated by a small number of fields, all of them in very shallow waters. Chief amongst these were the two fields that, as of 2000, have been the greatest GOM producers of all, in terms of cumulative oil output (BM002 and WD030, both located in about 50 feet of water). In 1970, there were nine fields producing more than 40 MBD, but by 1980 there was only one field in the whole shallow water sub-province whose output exceeded this mark (EI330). Throughout this decade, there was a major increase in the number of producing fields and in the spread of those fields across the area, which was accompanied by a strong decline in the productive capacity of key fields.

As far as oil was concerned, the performance of fields found during the 1980s was particularly disappointing: by 2000, for instance, 1980-vintage fields were producing less oil than either those of the 1970s or the 1960s, and only slightly more than the residual production of fields commissioned back in the 1950s (Figure 3.9). Output at natural gas fields discovered during the 1980s held up much better, by contrast (Figure 3.10). With the gradual dissipation of the 1986 price crisis hangover, the number and spread of shallow water fields in GOM again increased markedly. Even so, in terms of production, no single shallow water field recorded an output of 40 MBD and, indeed, none even reached the 30 MBD mark. Production became very highly diversified across fields: the main cluster of major fields tended to produce 10–20 MBD at best, with the loss in the maximum productive capacity of larger fields being compensated – albeit not entirely – by the sheer increase in the number of smaller fields.

The onset of the deepwater boom witnessed a reversal of roles for crude oil and natural gas production. As the former surged in response to the commissioning of large deepwater fields, the latter languished and then began to decline quite rapidly. Indeed, the performance of 1990s vintage gas fields has mirrored that of 1980s oilfields, and this

Table 3.2: Thirty Largest GOM Shallow Water Fields by Rank Order, Based on Proved BOE Reserves as of 2002

Rank	*Field name*	*Discovery year*	*Water depth (feet)*	*Field type*	*Field Gas/Oil Ratio (SCF/STB)*	*Proved reserves*		*Cumulative production through 2000*		*Remaining proved reserves*	
						Oil (MMB)	*Gas (MMBOE)*	*Oil (MMB)*	*Gas (MMBOE)*	*Oil (MMB)*	*Gas (MMBOE)*
1	EI330	1971	246	O	4,297	425.0	325.0	402.1	313.6	22.9	11.4
2	WD030	1949	49	O	1,488	572.3	151.5	547.9	142.6	24.4	8.9
3	GI043	1956	139	O	4,345	367.5	284.2	353.2	265.1	14.3	19.1
4	BM002	1949	50	O	1,057	523.9	98.5	513.2	93.4	10.7	5.1
5	TS000	1958	13	G	85,126	37.3	564.8	36.4	556.1	0.9	8.8
6	VR014	1956	26	G	64,373	48.1	550.5	47.7	539.3	0.3	11.3
7	MP041	1956	42	O	5,665	265.0	267.1	242.0	248.2	23.0	18.8
8	VR039	1948	38	G	82,271	31.7	463.7	30.6	447.3	1.1	16.4
9	SS208	1960	103	O	6,362	218.2	247.0	211.1	232.6	7.1	14.4
10	WD073	1962	177	O	2,639	270.0	126.8	253.2	107.2	16.9	19.6
11	GI016	1948	54	O	1,275	299.3	67.9	294.6	66.0	4.7	1.9
12	SP061	1967	220	O	1,929	262.7	90.2	251.6	87.1	11.2	3.1
13	EI238	1964	146	G	16,850	86.6	259.8	76.6	234.7	10.0	25.1
14	SP089	1969	425	O	4,415	193.0	151.6	182.6	132.4	10.5	19.2
15	ST172	1962	98	G	157,847	11.6	324.8	10.4	314.0	1.2	10.8
16	WC180	1961	49	G	138,546	13.0	320.5	12.4	307.2	0.6	13.3
17	ST135	1956	130	O	4,956	172.2	151.8	161.7	96.1	10.5	55.7
18	ST021	1957	46	O	1,640	245.1	71.5	239.1	68.4	6.0	3.1
19	SM048	1961	100	G	56,229	28.4	284.2	27.1	264.6	1.3	19.6
20	EI292	1964	211	G	85,509	19.1	290.2	17.6	282.7	1.4	7.4

Table 3.2: *Continued*

Rank	*Field name*	*Discovery year*	*Water depth (feet)*	*Field type*	*Field Gas/Oil Ratio (SCF/STB)*	*Proved reserves*		*Cumulative production through 2000*		*Remaining proved reserves*	
						Oil (MMB)	*Gas (MMBOE)*	*Oil (MMB)*	*Gas (MMBOE)*	*Oil (MMB)*	*Gas (MMBOE)*
21	EC271	1971	171	G	19,257	68.9	236.1	65.9	229.4	3.0	6.7
22	EC064	1957	49	G	57,535	26.9	275.6	26.1	270.3	0.9	5.3
23	ST176	1963	127	G	14,990	81.8	218.3	76.4	188.6	5.4	29.6
24	SS169	1960	63	O	5,288	152.9	143.8	143.4	135.8	9.4	8.0
25	WC587	1971	210	G	119,223	13.1	278.0	12.5	267.4	0.6	10.5
26	SP027	1954	63	O	5,300	149.3	140.8	147.4	131.9	1.8	8.8
27	SS176	1956	100	G	20,588	62.0	227.0	60.2	219.8	1.7	7.2
28	WD079	1966	125	O	3,810	162.6	110.2	159.3	107.6	3.3	2.6
29	EI296	1971	213	G	69,270	20.4	252.0	20.2	247.6	0.3	4.3
30	WC192	1954	57	G	60,972	106.3	243.5	20.8	226.4	1.6	17.0
TOTAL						4,934.2	7,216.8	4,643.5	6,823.6	206.8	393.2

Source: MMS

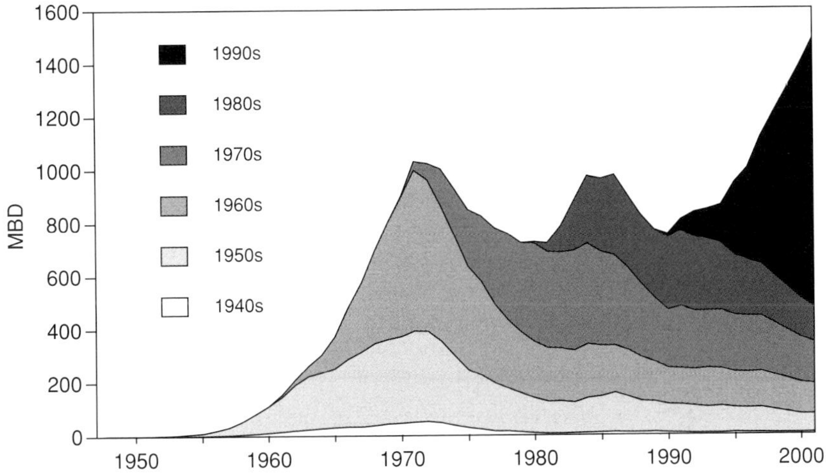

Source: MMS

Figure 3.9: GOM Federal OCS Crude Oil Production Profile, by Decade, 1947–2001

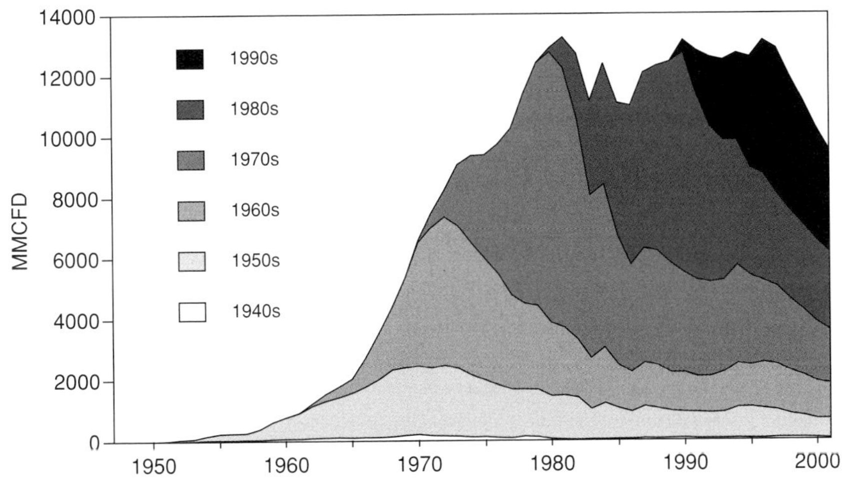

Source: MMS

Figure 3.10: GOM Federal OCS Natural Gas Production Profile, by Decade, 1947–2001

is a major cause of concern for the future health of the US natural gas market, as we shall explain later in greater detail.

As a general rule, the larger shallow water gas fields followed quite a leisurely path to their production peak, as illustrated by Figure 3.11 (which plots the production profile of the VR014 field, which at 3.1 TCF of reserves is the largest gas field ever found in GOM). Output at such fields throughout much of the 1970s was choked back because of the distortions derived from the regulatory regime on inter-state commerce of natural gas. These distortions translated into supply shortages in markets dependent on inter-state flows for their supplies. The regulation-induced shortages were supposed to be addressed by the Natural Gas Policy Act (NGPA) and the Powerplant and Industrial Fuel Use Act (PIFUA), both enacted in 1978. In reality, this new legislation only muddied the waters further chiefly because price caps for the categories of gas subject to escalators related with the price of other fuels ended up exceeding the market price for gas by a considerable margin. 'This', as the National Petroleum Council (NPC) explains, 'resulted in high reserve additions, while at the same time ... high prices were having a dampening effect on demand. By the early 1980s, the shortage ... had been replaced with a surplus.'[46] The majority of the large GOM offshore fields (like VR014) peaked around 1981, and they made a key contribution to the formation of the so-called 'gas bubble' (which began to deflate – albeit very slowly – after the repeal of PIFUA in 1987).

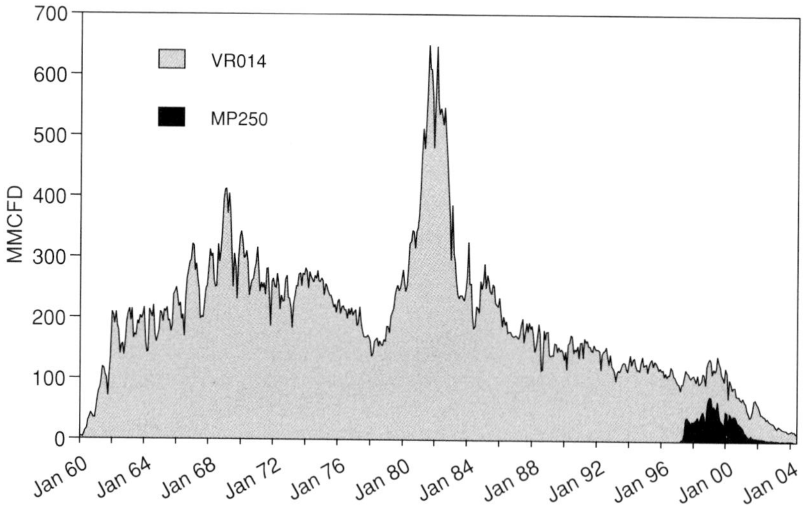

Source: MMS

Figure 3.11: Production Profile of VR014 Field, 1960–2004

The contrast between the VR014 type of production profile, on the one hand, and that of later vintage small gas fields (exemplified in Figure 3.12 by the MP250 field, discovered in 1997 and with 15.5 BCF worth of gas reserves), on the other, is quite striking (MP250 output is a piffle in comparison to VR014). The production profile of late vintage oilfields (exemplified in Figure 3.13 by the EW910 field, discovered in

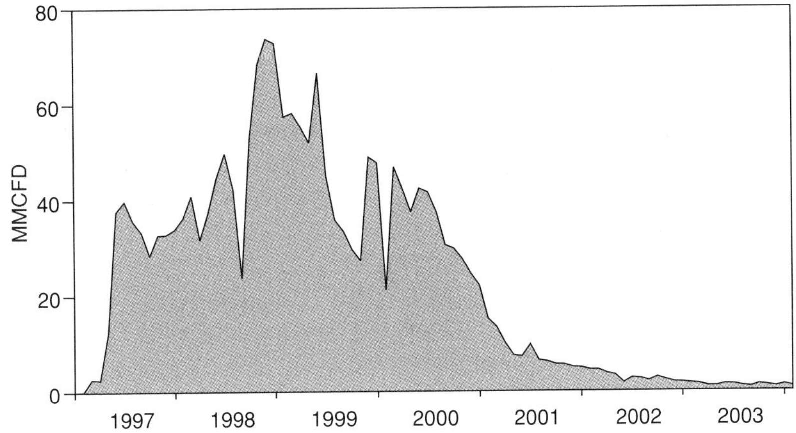

Source: MMS

Figure 3.12: Production Profile of MP250 Field, 1997–2004

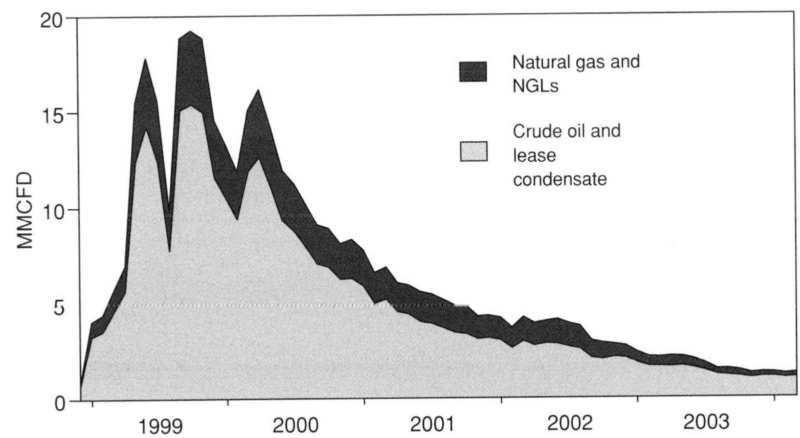

Source: MMS

Figure 3.13: Production Profile of EW910 Field, 1998–2004

1996 and with 24.4 MMBOE of reserves) closely resembles that of small late vintage gas fields.

The larger oilfields found in the shallow water province display a production profile that is very different in turn from the one characteristic of the deepwater fields responsible for the recent resurgence in GOM production. Figures 3.14 and 3.15 show the production profile

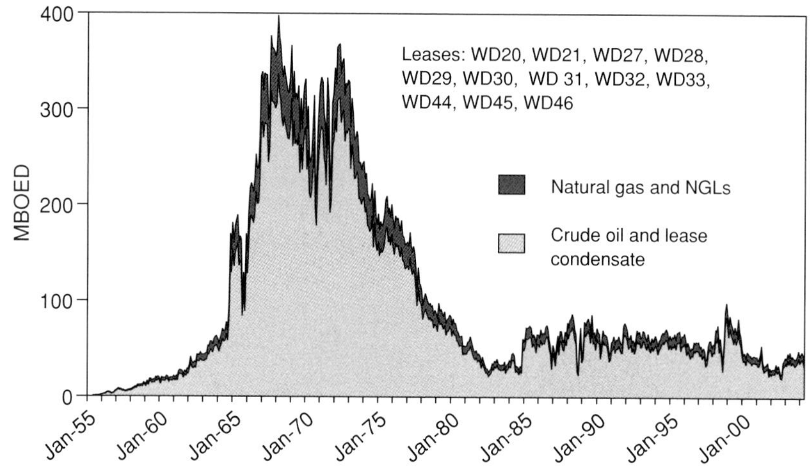

Source: MMS

Figure 3.14: Production Profile of WD30 Field, 1955–2004

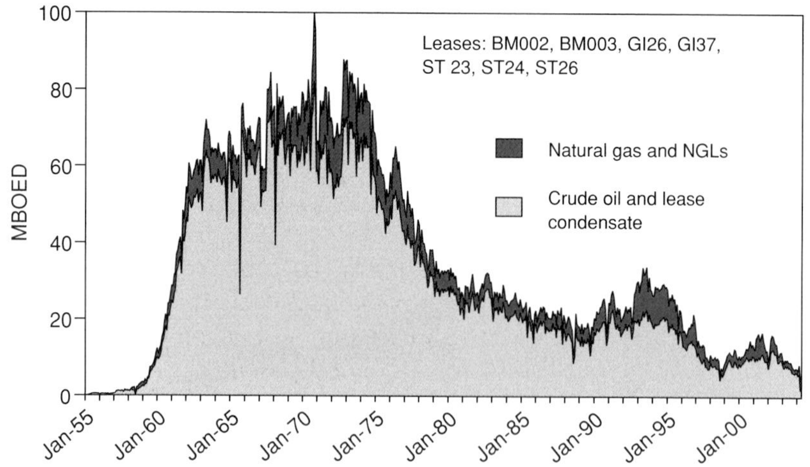

Source: MMS

Figure 3.15: Production Profile of BM002 Field, 1955–2004

of the WD30 and BM002 fields respectively. As can be appreciated, these fields took around 15 years to peak, and they then underwent a period of ten years of quite rapid output decline, before settling into a long period of slower decline.

The third largest cumulative oil producer to date, EI330, shows an overall profile that is more akin to that of modern deepwater fields: a more rapid ramping up to a peak, followed by a swifter decline and what is likely to be a relatively shorter period of stabilisation leading to eventual abandonment (Figure 3.16). During this latter period, though, output at the EI330 field underwent a most unusual recovery that culminated in a second production peak (obviously at a much lower level than the first peak) in late 1995.

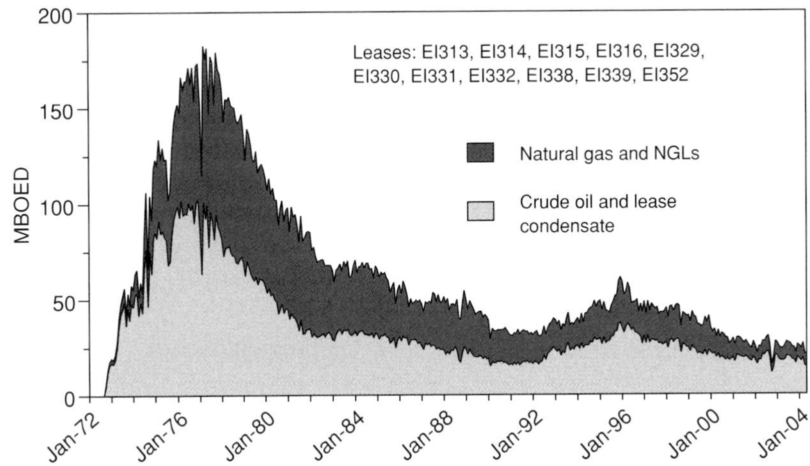

Source: MMS

Figure 3.16: Production Profile of EI330 Field, 1972–2004

This pleasant surprise came on the back of indications that the remaining estimated reserves in the field were not depleting quite as fast as would have been predicted on the basis of recorded production rates.[47] Even more remarkably, geochemical evidence (as well as pressure, temperature and seismic amplitude anomalies) pointed towards the likelihood that at least some of the oil produced in the mid-1990s may not have been present in the field at the beginning of production in 1973. Indeed, samples of EI330 crude taken at various points in time by the Texas A&M Geochemical Study of Gulf Coast Oils showed that 'oils from shallow reservoirs at 4,200 feet and 5,200 feet were heavily biodegraded in 1972, but the 1984 oils were less heavily biodegraded than either the 1972 or the 1988 oils *from the same perforation depths in the same wells*'.[48] In short, the behaviour of EI330 suggested 'overproduction caused by the migration of new hydrocarbons from a deeper source region into these shallow reservoirs even as they are being produced'.[49]

The phenomenon of potential hydrocarbon migration from undiscovered hydrocarbon columns buried within geopressured turbidites in GOM was held to be of such import that the EI330 field became the focus of a major study sponsored by the DOE in the mid-1990s, as part of its Advanced Oil Recovery Programme. The study had a budget of USD 20 million, a tidy sum that could have bankrolled an exploration campaign by a small E&P company.[50] The promoters of the study posed the hypothesis that the migration process might also be taking place at other locations, in which case significant new deep reserves (20 billion barrels plus, it was hoped) could be waiting to be tapped by the relatively simple expedient of finding their migration pathways, as opposed to having to drill through thousands of feet of salt and sediments. Rather hastily, some observers concluded that this novel way of exploiting oil resources might 'arrest if not reverse the production decline within US borders'.[51] Others drew even more adventurous conclusions, seeing in the refilling of producible reservoirs at EI330 nothing less than a confirmation of the theory of the inorganic origin of petroleum.[52]

The Dynamic Enhanced Recovery Technologies study did identify the pathways, fault zones and possible mechanisms for oil migration into EI330 reservoirs. It also showed that 'oil and gas [were] recoverable from the fault zone … and would flow under pressure'. It did not, however, find a way 'to make the fault economically producible'[53] and, therefore, it could offer no viable alternative to the colossal expense of deep drilling. Had the project been able to fulfil some of the more sanguine expectations that some of its promoters harboured in this regard, or had the oddities of EI330 proved that oil is in fact a renewable resource of sorts, then this field would have a reasonable claim to be perhaps the most significant in the history of the industry. As things have turned out, however, EI330's main claims to fame are that it is the largest Pleistocene reservoir ever found, on the one hand, and that it is the only one among 2200 fields with reserves greater than 100 MMB where a significant decrease in the natural decline rate has ever been reliably attested, on the other.[54] But, given that its replenishment rate appears to be diminishing rapidly, it is likely that EI330's next claim to fame will be to become the largest abandoned Pleistocene discovery.

3.3 Outlook

The GOM shallow water province has been explored, drilled and exploited almost literally to the point of exhaustion. Out of the twenty GOM fields with the most remaining reserves as of year-end 2000, for instance, only four lay in shallow water: GI116 (85 MMBOE); SP089 (70 MMBOE); MO823 (64 MMBOED) and ST190 (53 MMBOE). At that point, their reserves represented only around 11 percent of the remaining reserves in this group of fields, and 4 percent of the total proved remaining reserves in GOM. Another good indication of the maturity of the province is its creaming curve, which displays an unmistakably asymptotic profile (Figure 3.17). Equally revealing is the curve plotting the mean field size of new discoveries in the province (Figure 3.18).

Given the astonishing number of wells drilled in GOM shallow waters since 1953 (around 30,000 more than the total wells drilled in the British, Danish and Norwegian sectors of the North Sea to 2004), it is remarkable that petroleum is still being found in paying quantities here. Two factors explain why GOM has been able to sustain a much more intensive exploration effort for a longer stretch of time when compared to other provinces. On the one hand, the sediments are thick

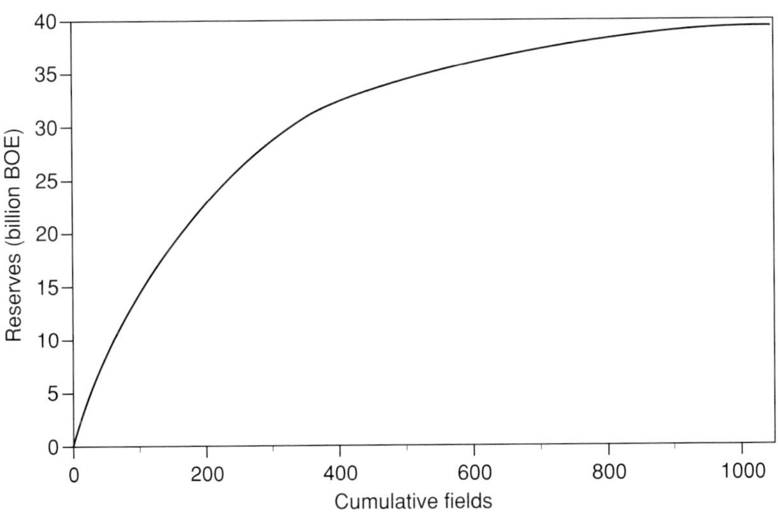

Source: MMS

Figure 3.17: Creaming Curve for Fields in the GOM Shallow Water Sub-province as of 2000

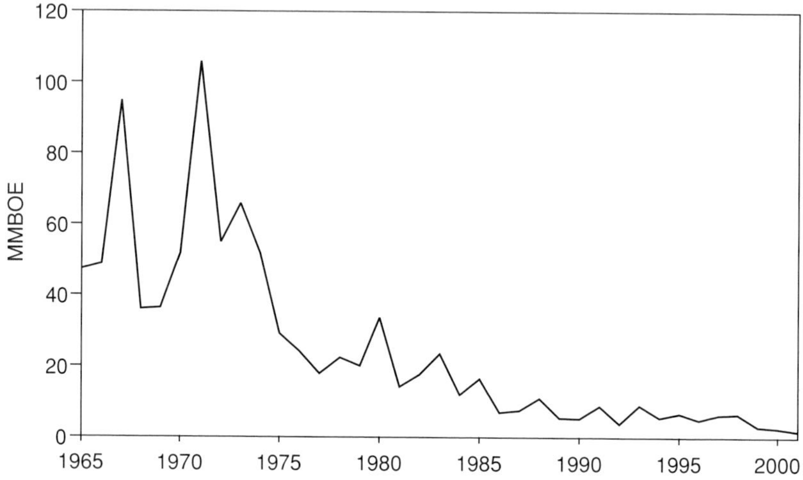

Source: MMS

Figure 3.18: Mean Field Size of New Discoveries in the GOM Federal OCS Shallow Water, 1965–2001

and clastic, with ample reservoirs. On the other hand, the movement of salt has created a prodigious number of structural traps.

As mentioned above, the introduction of new seismic technology has had a very salutary effect on GOM shallow water reserve figures, by allowing for the identification and tapping of numerous small pools which had been by-passed in earlier, less sophisticated exploration efforts. In recent years, around three-quarters of the reserve growth has come from additions to older fields, as opposed to new reservoirs of exploratory significance in older fields or reserves from new fields.[55] In contraposition to trends seen in other major provinces, though, reserve additions in giant fields have been relatively small. The explanation behind this is that the magnitude of these fields was recognised at a very early stage in their development, and they were therefore treated as core assets by their operators, undergoing several cycles of re-evaluation and renewed exploration. The intensity of their early exploitation left relatively few reserves to be added later on.

Table 3.3 shows the remaining proved reserves in the GOM shallow water, by administrative subdivision. Richard Nehring considers that there is a reasonably high probability that at least 8000 MMBOE of additional shallow water reserves will be found, and a very low probability that such future reserve additions will reach 17,000 MMBOE.[56] Nehring also considers that, in coming years, the annual reserve addition rate is likely to average 700–800 MMBOE (with natural gas making up as

Table 3.3: Estimated Oil and Gas Reserves in the GOM Federal OCS Shallow Water, as of December 31, 2002

	Number of fields					
Area	*Proved active producing*	*Proved active nonproducing*	*Proved expired depleted*	*Unproved Active*	*Unproved Studied*	*Expired nonproducing*
Western Planning Area						
Brazos	23	6	9	0	0	2
Galveston	22	5	18	0	0	3
High Island and Sabine Pass	72	18	31	2	2	14
Matagorda Island	23	2	3	0	0	3
Mustang Island	15	1	11	1	1	6
N.& S.Padre Island	6	1	4	0	0	1
Western Planning Area Subtotal	161	33	76	3	3	29
Central Planning Area						
Chandeleur	6	1	3	0	0	0
East Cameron	53	4	10	0	0	0
Eugene Island	60	5	10	3	3	7
Grand Isle	16	3	2	0	0	1
Main Pass and Breton Sound	57	8	13	3	3	6
Mobile Bay	18	4	5	0	0	3
Ship Shoal	48	4	8	2	2	3
South Marsh Island	40	4	5	0	0	0
South Pass	11	1	1	1	1	0
South Pelto	9	0	0	0	0	0
South Timbalier	43	4	6	4	4	2
Vermilion	64	7	12	0	0	3
West Cameron and Sabine Pass	76	8	19	2	2	4
West Delta	21	1	2	0	0	2
Central Planning Area Subtotal	522	54	96	15	15	31
TOTAL	683	87	172	18	18	60

Reserves: oil in MMB at 60 °F and 1 atmosphere; gas in BCF at 60 °F and 15.025 psia.
Cumulative production: oil in MMB; gas in BCF

Source: MMS

Table 3.3: *Continued*

Area	*Proved reserves*		*Cumulative production through 2002*		*Remaining proved reserves*	
	Oil	*Gas*	*Oil*	*Gas*	*Oil*	*Gas*
Western Planning Area						
Brazos	11	3,579	9	3,086	2	493
Galveston	55	2,041	47	1,794	8	247
High Island and Sabine Pass	376	14,680	346	13,796	30	884
Matagorda Island	26	5,615	22	4,702	4	913
Mustang Island	7	1,995	5	1,602	2	393
N.& S.Padre Island	0	523	0	475	0	48
Western Planning Area Subtotal	475	28,433	429	25,455	46	2,978
Central Planning Area						
Chandeleur	0	346	0	337	0	9
East Cameron	327	10,505	308	10,036	19	469
Eugene Island	1,603	19,042	1,498	17,808	105	1,234
Grand Isle	965	4,782	923	4,411	42	371
Main Pass and Breton Sound	1,092	6,565	943	5,684	149	881
Mobile Bay	0	2,111	0	1,503	0	608
Ship Shoal	1,354	11,902	1,277	11,173	77	729
South Marsh Island	900	14,040	814	13,231	86	809
South Pass	1,076	4,357	1,023	4,043	53	314
South Pelto	157	1,151	139	943	18	208
South Timbalier	1,495	10,043	1,392	8,518	103	1,525
Vermilion	543	16,208	494	15,283	49	925
West Cameron and Sabine Pass	207	19,625	185	17,993	22	1,632
West Delta	1,396	5,591	1,309	5,099	87	492
Central Planning Area Subtotal	11,115	126,268	10,305	116,062	810	10,206
TOTAL	11,590	154,701	10,734	141,517	856	13,184

Reserves: oil in MMB at 60 °F and 1 atmosphere; gas in BCF at 60 °F and 15.025 psia.
Cumulative production: oil in MMB; gas in BCF

Source: MMS

much as 75 percent of this total), with anything from half to three-quarters coming from growth in existing major reservoirs of known producing fields located for the most part offshore Louisiana.

One could argue that Nehring's finding rate and estimates of undiscovered reserves (which include deep gas) are optimistic. What is indisputable, though, is that most of the future shallow water reserve additions will be made by independent oil companies. These firms, which account for most of the 150 or so operators currently active in the GOM region, already do most of the drilling, record most of the discoveries and build most of the offshore platforms in the sub-province.

Independents played a crucial role in the development of GOM shallow water resources, accounting for 60 percent of total bonus payments for shallow water acreage and slightly over 50 percent of cumulative GOM shallow water output.[57] Largely as a result of their presence, the degree of concentration prevailing in the shallow water province has always been modest (Table 3.4),[58] much lower than in other offshore provinces outside the USA although never quite as low as it is in the ultra-atomised onshore US states like Texas or Louisiana. Indeed, no single player has had a dominant presence throughout the shallow water province at any point in its development. Moreover, the jockeying for position by large and small firms across the sub-province has been quite intense throughout the years. This, in turn, is indicative of low barriers to entry and exit, and the resulting vigorous competition

Table 3.4: Concentration Indices* for GOM Hydrocarbons Production, 1947–2005

Year	*Crude oil and condensate*	*Natural gas and NGLs*
1947–1995	9.50	19.29
1996	13.17	25.03
1997	11.25	25.08
1998	8.29	19.97
1999	8.74	17.15
2000	7.45	17.30
2001	7.24	16.14
2002	6.61	15.29
2003	7.29	18.22
2004	8.11	20.53
2005	7.89	18.20

* Inverse Herfindahl Index

Source: MMS

that has characterised the market for shallow water leases throughout its history.

Of late, the relative share of independents in the rapidly decreasing shallow water output has grown even more, as major oil companies drastically re-dimensioned the scale of their operations in this sub-province from 1996 onwards. Despite the rise in the GOM-wide level of concentration that this development has entailed, most analysts have interpreted it as being unambiguously positive for smaller independent companies. The following lines, penned at a time of rapidly rising prices, distil the conventional view on the matter:

> With the price of oil around $30/bbl, now is an excellent time for producers to invest in shelf ventures to generate even more cash to grow their companies. To milk this cash cow, they will have to have a three-legged stool: some money to buy the old fields that the majors are selling off, some 3D seismic data and a talented, multidisciplinary asset team. Then they will need a pail, which in this case is a way to tap into the existing infrastructure to fully deplete these shelf reservoirs. Once they have the stool and the pail, they can make money 'til the cows come home.[59]

Oil prices have nearly doubled since this analysis was published, so one would be excused for thinking that the prospects of small E&P outfits would be so much the brighter. Unfortunately, this rosy view of their future neglects to mention one important detail; namely, that there is every indication that the aforementioned cows may be just about to set off for the abattoir.[60]

NOTES

1 As Kerr-McGee (the company that is now remembered as the original GOM offshore pioneer) put it: 'We decided to explore the areas where the really potential prolific production might be – salt domes – the good ones on land were gone, but we could move out into the shallow water and, in effect, get into a virgin area where we could find the real class-one type salt dome prospect' (Ezell 1979: 154).

2 Pratt (2004: 23) points out that 'shifting ocean sediments caused by earthquakes had been known to break telephone cables on the ocean floor, and as early as 1950, oceanographic consultants had studied the possibility that unburied offshore pipelines might move during hurricanes'. However, the designers of oil platforms 'had not appreciated that, under certain conditions, mudslides might pose catastrophic threats to platforms.' This particular phenomenon was first encountered after the passage of Hurricane Camille (a '400-year' storm), in August 1969. The hurricane produced some waves that were 70–75 feet high (when conventional wisdom had it that hurricane

waves would seldom if ever, exceed a height of 20 feet). The hurricane caused USD 100 million worth of damage, and destroyed a number of facilities, including three state-of-the-art platforms (one of which, in South Pass block SP70, was only five months old at the time, and held the world's depth record at 300 feet of water). These platforms had been installed by Shell, universally acknowledged as the leading company in offshore design at the time. The platforms had been designed to withstand 100-year waves, but mudslides caused by the storm lowered the ocean floor and effectively placed them in deeper waters, whereupon the structures toppled over.

3 Pratt, Priest and Castaneda 1997: 13.

4 *Ibid.*

5 The Creole platform, a triumph of ingenuity for its time (1937–8), rested in 15 feet of water. It was an artificial island of sorts, supported by 300 piles. Despite its imposing size (55 x 100 metres), it was seen – correctly – as being too flimsy for living quarters to be sited there, especially given the lack of speedy and reliable maritime transportation and a hurricane warning system; thus, its workforce had to undertake a 26-mile round trip every day, always in the hope that rough seas would not prevent their debarkment (an expensive eventuality, since the wells could not produce without a crew). At its peak, Creole produced around 4 million barrels of oil per year. On 18 March 1938 (on the same day that the Mexican government decreed the expropriation of US and British oil properties in Mexico and the first successful discovery well in Saudi Arabia came on stream), the crew in Creole completed the first successful GOM offshore well (albeit still within sight of land). Then, in 1940, the Creole platform became the first offshore structure in the Gulf to survive a hurricane (although a rather small one).

6 Pratt, Priest and Castaneda 1997: 7.

7 Pratt 2004: 15.

8 Gramling 1995: 55.

9 Pratt 2004: 11.

10 More than 300 of these leases were located beyond the 3-mile boundary limit. A further 50 straddled the state–Federal boundary line.

11 Pratt, Priest and Castaneda 1997: 30.

12 Priest 2004: 33. In addition, the majors were responsible for over 75 percent and 80 percent of wells drilled and discoveries, respectively, in Texas and Louisiana state waters.

13 See Bartley 1953; Fitzgerald 2001; Boué 2002. According to sources quoted by Christopher (1953: 24), 900,000 words were spoken during the five weeks that the US Senate spent debating the bill that eventually became the Submerged Lands Act.

14 Kreidler 1997: 99.

15 Oil companies and oilmen as well as individuals otherwise closely connected to the industry had been Eisenhower's prime source of funding in the 1952 presidential campaign. The Eisenhower administration, in the person of Secretary of Interior Douglas McKay, went out of its way to

repay this support by greatly expediting private access to lands in Federal lands. Such was his zeal that a Congressional investigation into the matter ensued (Strohmeyer 1993: 34). The passage of both SLA and OCSLA is consistent with the desire on the part of this administration to obtain more highly prospective acreage that it could make available to oil industry supporters at a pace that suited them.

16 Out of the 90 tracts leased in this first ever sale, 26 remain active and continue producing oil and gas. Throughout their lifetime, those 26 leases have produced 508 MMB of crude and 1943 MMCF of natural gas.

17 Pratt 2004: 22.

18 Priest 2004: 42.

19 *Ibid.*

20 The president of the American Association of Oil Well Drilling Contractors wrote in 1959: 'the rapid rise and correspondingly rapid decline in offshore drilling operations in the Gulf of Mexico is one of the most surprising phenomena which has occurred in the oil business in many years' (*ibid.*).

21 Around 26 percent of the wildcat wells drilled in the USGC offshore until 1956 were productive, compared to the US onshore average at the time of 11 percent. Indeed, oil companies discovered an average of nearly 38 million barrels of oil for every wildcat well they drilled, a success rate that completely eclipsed that of US onshore fields (*ibid.*: 33).

22 The market-demand pro-rationing practices of Louisiana and Texas were applied in the GOM Federal OCS until 1970 inclusive (McDonald 1979: 178). The production allowable for a 10,000 ft. deep well offshore (242 barrels) was nearly double the allowable for a comparable well onshore (132 barrels). By encouraging a greater spacing of development wells, this reduced field development costs in a way that compensated for the higher costs of individual wells (Pratt, Priest and Castaneda 1997: 39).

23 For more details, see Gramling 1995: 53–85; Pratt, Priest and Castaneda 1997: 34–70.

24 Anderson 1984: 148.

25 Kreidler 1997: 197.

26 *Ibid.*: v.

27 *Ibid.*: 205.

28 *Ibid.*: 203–4.

29 *O&GJ*, 1 April 1963: 78–9. The largest bonus outlays during the 1962 sale, for instance, came from SONJ affiliate Humble (USD 63.1 million), Gulf (USD 46.6 million), and Shell (USD 45.5 million). The largest outlay by an independent came from Tenneco (USD 43.3 million).

30 The account of the 1960 and 1962 sales can be found in Priest 2004.

31 In actual fact, it took two days to read the bids because the sale was split (Priest *ibid*: 47). On the first day, 401 tracts were offered and of these, 206 were leased for cash bonuses totalling over USD 177 million. On the second day, 410 tracts were offered and 195 were leased for cash bonuses totalling USD 269 million. Notwithstanding the fact that the average bid received for a tract offered during the second day was higher than the average for

a tract offered during the first, the idea of holding this type of split sale as a means of increasing government revenue did not take hold in DOI, and the two-day procedure was never to be used again.

32 As Priest (2004: 50) points out, the results of the 1962 lease sale alerted the Federal government to the fact that the OCS leasing programme, managed by no more than 30 employees (some of whom did not even devote their full time to it), had taken in more money in a single sale than the combined total of all the timber sales in Oregon and California, plus the proceeds from onshore mineral leasing for that year. The time would come when the signature bonus paid for *individual* OCS blocks would exceed the total from timber sales and onshore leasing.

33 Gramling 1995: 72. The rate of technological progress and change was so rapid that, at the time, it was remarked 'that mobile platforms termed revolutionary only a year ago, may already be considered conventional' (quoted by Veldman and Lagers 1997: 56).

34 Out of the 54 giant fields discovered in the USA between the end of the Second World War and 1970, 23 lay in the GOM Federal OCS (McKellar 1998).

35 Out of the 420 leases assigned after the 1962 sale, 252 (60 percent) were in production by 1969 (the figure for the previous four sales was 178 productive leases out of 410 assigned). As of 1968, 14 of the 62 giant fields discovered in the United States were offshore Louisiana, and 11 of those 14 lay either wholly or partially within federally administered areas.

36 Pratt, Priest and Castaneda 1997: 91.

37 Lore 1992: 161.

38 *Ibid.*: 160–1.

39 Put somewhat simplistically, two-dimensional seismic generates images resembling vertical slices through the earth, whereas three-dimensional seismic generates images that resemble a cube cut from the earth.

40 Pursell 1998: 10.

41 *Hart's E&P*, April 2001: 49.

42 EIA 2000: 4.

43 NPC 2003, v. II: 155.

44 See Pursell 1998: 11.

45 Within the shallow water province, 31 gas fields with reserves of more than 1 TCF have been found (most of which are non-associated gas).

46 NPC 2003, vol. III: II–11.

47 Anderson 1993: 88.

48 *Ibid.*: 90; italics ours. Also, light gasoline ratios in 1972 vintage oils were generally greater than those in 1985 vintage oils, which in turn were smaller than those found in 1998 vintage oils.

49 *Ibid.*: 88

50 Indeed, the study was not unlike an exploration drilling effort. It had the unique feature of offering academics the chance to test models and visualisation results at first hand, in the field, through the location and drilling of wells 'to verify that … data interpretations are correct and to discover a

mechanism to extract hydrocarbons from ... migration streams efficiently' (*ibid.*: 87). A more complete discussion of the drilling programme and its modest results is found in Anderson *et al.* 1994.

51 *Ibid.*: 87.

52 This theory was originally put forward by Soviet academics (see Porfir'yev 1974), but has recently been given a new lease of life by Gold (1999). It has never been taken seriously in Western academic circles, although some observers (Odell 2001: 199) still believe that its dismissal might have more to do with the stigma of its Soviet pedigree than with its inherent unsoundness.

53 Anderson *et al.* 1994: 103.

54 Laherrère 2003: 239.

55 *AAPG Explorer*, October 2000. From 1983 to 1998, for instance, the average reserve additions per field from older fields were 27 percent larger than the average size of new field discoveries (a margin that increases to 75 percent if the very numerous negative revisions to older fields are not taken into consideration). Both the average size and the number of new field discoveries declined significantly between 1991 and 1998, compared to the previous seven years. The decline in the average size of new discoveries was nearly universal across all areas, with the exception of the East Louisiana area, where the average size of new discoveries actually doubled.

56 *AAPG Explorer*, October 2000.

57 Boué 2002: 78.

58 The Inverse Herfindahl Index is an indicator of the number of equal-sized firms active in a market or industry.

59 *Hart's E&P* April 2001: 71.

60 As MMS itself points out, 'in the absence of primary lease term extensions, all active [i.e. leased out but non-producing] shallow water leases will expire before 2008' (Baud *et al.* 2002: 44).

CHAPTER 4

THE SHALLOW WATER SUBSALT

The year 1993 is now universally recalled as having marked the beginning of the revival of GOM as a vibrant petroleum province. What is not so well remembered is that this recovery was seen as resting on two quite distinct pillars: increased production in the deepwater province, on the one hand, new production from the shallow subsalt province, on the other hand. High hopes were pinned on the latter play in the wake of a string of significant discoveries,[1] but reality has not matched hopes.

4.1 Antecedents

For a long time, salt bodies in the GOM were thought to be rooted in extensive and incredibly thick salt deposited during Jurassic times. Thus, GOM oil and gas wells were generally stopped as soon as the drill bit encountered salt, because it was assumed that there could be no oil-bearing sediments below it. By the early 1980s, though, advances in geological science had gradually made it clear to the oil industry that a large proportion of the sediments in the GOM region, more than 35,000 square miles in extension, was actually covered not by salt sheets but by colossal tabular salt diapirs (tongues, nappes, canopies) that had been extruded from underlying sheets (hence their designation of 'allochthonous').[2]

As Table 4.1 indicates, Placid Oil drilled the first genuine subsalt well in Ship Shoal SS366 block in 1983. This well penetrated only 295 feet of subsalt sediments, but it nevertheless encountered three separate salt intervals. Three years later, after a further handful of subsalt wells had been drilled by a variety of companies, a Diamond Shamrock subsalt well in South Marsh Island block SMI200 encountered a massive interval (800 feet net thickness) of reservoir-quality sandstones (with porosities in excess of 30 percent and permeabilities approaching 2000 millidarcies), which unfortunately proved to be water wet. This well provided the industry with decisive evidence that excellent reservoir quality sands could be found underneath salt bodies. At that point, though, oil companies that found themselves drilling a subsalt well still did so in

Table 4.1: Shallow Subsalt Wells Drilled in the GOM Federal OCS, 1983–2004

Date	*Area*	*Block*	*TVD (feet)*	*Water depth (feet)*	*Operator*	*Prospect name*
10/18/83	Ship Shoal	366	8,203	453	Placid	–
4/18/84	Garden Banks	171	10,597	670	Marathon	–
5/27/84	Green Canyon	98	13,159	853	Conoco	–
9/12/84	West Cameron	505	18,500	138	Gulf	–
9/12/85	High Island	A 374	15,000	362	Mobil	–
1/29/86	South Marsh Island	200	13,500	475	Diamond Shamrock	–
1/7/87	Vermilion	412	10,496	471	Mobil	–
12/18/87	Vermilion	356	17,000	265	Amoco	–
3/20/92	Garden Banks	165	18,000	724	Chevron	–
9/4/93	Ship Shoal	349	16,563	370	Phillips	Mahogany
10/10/93	South Timbalier	260	16,611	372	Phillips	–
11/17/93	South Marsh Island	169	18,020	288	Amoco	Mattaponi
5/23/94	Vermilion	349	16,146	237	Anadarko	Mesquite
6/20/94	Ship Shoal	349	19,101	375	Phillips	Mahogany
6/20/94	Garden Banks	128	18,454	705	Shell	Enchilada/Elmer
8/6/94	Ship Shoal	360	19,000	393	Unocal	Rhino
8/27/94	Ship Shoal	250	17,750	190	Japex	–
9/21/94	Ship Shoal	368	17,500	457	Amerada Hess	Citation
10/6/94	South Timbalier	289	18,034	398	CNG	Cypress
4/12/95	Ship Shoal	359	19,665	375	Phillips	Mahogany
5/13/95	Garden Banks	127	14,730	630	Shell	Enchilada/Elmer
6/30/95	Vermilion	308	20,399	200	Amoco	South Ana
2/16/96	Ship Shoal	361	16,163	405	Phillips	Agate
3/9/96	Ship Shoal	359	19,094	372	Phillips	Mahogany
3/19/96	Ship Shoal	337	17,851	295	Phillips	Alexandrite
5/23/96	South Marsh Island	97	9,643	181	Pennzoil	–
8/21/96	Vermilion	375	16,856	318	Anadarko	Monazite
10/29/96	Ship Shoal	350	16,422	310	Vastar Resources	–
11/14/96	Garden Banks	128	16,535	663	Shell	Enchilada/Elmer
1/30/97	Ship Shoal	357	20,610	420	Amerada Hess/Oryx	–
7/28/97	South Timbalier	299	17,540	289	Challenger	–
10/18/97	Ship Shoal	359	16,358	370	W & T Offshore	–
3/25/98	Garden Banks	127	16,909	633	Shell	Enchilada/Elmer
5/31/98	Eugene Island	346	11,833	314	Anadarko	Tanzanite
10/4/98	Grand Isle	116	21,600	323	Anadarko	Hickory
11/15/98	Garden Banks	128	17,197	633	Shell	Enchilada/Elmer
12/20/98	Ship Shoal	359	18,757	372	Phillips	Mahogany
2/17/99	Grand Isle	110	21,540	323	Anadarko	Hickory
5/12/99	Grand Isle	72	9,400	116	Walter Oil and Gas	–
8/4/99	Eugene Island	346	15,755	318	Anadarko	Tanzanite
8/6/99	East Cameron	347	15,814	296	Anadarko	–
9/6/99	East Cameron	347	14,313	298	Anadarko	–
3/28/00	Grand Isle	116	21,475	326	Anadarko	Hickory
5/1/00	Grand Isle	106	19,204	338	BHP Billiton	–
7/6/00	Grand Isle	116	18,000	323	Anadarko	Hickory
9/28/00	Grand Isle	111	21,057	320	Anadarko	Hickory
11/3/00	Grand Isle	110	21,269	320	Anadarko	Hickory
3/3/01	Garden Banks	272	22,047	560	McMOran	–
4/20/01	Ship Shoal	359	18,678	370	W & T Offshore	–
5/20/01	Ship Shoal	359	18,813	370	W & T Offshore	–
7/12/01	East Cameron	185	12,409	93	Remington O&G	–
11/25/01	East Cameron	179	11,670	95	Remington O&G	–
12/11/01	East Cameron	179	12,226	95	Remington O&G	–
12/29/01	South Timbalier	308	19,884	484	Anadarko	Tarantula
7/25/02	Grand Isle	116	16,620	325	Anadarko	Hickory
10/10/02	Grand Isle	110	22,204	325	Anadarko	Hickory
12/5/02	South Timbalier	308	20,960	484	Anadarko	Tarantula

Source: MMS

the main because they thought that the anomalous seismic reflections given off by salt were caused by the presence of hydrocarbons.

The late 1980s saw the drilling of a handful of subsalt wells, at depths ranging between 140 to 670 feet. Lackadaisical as this effort might appear, in retrospect it can be seen that it amounted to a radical about-face in the industry's attitude towards subsalt drilling, as companies singled out salt accumulations as prospective targets, instead of merely encountering them while looking for something else. By 1990, one of these more 'intentional' wells (drilled by Exxon only a fortnight after Shell had announced its major Mars deepwater discovery) struck subsalt oil for the first time in the history of GOM exploration, and in what appeared to be reasonable quantities (reserve estimates ranged between 50 to 200 MMBOE). Unfortunately, this prospect (nicknamed Mickey) lay beneath 4300 feet of water (to say nothing of 3000 feet of salt), which made its development a very expensive proposition (by the end of 2005 it had yet to receive a commercial development decision, extraordinarily high oil prices notwithstanding). Two years later, Chevron drilled a well in Garden Banks block GB165 that, in the eyes of the geological fraternity, was of 'major historical significance' because it 'demonstrated that unprecedented thicknesses of salt could be drilled with continued penetration of a highly prospective underlying clastic section'.[3] The well (which penetrated 6950 feet of salt, and tested approximately 5150 feet of subsalt section at a depth of more than 15,000 feet) would have been even more significant had the prospective underlying sediments harboured any oil but, alas, they did not. Thus, in exchange for a vast expenditure, Chevron was left with the scant consolation of having secured a very minor footnote in the annals of offshore petroleum prospecting.

4.2 Breakthrough?

The persistence of the American oil industry with the shallow subsalt play did not pay off until 1993, when Phillips Petroleum finally struck oil in Ship Shoal block SS349 (a prospect later baptised as Mahogany). Mahogany, with a peak production rate of 33 MBD of oil and 10 MMCFD of gas, became the first commercial subsalt oil development in GOM when it came on stream in December 1996.[4] Just as importantly, it gave rise to a frenzy of activity in the shallow subsalt province: in the four years after the Mahogany discovery, 15 companies drilled nearly 30 wildcat subsalt wells, and made 11 discoveries.[5]

In the lease sales held after the Mahogany discovery, values for leases located in proximity to subsalt prospects soared. Ship Shoal Addition block SS337, located updip from Mahogany, was leased in 1994 for USD 40 million (a sum equivalent to 50 percent of the total bonus payments received by MMS less than two years before!). The number of bids that this block received – nine – was also without precedent in the acreage auctions held since 1983.[6] Nor was it alone in attracting so much attention: there were seven bids for Vermillion South block VE375, six bids for East Cameron block EC 357 and five for Vermillion South block VE295. None of the latter blocks came close to reaching the auction value of USD 7000 per acre that Anadarko paid for SS337, however. This figure, a record for a post-1983 auction, conveys well the desperation underlying the efforts of second-tier majors like Phillips and Amoco or first-tier independents like Anadarko to redress the damage suffered by their E&P portfolio by their having effectively sidelined themselves from the GOM deepwater, which with hindsight came to be appreciated as a most prospective exploration play. It is also illustrative of the risks inherent in offshore exploration: the SS337 block harboured a structure covering 3500 acres, but Anadarko and its partners in the venture (Phillips, Amoco) ended up by spending USD 11 million on a wildcat well that could not find hydrocarbons in commercially exploitable amounts.

Between 1994 and 1996, success rates in the shallow subsalt province were reasonably attractive (approximating 40 percent) but, nonetheless, oil companies soon began to realise that 'structural complexities, seismic uncertainties, and drilling difficulties associated with subsalt exploration made the play very high risk'.[7] Thus, by mid-1997, the *Oil and Gas Journal* was reporting that 'the euphoria touched off by a string of early discoveries [had] given way to a more subdued pragmatism, as operators [struggled to] cope with some very real challenges'.[8] The first of these had to do with the technical complications that companies encountered in drilling through shallow formations deformed by the migration of salt: lost circulation of drilling mud due to faulting and rubble zones, excessive wear and tear of equipment, corrosion due to carbon dioxide and hydrogen sulphide, high downhole temperatures and abnormal pore pressures, to name but the most important.[9] For instance, even though the wildcat in Anadarko's Teak prospect was sunk at a time when the industry had already gone some way up the learning curve, the (unsuccessful) well ended up by costing USD 24 million instead of the envisaged USD 12 million, chiefly because drilling it consumed eight months of rig time (plus a further two for testing), instead of the 3–4 months it was supposed to have taken.

Second in the list of problems was the interpretation of subsalt seismic data, a task that notwithstanding the advent of advanced (and very dear) supercomputing capabilities was fraught with great difficulties. Even today, imaging a relatively uncomplicated subsalt structure may take as much as three years; imaging less straightforward structures can easily consume 50 percent more time. The pitfalls of subsalt imaging can be put into stark relief by the travails of Anadarko/BHP at their Monazite prospect in Vermillion block VE375, where they sank a well that revealed multiple pay zones but which had to be plugged and abandoned due to unforeseen problems related to the nature of the sediments penetrated by the well.

The third challenge stemmed from the expense associated with the great depths to which susbalt wells had to be drilled. Figure 4.1 shows

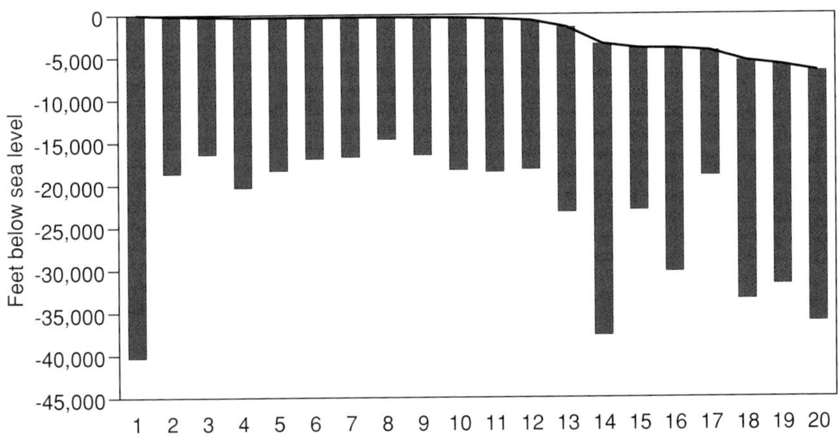

	Name	*Operator*		*Name*	*Operator*
1	Kola (USSR–Russia)		11	Tarantula	Anadarko
2	West Cameron 505	Gulf Oil	12	Enchilada	Shell
3	Mesquite	Anadarko Phillips	13	Conger	Amerada Hess
4	Lion/Emerald	BHP	14	Knotty Head	Chevron
5	Mattaponi	Amoco	15	Gemini	Chevron
6	Teak	Anadarko	16	Tahiti	Chevron
7	Kingfisher	Vastar	17	Mickey	Exxon
8	Tanzanite	Anadarko	18	Catahoula	Texaco
9	Monazite	Anadarko	19	Thunder Horse	BP
10	Mahogany	Phillips	20	Loyal	Texaco

Sources: MMS, O&GJ, BP 2001

Figure 4.1: Water Depth and True Vertical Depth for Selected GOM Subsalt Wells (Shallow and Deep Water)

the true vertical depth achieved by a representative sample of subsalt wells, most of them located in shallow waters. To bring the depth figures on this graph into context, one need only recall that the average depth of all US onshore wells between 1987 and 1998 was only 4900 feet.[10] The depths of most of the wells on the graph are similar to those of deep gas wells in the onshore Permian basin and significantly exceed typical values for wells in Mexico's onshore Reforma basin, which for a long time after its discovery in the early 1970s was seen as the most celebrated example of a very deep oil-producing province (the producing horizons of the supergiant A.J.Bermúdez field, for instance, are located at a depth of 14,000 feet). Indeed, the depths reached by some GOM subsalt wells bear comparison to the deepest drilling depth ever attained (a geophysical research hole drilled in Russia), particularly when one thinks that the latter hole was drilled onshore, in an admittedly inhospitable place (the Kola peninsula) but over a leisurely twenty-year interval.[11]

One final (albeit relatively minor) problem was the quality of the crude found in shallow subsalt wells. This turned out to be significantly lower than that of crude extracted from 'traditional' shallow water leases, and even that of most deepwater leases (Mahogany lease crude, for instance, has an API gravity of 25° and a sulphur content of 1.9 percent). In the end, this meant increased transportation costs for crudes from most shallow subsalt strikes, because they had to be taken to market through pipelines like Poseidon, dedicated to gather lower quality common streams from deepwater areas. Had the quality of shallow subsalt crudes been better, they might have been able to access pipelines dedicated to lighter crude streams (like the Eugene Island Pipeline System), which had significant amounts of spare capacity at their disposal.

The effect of some of the factors mentioned above on costs in the shallow subsalt province was quite significant. Drilling costs were most heavily affected: the pioneering wells in the province set their sponsors back a steep USD 35 million. This figure has been quickly slashed (by as much as half), but even after these improvements, shallow subsalt wells remain much more expensive to drill than traditional shallow water wells and only slightly less expensive than wells in the more amenable deepwater areas. Thus, as the list of subsalt disappointments began to lengthen, industry enthusiasm for the shallow subsalt play began to wane, and acreage prices suffered accordingly (in Lease Sale 152, held in May 1995, bidding for tracts in subsalt hot spots was distinctly underwhelming, with Anadarko – the self-designated leader of the subsalt pack[12] – buying only one block, and that in partnership with Marathon).

4.3 A False Dawn

Although drilling activity in the shallow subsalt resumed after the enforced hiatus brought about by the 1998 price crisis, it is nowhere near as healthy as it used to be before 1996. At its peak, hydrocarbons output in this sub-province reached 110 MBOE during June 2001 (Figure 4.2). The peak production months for oil and natural gas were May 2001 (37 MBD) and June 2001 (74 MBOED), respectively. Overall, the most important cumulative producer in the sub-province is Hickory, at 29.7 MMBOE, followed by Enchilada at 21 MMBOE and Mahogany at 18 MMBOE. Output at the first two of these fields has been mostly natural gas and NGLs (76 and 80 percent, respectively), while that of Mahogany has consisted mainly of oil and condensate (64 percent of the total). These peak production figures are unlikely to be surpassed any time soon because the oil industry as a whole seems to have become disenchanted with the shallow subsalt play.

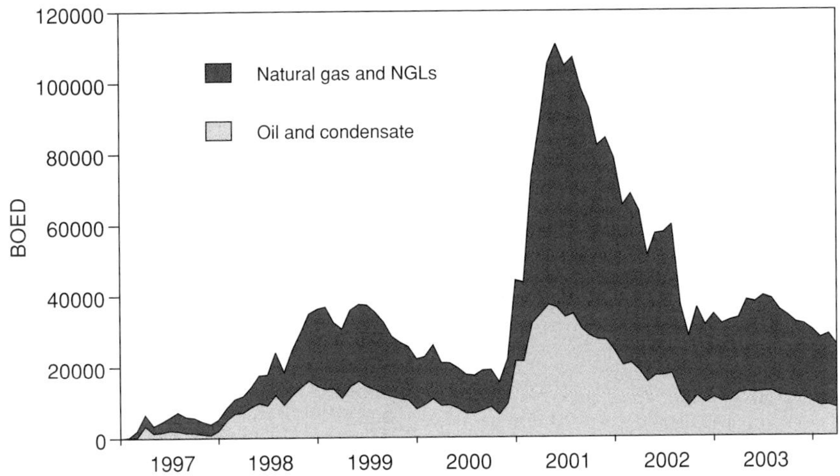

Source: MMS

Figure 4.2: Monthly Hydrocarbons Production in the GOM Federal OCS Shallow Water Subsalt Province, 1997–2004

Post-1998, Anadarko and ConocoPhillips were still the most prominent shallow subsalt operators. Indeed, Anadarko in particular ramped up its exposure to subsalt acreage through a 1999 agreement with Texaco giving it operatorship rights on 82 blocks, and a 2002 joint venture with BP giving it the option to earn a 33–66 percent working interest in 95 blocks in the Garden Banks and Keathley Canyon areas, most of them harbouring subsalt objectives. The company set itself the target of drilling 4–5 subsalt wildcats a year, and largely fulfilled this plan.[13] These actions suggest not only that Anadarko's post-1998 exploration strategy was essentially unchanged, but also that our assertion regarding the apparent loss of interest by the industry in the shallow subsalt might be wide of the mark. However, the fact is that Anadarko's strategy underwent a radical reorientation, with a shift in focus from the shallow subsalt to the deepwater (all the subsalt blocks involved in the agreement with BP lie at water depths of between 3000 and 6000 feet). The reason behind this was that a spate of very large subsalt deepwater discoveries (Thunder Horse, Neptune, Atlantis, Mad Dog) convinced the company that it could not afford to ignore the deepwater any more, and that its subsalt expertise could be put to more profitable use there.

Anadarko admitted that it had been late getting into the deepwater action because it did not believe either that 'economic production technology would develop as fast as it did' or that 'the [deepwater] sands would be as thick'.[14] The company's acknowledgement that it had been wrong on both these counts encapsulates the reason why the shallow subsalt has been relegated to the back stage. Simply put, while upstream activities in the shallow subsalt and the deepwater share a number of disadvantages – very high drilling costs, seismic interpretation fraught with complications, high leasing costs in certain areas – the former have been found lacking in the departments that make the latter such a powerful magnet for exploration budgets: size of finds and productivity of wells. In other words, Anadarko ceased to believe that the shallow subsalt had the potential to be the engine of the company's growth, and it even took to referring to its newest developments in the province (i.e. Tarantula) as bread-and-butter projects undertaken chiefly in the interests of having a balanced portfolio of riskier (i.e. deepwater) and more tractable properties. Anadarko never managed to achieve this balance, though. During August 2004, as part of a far-ranging divestiture programme, the company sold off its GOM exploration portfolio to Apache and Morgan Stanley, for USD 1.3 billion.[15] This denouement to the company's checkered history in the Gulf poignantly, by the way, illustrates the unacceptably high costs that many large independents and second-tier majors ended up paying as a result of their efforts to play catch up (in very disadvantaged circumstances) with the first movers in the deepwater sub-province. Why these companies had to play catch up in the first place is a critically important question, but one that will be tackled in a subsequent chapter.

What does the future hold for the shallow subsalt province? With its traditional (and many might say misplaced and even unrepressed) optimism, MMS announced at one point that it saw a resource potential of as much as 6–16 TCF of natural gas in the subsalt. ConocoPhillips, for its part, still considers that there remain to be found at least 1 billion barrels of oil.[16] However, the near-term development of these resources cannot be taken for granted, especially in the wake of Anadarko's high profile retrenchment in the shallow water GOM. As an *O&GJ* article written when the shallow subsalt drive was beginning to lose momentum put it: 'operators [are] stress[ing] caution will be the future watchword for the ... subsalt play, as they seek answers to the basic questions: Where can substantial reserves be found, and how economically can they be brought on stream'.[17] Larger companies seem to have made up their mind with regards to both of these questions, and their strategic compasses are pointing firmly away from the shallow subsalt, and towards the deepwater province. Smaller companies, for their part, have turned their sights back to the shallow waters, but not in order to drill subsalt projects. These companies are once again on the lookout for gas fields of a very special kind.

NOTES

1 In 1994, *Platt's* wrote that 'it is evident to all that the subsalt will be one of the driving forces in the region for the foreseeable future' (*PON*, 4 April 1994: 1).

2 Salt is a crystal and therefore is not compacted, so it retains its low density of 2.1g/cm^3 after burial. At depth, sediments surrounding the salt tend to compact and become denser than the salt. The salt, in contrast, will tend to flow, just like a glacier (which, incidentally, is also formed by crystals). The specific gravity of salt is less than that of rock, so salt will tend to move upwards if overlying sediments do not offer significant resistance or if extensional faulting in these sediments develops, in a flow that is plastic in nature and takes place over eons. Even if the sediments resist, though, salt may often push through, creating faults in the process.

3 Montgomery and Moore 1997: 875.

4 Camp and McGuire 1997.

5 In 1994, Shell Offshore, Pennzoil, and Amerada Hess discovered a field thought to contain 400 BCF of gas and 25 MMB of condensate in Garden Banks block GB 128. The field, named Enchilada, was brought on line in July of 1998. In 1996, Texaco and Chevron discovered the Gemini field (estimated reserves: 250–300 BCF of gas and 3–4 MMB of condensate) in Mississippi Canyon block MC292. In 1996, Anadarko and Phillips discovered the Agate field in Ship Shoal block SS361. In 1997, Amerada Hess discovered the Conger field in Garden Banks block GB216, developed

through a subsea tieback to the Baldpate platform. In July 1998, Anadarko announced a subsalt discovery (Tanzanite) at its Eugene Island block EI346, thought to contain 140 MMBOE. In that same year, Anadarko announced the discovery of the Hickory field in Grand Isle block GI116. The discovery well for this field, estimated to contain 400 MMBOE, was drilled to a total depth of 21,600 feet, and penetrated one of the thickest sections of salt ever drilled in GOM (8000 feet).

6 In auctions held under areawide leasing rules (1983–2002), offered blocks have attracted an average of only 1.4 bids per block.

7 Bascle, Nixon and Ross 2001: 44. The first major setback in the subsalt play came in the form of the Anadarko/Phillips Mesquite prospect, drilled in Vermilion block VE349 to a depth of 16,146 feet.

8 *O&GJ*, 14 July 1997: 21.

9 Ideally, salt must be drilled through in a relatively slow and controlled manner because vibrations within the salt can damage the bottom hole tool assembly. However, since salt tends to flow like plastic under conditions of high temperature and pressure, salt sections have to be drilled as quickly as is feasible. Otherwise, the vertical and lateral movement of salt can reduce the wellbore gauge in an open hole. Drillers have to pay close attention to drilling fluid weight when drilling through the lower portion of a salt section because, often, just underneath, there will be found a section of unconsolidated sediments (commonly referred to as 'gumbo') that can quickly absorb large amounts of drilling fluid when an over-balanced condition exists (in effect, the well 'blows into' the formation). But if an under-balanced condition is allowed to develop and the fluids in the unconsolidated zone are highly pressured, this may lead to formation flow into the wellbore (raising the risk of a 'blow out'). Quite apart from the difficulty in achieving this balancing act, drilling through great thicknesses of salt requires the use of very expensive special drilling muds that prevent water loss to the formation, on the one hand, but at the same time prevent the wellbore wall from being dissolved by undersaturated fluids. Additional drilling costs are also incurred because subsalt wells require an extra string of heavy-walled, intermediate casing set through the salt in order to counteract the forces created by the tendency of salt to flow. Setting casing can become a problem if salt ledges or washout zones occur in the borehole. Salt ledges can hang up centralisers, while washout zones can prevent good cement bonding between the casing and the formation, leading to nonuniform loading on the casing. When casing strings have been cemented in place for long periods of time, salt creep can bend, stretch, and shear them. In subsalt producing wells, casing and production tubing through the salt interval can shift significantly in a lateral direction. This lateral movement can create significant problems for well workovers.

10 Gerking, Kunce, Kerkvliet and Morgan 2000: 142.

11 This geophysical research hole was drilled between 1970 and 1994.

12 At that time, Anadarko had 60 tracts with subsalt prospects.

13 *Hart's E&P*, November 1999: 20.

14 *Ibid.*
15 *PON* 23 August 2004: 1–5.
16 Natural gas accounts for about 60 percent of the shallow subsalt hydrocarbons discovered to date.
17 *O&GJ*, 14 July 1997: 21.

CHAPTER 5

THE DEEP GAS PROVINCE

The deep gas sub-province is the only one within GOM whose spatial, temporal and economic boundaries can be demarcated with complete precision (it includes any productive gas well drilled in a lease lying in up to 650 feet of water and wholly west of 87° 30' West Longitude,[1] *drilled* after March 2003 to a true vertical depth of at least 15,000 feet below sea level, and achieving production before 2009[2]). These limits were fixed by administrative fiat, in response to a readily identifiable problem (namely the serious and widening imbalance between US natural gas supply and demand), with a clear objective in mind (using royalty relief to accelerate the exploration and production of non-deepwater gas in GOM), and taking into consideration some very specific political constraints (namely the impossibility of granting across-the-board royalty relief to all gas-producing leases in the GOM shallow water without triggering the pay-as-you-go provisions of the Budget Enforcement Act).

5.1 Antecedents

At the end of the 1990s, US demand growth rates for gas were expected to accelerate sharply (chiefly because 95–98 percent of the electricity generating capacity to be installed up to 2010 was supposed to be gas-fired, as a result of the long-delayed impact of Clean Air Act requirements enacted in the early 1990s). According to an extremely influential report that the National Petroleum Council (NPC) put in the public domain in 1999, total gas demand was expected to grow to 29–30 TCF by 2010 (compared to the 22.5 TCF registered in 2000) and 35 TCF by 2025, a more than twofold increase over the 16.2 TCF recorded when it reached its nadir in 1986.[3] At the same time, gas availability (US domestic output plus pipeline imports from the Western Canada Sedimentary basin) was forecast to experience slight but steady year-on-year declines, until such time as new supplies from the GOM deepwater province, Alaska and the McKenzie Valley were available to turn the tide. Unfortunately, even according to the most optimistic scenarios,[4] these new sources of gas would not come on stream before

2006, which left US energy policymakers staring down the barrel of a modest supply gap for the years 2000–2006.

A mild sense of foreboding in the US political establishment must have turned to near panic after the extreme price spike of January 2001, during which natural gas prices temporarily reached USD 10/MMBTU (a year-on-year increase of USD 7.50/MMBTU).

This event led policymakers to conclude that, 'without dramatic change in exploration and development patterns, production from the GOM may not be able to meet the expected share of future natural gas supply needed ... to meet growing demand'.[5] Furthermore, and more or less concurrently with this price spike, other complications began to loom large on the horizon. Firstly, there was the realisation that the convoluted approval process for long distance pipelines in both the USA and Canada would inevitably take on-stream dates for key projects significantly beyond 2006 (indeed, at the time of writing, there appears to be only the remotest chance that Alaskan gas may reach consumers in the Lower 48 through a pipeline before 2016). Secondly, there was the strong growth in Mexican natural gas net imports, which are sourced from southern Texas.[6] Finally, there was an unexpectedly modest expansion in Canadian export availabilities (the result among other things of rapid output decline in the much-vaunted Ladyfern field in British Columbia[7]), which by 2002 actually deteriorated into the first ever annual gas production decline for the Western Canada sedimentary basin. This was by far the most worrying development of all, as Canadian gas imports had satisfied three-quarters of the total incremental US gas demand over the 1986–1999 period (during this time, Canadian production grew twofold and exports to the USA grew fourfold).[8] Moreover, Canadian demand for natural gas was expected to grow strongly, not least because of the expansion of production in the Alberta oil sands.[9]

Throughout energy industry and government circles in the USA, there was a clear understanding that the factor making the greatest contribution to the overall tightness in the natural gas market was the accelerating pace of reserve depletion in the GOM shallow water province (proven GOM gas reserves declined from 46 TCF in 1986 to 24 TCF by 1999), and the precipitous decline in GOM shallow water output (from 4.8 TCF in 1997 to 3.4 TCF by 2002). Industry officials pointed out that there was a silver lining to this particular cloud; namely, that only a small fraction (around 5 percent) of GOM wells had ever breached the 15,000 foot mark, a depth beyond which enormous volumes of sedimentary rocks – which MMS saw as containing anything between 5 and 20 TCF of recoverable gas reserves (with the likeliest figure being put at around 10 TCF[10]) – lay largely unexplored. Indeed, up to 1998 inclusive, a total of 724 producing deep gas wells had been drilled in the GOM Federal OCS (out of a US-wide total of 1097 wells drilled to 15,000 feet or more), with 23 of these drilled to depths of 20,000 feet or more (Table 5.1).[11]

Table 5.1: Number of Deep Producing Wells in the GOM Federal OCS to 1998

Depth interval (thousands of feet)	*Total wells*	*Gas wells*
15–16	502	316
16–17	302	204
17–18	163	118
18–19	64	41
19–20	33	22
20–21	12	6
21–22	11	7
22–23	7	7
23–24	3	3
TOTALS	1,097	724

Source: Dyman and Cook 2001: 6.

Opening up this apparently uncharted frontier located on the industry's doorstep seemed to offer the best short-term opportunity for achieving the large reserve additions and flow rates necessary to increase the volume of gas production from the GOM Federal OCS during the period 2001 through 2006, not least because the abundance of shallow water platforms, production and separation facilities and pipelines would make it possible for incremental 'deep gas' to reach the market very quickly. The industry told MMS that all the agency would have to do in order to expedite the pace of deep gas development would be to provide 'certainty and stability, and incentives that are predictable and transparent'.[12] Moreover, MMS would not find it politically difficult to grant such incentives, because the fact that no gas was being produced at the time from deep wells in the GOM Federal OCS outside of the Mobile Bay administrative division meant that the royalty holiday would not impact royalty income adversely, which meant in turn that it would not run afoul of the pay-as-you-go provisions of the Budget Enforcement Act that allow a point of order to be made against any legislation that either increases federal mandatory spending or reduces mandatory offsetting receipts. At the same time, the fact that *some* deep gas was already being produced in Mobile Bay made the granting of fiscal incentives to search for and produce more such gas sound an even more attractive proposition.

5.2 First Taste of Difficult Deep Gas: the Offshore Norphlet Formation

The American oil industry drilled around 300 deep offshore wells between 1950 and the late 1970s. Around half of this total was drilled during the decade of the 1970s, spurred by the price incentives that the regulatory framework then in force gave to gas produced from deep wells. Then, during the early 1980s, some companies managed to find highly prospective deep sediments offshore Alabama, in both state and Federal waters. Between 1981 and 1984, a total of 17 offshore wells were spudded in the deep Jurassic Norphlet trend, a formation well known from onshore deep wells. Despite the crudeness of the 2-D seismic technology available at that time, no less than 13 of these wells became significant gas discoveries. Indeed, these deep wells uncovered around 2.5 TCF of reserves in 24 pools, for an average discovery size of 105 BCF per pool (by way of comparison, GOM deep gas wells drilled outside the Norphlet trend during this period discovered 479 deep gas pools with 7.5 TCF of reserves, for an average of 15.7 BCF per pool).

The offshore Norphlet formation was discovered by Mobil in 1978, after a particularly long and painful gestation period (the company had leased four tracts in Alabama state waters back in 1969 but only managed to start drilling nine years later, after two appearances in Federal courts and one in state courts, and the posting of a USD 55 million bond meant to ensure that Mobil would discharge no effluents whatsoever in the waters of Mobile Bay[13]). The discovery had a fortuitous aspect to it, in that it was only made because Mobil decided to carry on drilling for an additional mile after it had reached its original depth target (as a result of this, the well ended up costing Mobil USD 20 million, instead of the initially estimated USD 7 million).

Mobil's extraordinary patience with its Alabama leases, and its resolve in the face of mounting drilling costs, reaped a handsome reward: the discovery well of the Mary Ann field struck up to 283 feet of net play (average 100 feet) in rocks that were both highly porous (up to 20 percent) and permeable (1 millidarcy). The gas found in the Mary Ann field had a very high (9 percent) hydrogen sulphide content, an unfortunate characteristic shared to a greater or lesser extent by all Norphlet fields found since, and which places significant processing challenges on the developers of these resources.[14] Meeting these challenges proved very costly. Indeed, drilling and processing costs vastly exceeded original expectations: most onshore Norphlet fields are only slightly over-pressured,[15] so the pressures encountered offshore probably took the industry by surprise. However, none of this was seen by the industry as an insurmountable problem because, just as the first Norphlet fields were being developed, the regulated price of deep gas under NGPA Section 107 ranged between USD 7–9/MMCF. Unfortunately, by the time the first wells finally came on stream, the newly deregulated natural gas prices had fallen to only USD 2.5/MMCF.[16] Ultimately, this made most of the Norphlet discoveries 'uneconomic to drill and produce' but, in the opinion of the MMS, they were produced nevertheless, 'probably to meet long-term gas [supply] contracts' (Figure 5.1).[17]

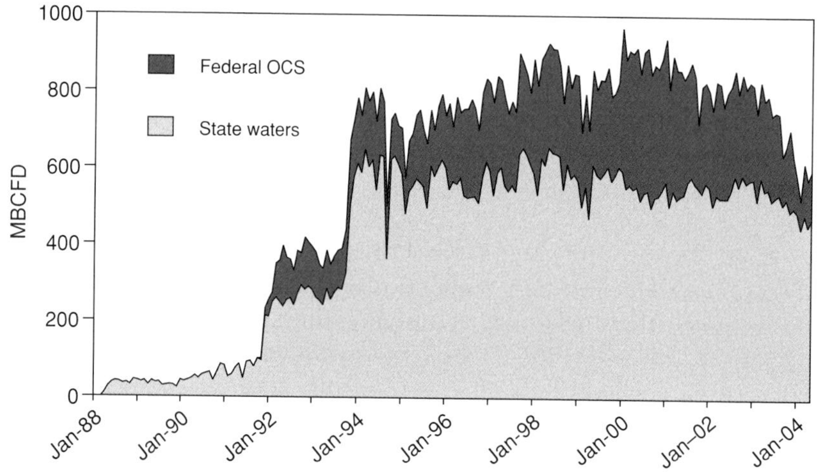

Sources: ASOGB, MMS

Figure 5.1: Monthly Gross Gas Production from Offshore Deep Wells in the Norphlet Formation, 1988–2004

At the time of its discovery, the Norphlet formation was hailed as a potentially colossal play, extending east from Louisiana across Mississippi and Alabama and eastwards all the way to the Destin Dome area offshore the Florida Panhandle. A high E&P official at Exxon, a company not associated with pie in the sky fantasies, once waxed lyrical in print about how the resources of the 'offshore Norphlet trend may someday surpass Prudhoe Bay's 26 TCF'.[18] Exxon, moreover, put its money where its mouth was: prior to its merger with Mobil, this company accounted for around 40 percent of the cumulative industry expenditure in Norphlet leases in both Alabama and Federal waters. Exxon's total exploration and development expenditure in the offshore Norphlet comes to around USD 1.2 billion (USD 600 million in processing facilities and 600 million in wells). This is equivalent to around 30 percent of the USD 4 billion plus that the oil industry has dedicated since 1981 to developing, treating and bringing to market these reserves.[19] Mobil's Norphlet development expenditures, in turn, can be estimated at *circa* USD 1.6 billion. Hence, ExxonMobil's combined share of Norphlet development expenditure could be as much as 70 percent of the total.[20]

According to Wade, Plater and Kelley, ExxonMobil's very high profile in offshore Norphlet operations is a reflection of the fact that 'the frontier nature of the Norphlet geology and associated production engineering and technology challenges in many ways parallel the equally daunting but different challenges of Alaska's harsh, remote locale. Only a few of the major oil and gas companies could afford the billions of dollars and years of lead times necessary to bring the Norphlet into production.'[21] This parallel between the hostility of the environment in Alaska and the offshore Norphlet is overstated, even if it is undeniable that the latter region was far from hospitable (not least in terms of permitting and regulation). It is also true that the large majors were on the whole better equipped to deal with the Norphlet's challenges than their smaller peers, although both Unocal and Murphy tried their luck at the play (the latter with a conspicuous lack of success). Nowadays, though, thanks to the march of technological progress (especially the coming of age of 3-D seismic) and the reduction of drilling costs in real terms, the 15,000 feet depth offshore no longer constitutes a threshold beyond which only major companies dare or may venture. As explained more fully below, that is the key difference between the 1980s-vintage Norphlet offshore play, on the one hand, and the GOM-wide deep gas play in its current incarnation, on the other.

5.3 A Predictable Policy Response

In light of the traditional effectiveness of US oil industry lobbying and the reasonably encouraging finds in the Norphlet formation, and given that the US government could give fiscal incentives to deep gas without paying much of a political price (at a time when the US gas supply crunch was worsening perceptibly and independents had access to the technical means to tackle deep gas wells), it hardly came as a surprise when MMS announced in 2001 that, starting with Lease Sale 178, it would extend royalty relief to the first 20 BCF of deep gas output from shallow water leases. The royalty holiday would be available so long as production began no more than five years after lease assignment or before 2006, and the average price of gas did not exceed the threshold of USD 5/MMBTU[22] for a whole calendar year. At a price of USD 3.50/MMBTU, it was estimated that this royalty waiver would be worth about USD 12 million to a producer, a figure at the low end of the USD 9 to 23 million that MMS estimated drilling a deep well in GOM would cost.

These royalty waiver provisions were applicable to all 1240 GOM shallow water leases issued from 2001 onwards. However, at least 60 percent of the GOM deep gas resources were seen as lying under the 2400 active leases that had been issued before this date. Thus, in 2004, MMS significantly extended the scope of the royalty incentive programme to include such leases (all slated to expire before 2008[23]), and made its overall provisions considerably more attractive. The final MMS deep gas royalty relief rule (applicable after April 2004) granted royalty suspensions on the first 15 BCF of gas from wells drilled between 15,000 and 18,000 feet, and on the first 25 BCF for wells drilled beyond 18,000 feet. Supplementary dry hole incentives of 5 BCF (a maximum of two per lease), applicable to any future production of gas or oil from any drilling depth, were allowed for leases where operators drilled to a target reservoir at 18,000 feet or deeper.[24] Furthermore, the threshold price was raised by 70 percent, to USD 9.34/MMBTU (2004 dollars) for a whole calendar year, adjustable for inflation. Lessees holding undrilled blocks qualifying for deep gas royalty relief as per the MMS rule of March 2001 were given the option of replacing the existing deep gas relief scheme with the new terms (as it was, only three productive leases ever availed themselves of the 2001 deep gas royalty waiver incentives).[25]

The new provisions meant that leases where ultra-deep (18,000+ feet) wells were drilled could potentially produce as much as 35 BCF on a royalty-free basis (10 BCF earned from drilling up to two unsuccessful wells, plus 25 BCF from a successful well). The maximum royalty-free volume available for a lease with wells drilled to less than 18,000 feet, in contrast, would be 15 BCF.[26] MMS calculated that the average anticipated royalty savings for wells drilled to at least 18,000 feet, at 2003 natural gas prices, would be around USD 20 million.[27]

5.4 The Rationale for Deep Gas Royalty Relief

On the face of it, there are good reasons to think that incentives such as those described above can provide a decisive push for deep gas production. There is, after all, no doubt that deep gas is an expensive and technically challenging undertaking. Reservoir quality for deep strata is lower than that of younger strata (due to the depths and pressures), which means that amplitude anomalies are less clear and 'take new data, new techniques, more work with analogues and a lot of modelling to unravel'.[28] In addition, it is difficult to get seismic energy deep enough to illuminate the traps,[29] and the sinking of deep gas wells, just as in the Norphlet formation, continues to be a far from straightforward affair.

Deep gas wells in the GOM area are very demanding because drillers encounter (far more often than they do elsewhere) a combination of 'overpressured zones, high bottom-hole pressures and corrosive reservoir fluids [that] require special practices'.[30] As an official from an independent oil company put it, 'we partnered with a major oil company to drill a well … in the deep shelf that went to about 21,000 feet … [and] the major's engineers made the comment to us that this was the hottest, highest pressured well they had ever drilled … There is significant mechanical risk every time you drill one of these wells.'[31] In addition to this, GOM deep gas production (outside the Mobile Bay area) also requires that companies tap very high volume wells with infrastructure designed for comparatively modest flow rates. And while it may be true that a run-of-the-mill jackup rig may be able to manage the shallower water depths associated to deep gas operations without any trouble, these mechanical risks mean that 'much of the equipment on the deck of that jackup is likely to be substandard when it comes to drilling deep wells', to the tune of approximately USD 40 million (a figure that does not include the nearly USD 20 million that a jackup operator would lose in day rates for the time the rig would spend in the shipyard being upgraded).[32]

Just as in the shallow subsalt, the learning curve for companies entering the deep gas province has been steep, and drilling costs of early deep gas wells were quite high: for instance, El Paso Production's first deep gas well (drilled in Vermillion block VE47 in 1999) cost the company USD 37 million (and, to add insult to injury, it proved dry).[33] In real terms, this is not far removed from the costs faced by majors after the earliest stages of the development of the Norphlet play in Mobile Bay. Drilling costs were slashed very quickly,[34] although they remain on the high side if one compares them to traditional shallow water wells, at a total bill of around USD 8–20 million per well (with completion costs adding a further USD 3.5–4 million per well). Moreover, the long-term success rate in the play (which has been put at 25 percent, even though the overall success rate for wells drilled between 1998 and 2003 has only been 8 percent) [35] means that a company may very well have to spend USD 30–80 million on dry holes before encountering a commercial find. Nevertheless, such economics come to less than half of the costs that entrants to the early Norphlet play would have had to face, and that is why the deep gas play has come within the reach of even relatively small E&P outfits. At the moment, the financial means of such companies are only stretched to breaking point at extreme depth intervals – beyond 26,000 feet – where drilling a well may cost anything between USD 30 and 50 million,[36] and may easily take up to a year from spudding to total depth being reached.

The above complications notwithstanding, the deep gas play also has plenty of positive aspects to it. The same factors that affect source rock maturity in the play (depth, pressure) have tended to increase the amount of hydrocarbons in traps, thereby enhancing the reserve potential and pay thicknesses.[37] In the particular case of the Norphlet formation, porosity values are significantly higher than for comparable sandstones encountered elsewhere in the world.[38] Sustainable production rates are excellent: production rates of 20 MMCFD of gas are commonplace, and impressive average output figures of 80 MMCFD or more per well have been achieved encouragingly often.[39] Also, relatively large volumes of condensates have been encountered in many of the deep gas pools discovered to date.[40] As for dry deep gas reservoirs, operators can benefit from the fact that abandonment pressure can be kept quite low, which allows for high recovery rates (70–75 percent on average, and up to 90 percent if water intrusion can successfully be kept at bay).

Production test data accumulated thus far indicate that in the deep gas sub-province, reservoir quality does not decrease with depth. In fact, they strongly suggest that the deeper a drilling target is, the better the production rate achievable if any gas is found (Table 5.2).[41] The largest deep gas discovery thus far (found in El Paso's South Timbalier block ST204 in November 2000) contained 400–600 BCF of reserves, and reached a maximum daily output figure of 350 MMCFD of gas.[42] Finally, as has been said before, deep gas wells in the shallow water enjoy ready and cheap access to both transportation infrastructure and markets, and this means that 'a 50 million barrel [deep gas find] will make … more money in the shallow water than 200 million barrels will in deepwater'.[43] This assertion still holds, incidentally, even when

Table 5.2: Characteristics of Deep Gas Well Completions in the GOM Federal OCS, 2001–2

Depth interval (thousands of feet)	*Number of Completions in New Reservoirs*	*Average Maximum Well Test Rate (MMCFD)*
15–15.99	20	13.8
16–16.99	12	32.2
17+	13	44.8
TOTALS	45	27.7

Source: MMS 2003: 3.

royalty relief is not factored into the equation. And even though MMS has significantly *upgraded* its deep gas resource estimate to 55 TCF (more than doubling an earlier estimate of 20 TCF),[44] the easing of terms in the GOM deep gas province has continued apace. The latest instalments in this sweetening of terms saga are the raising of the threshold price for royalty relief (in the royalty relief provisions for deep gas wells included in the comprehensive energy bill approved by the US Senate in May 2005, threshold prices were not specified, although the idea of a price threshold was retained) and the authorisation for MMS to grant suspensions of operations (SOO) under certain circumstances to operators drilling wells targeting objectives at depths of 25,000 feet or more true vertical depth below the ocean surface. There are also moves afoot to extend the primary term of shallow water leases thought to harbour deep gas structures from the current five years to at least seven years and possibly more.[45]

All of these positive factors are surely very welcome for companies looking for deep gas, but they are eclipsed in importance by the degree to which the rapidly expanding US demand for natural gas has been outstripping supply (4.0 BCFD in 2003, by some reckonings[46]), and the effect that this has had on US natural gas prices. It is chiefly as a result of this trend, which shows no sign of abating in the near or medium term, that the deep gas province has become such a vital area of attention for oil companies, not least because the impending exhaustion of the traditional shallow water province has left many of the smallest among them with literally nowhere else to go.

The case of Pioneer Natural Resources (a company that emerged from the 1997 merger between Mesa Petroleum and Parker & Parsley) provides a good illustration not only of the growth options facing relatively small GOM players but also of the privileged place that deep gas now occupies within their growth strategies. Upon its formation,

Pioneer's management reached the conclusion that, even though international expansion of the company's operations was desirable, the bulk of its activity would still take place in the USA. Having examined all petroleum basins within the Lower 48, Pioneer concluded that the shallow water GOM had 'by far ... the most remaining potential'.[47] This potential, though, involved a very specific kind of prospect. As a Pioneer official explained, 'the [traditional] shallower plays on the shelf are very heavily drilled, so it seemed obvious to us *back in 1998* [i.e. three years *before* royalty relief was introduced for deep gas] that if we were going to play the shelf ... we would have to drill deep'. Thus, Pioneer 'built an inventory of deep gas prospects based on regional geologic work it initiated soon after its formation'.[48]

In this effort, Pioneer was by no means alone. Between 1999 and 2001, MMS assigned a total of 1371 shallow water leases, and most of these were obtained by parties who were intent on looking for deep gas. Indeed, lease sales held during these years witnessed the proportion of shallow water acreage as a proportion of total acreage leased once again exceeding the percentage of deepwater acreage leased by a comfortable margin (roughly 3 to 2). In contrast, during the 1995–8 timeframe, shallow acreage averaged only 40 percent of total acreage leased (with a low of 25 percent recorded in 1998).

The reason behind the interest in deep gas of companies such as Pioneer, *even in the absence of any fiscal incentive whatsoever*, is not hard to fathom: in the light of their exclusion from the deepwater province (due to factors explained elsewhere in this study), not to mention their inability to buy their way into the deepwater (*à la* BP), smaller GOM players see deep gas as their 'last chance saloon', and they have been prepared to face proportionately large upfront costs in order to make sure that they are not left out in the cold yet again. Thus, even a relatively small outfit like Spinnaker Exploration had managed to drill over fifty deep gas wells before royalty relief came into being, while El Paso had already sunk 24 deep gas wells (and drawn up plans for a further 41). It is important to note that, despite the widespread interest in deep gas among US oil companies, around two-thirds of the non-Norphlet identified remaining reserves are in the hands of only four players; namely Chevron, El Paso, Anadarko and Shell.

The density of transportation infrastructure in the shallow water province means that there is less scope for pipeline owners to expropriate surpluses generated by smaller players. Because of this, post-2000 OCS lease sales have witnessed some really spirited bidding for shallow water acreage, especially by smaller independents.[49] In Lease Sale 185, for instance, four of the ten highest bids received were placed on shallow water blocks, and two blocks attracted 18 bids between them, a phenomenon pretty much unseen since the netback crisis. Likewise, in August 2003, High Island block HI170 drew a total of 13 bids whose combined value (USD 111.5 million) came to 43 percent of the total monies exposed in all the 407 winning and losing bids that were placed on the 335 blocks offered in Lease Sale 187.[50] The winning bid (submitted by LLOG Exploration Offshore, a privately held E&P independent) came to USD 22.6 million, a sum rarely encountered these days outside hot areas in the deepwater province.

5.5 Can Deep Gas Royalty Relief Deliver?

As Figure 5.2 shows, between 2000 and 2002, GOM deep gas production – exclusive of output from Norphlet formation wells – increased by around 375 MMCFD (from 778 MMCFD to 1.15 BCFD). Given that deep gas output contributed around 12 percent of GOM shallow water production in 2002,[51] just how likely is it that the deep gas royalty relief initiative will make an appreciable difference in terms of the US supply/demand balance for natural gas during the critical 2005–2009 window?

The answer to this question, at first glance, appears to be: not likely. After all, even high gas prices, which have a much more dramatic

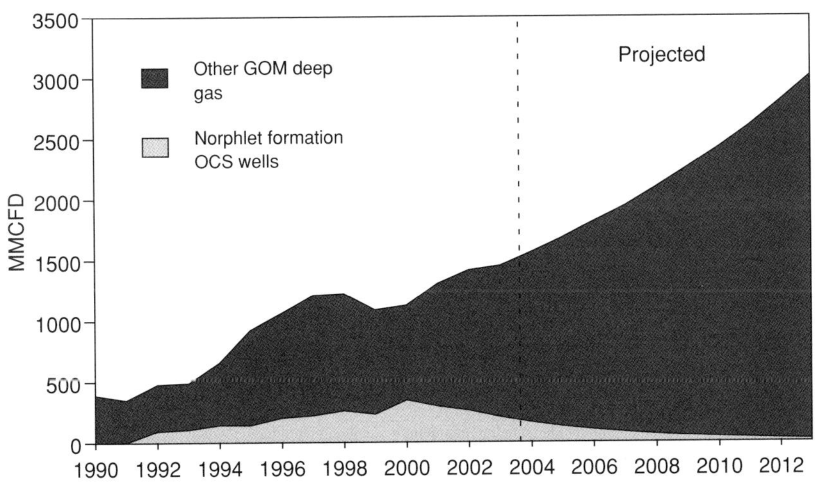

Source: MMS

Figure 5.2: Natural Gas Output from Deep Wells in the GOM Federal OCS Shallow Water, 1990–2013

effect on oil company cash flows than any royalty holiday, have failed to elicit a sufficiently large output response from the industry.[52] Indeed, the royalty relief sweetener has not even managed to coax companies to drill significantly more deep gas wells, although it is true that a total of 293 deep gas wells were drilled over the 1998–2003 period (with 49 wells drilled to at least 18,000 feet and nine to at least 20,000 feet).[53] Deep gas production outside the Norphlet trend continued to rise beyond that year but, according to consultants Wood Mackenzie, the 1.2 BCFD figure posted in 2003 represents a peak that will not be exceeded for some time,[54] despite the number of deep gas wells coming on stream (almost 70 in 2003 alone), and a couple of new significant discoveries.[55] This is a vision that is strikingly at variance with that of MMS, which is confidently predicting that deep gas output will continue expanding steadily and will reach a level of 3.01 BCFD by 2013.

MMS envisions that, over the 2003–2009 timeframe, the oil industry will drill a total of 62 wells to 18,000 feet or deeper, plus 71 wells to between 15,000 and 18,000 feet, on an annual basis. These figures represent an increase of 38 and 18 wells, respectively, over the number that MMS considers would have been drilled in the absence of royalty relief.[56] The output that these additional wells are expected to churn out over their active lives is respectable: 3.8 TCF of natural gas (out of which 1.5 TCF would be royalty free). Unfortunately, as can be appreciated in Figure 5.3, drilling activity over the 2000–2005 period has

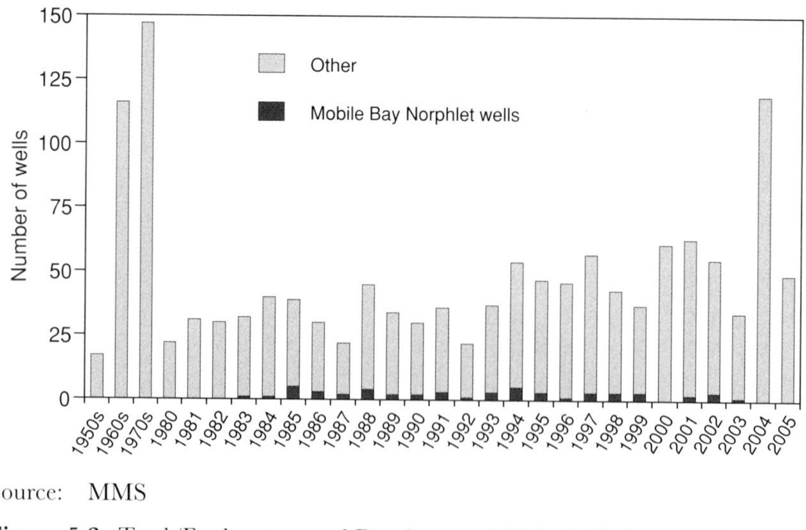

Source: MMS

Figure 5.3: Total (Exploratory and Development) Wells Drilled to a TVD of 15,000 Feet or More in Water Depths of 1,000 feet or less, in the GOM Federal OCS, 1950–2005

fallen well short of the MMS estimations (with the exception of 2004), despite very high gas prices (an element that should have prompted more rather than less drilling). Thus, incremental deep gas output is unlikely to compensate for the continued decline in traditional shallow water gas production (the MMS 2004–2013 production forecast posits that incremental deep gas production will compensate for 60 percent of the traditional shallow water decline over this period).

One should also point out that even if deep gas drilling activity were to conform to MMS expectations, the output decline rate in the GOM shallow water will not necessarily slow down, as deep gas fields are characterised by even quicker depletion profiles than those of the small traditional shallow water fields (see Figures 5.4 and 5.5, for the examples of ST204 and HI202 fields). On the strength of initial reservoir mapping and analyses of production and well test data for the 45 deep gas wells completed over the 2001–2 period, MMS reached the tentative conclusion that their production half-life (defined as the amount of time it takes for output in a well to decline by 50 percent from the initial maximum rate) would be 24 months.[57] Unfortunately, the production half-life of ST204 turned out to be only 15 months (i.e. 40 percent less). For its part, the production half-life at Spinnaker's HI202 field was an even shorter nine months, and this field ceased production altogether during February 2004. Thus, finding and developing more deep gas fields may very well accelerate the province-wide fast gas treadmill, especially since deep gas output is highly concentrated across a small number of fields: in 2002, for instance, the largest ten

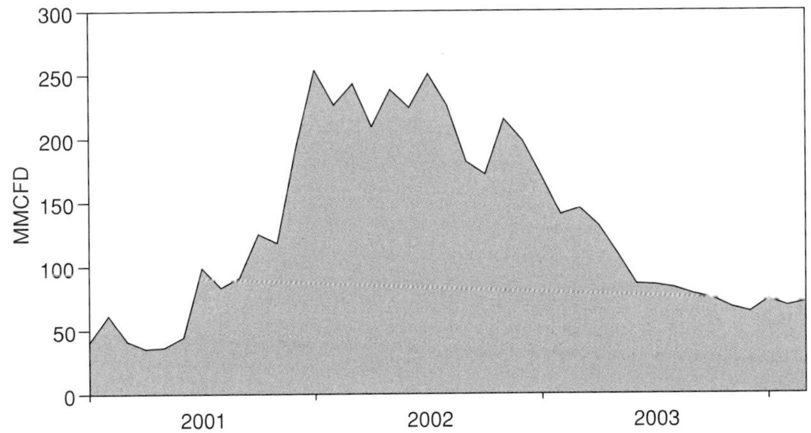

Source: MMS

Figure 5.4: Production Profile of ST204 Field, 2001–2004

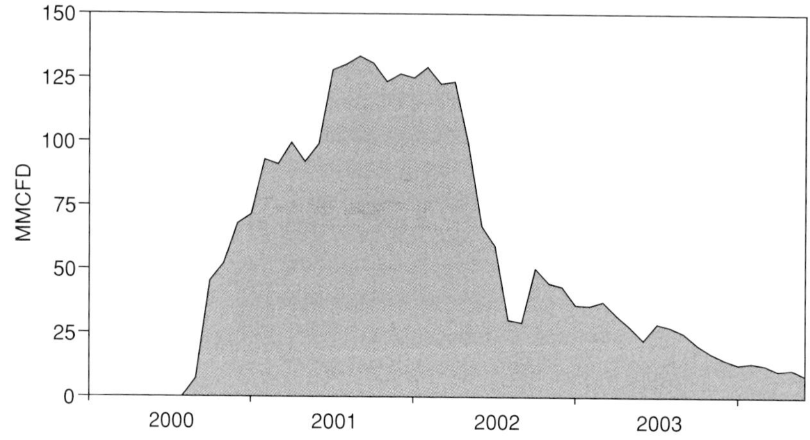

Source: MMS

Figure 5.5: Production Profile of HI202 Field, 2001–2004

deep gas fields were responsible for 55 percent of output in the sub-province (excluding the Norphlet formation), while the next 35 fields contributed with a further 32 percent.[58]

Contrasting the production profiles of the new breed of deep gas fields with those of their closely comparable predecessors in the Norphlet formation is both an instructive and worrisome exercise. Figure 5.6 plots the monthly output, from commencement of production, of two Norphlet trend fields in Alabama state waters (Mary Ann and Fairway) as well as three deep gas OCS fields (ST204, HI202 and Alex, in Brazos block BA19). The erratic production path for the Mary Ann field reveals why it has been said that 'keeping Norphlet wells producing at design rates is as difficult as finding the reservoirs four miles beneath the surface'.[59] Indeed, chronic mechanical and completion problems have plagued operators in the Norphlet, and the Mary Ann field has had its fair share of those. This explains why Mobil only achieved planned utilisation of the original processing facilities three years after the field came on stream.[60] In contrast, Shell's greater experience in dealing with high pressure, high temperature fields allowed it to ramp up production at the Fairway field much faster. This field was developed by means of five wells producing 40 MMCFD each (the best production figure achieved at the Mary Ann field was 35 MMCFD, in contrast). Nevertheless, Fairway still experienced major calcium fluoride and calcium carbonate scaling problems, which adversely impacted output until mid-2002, when Shell had to re-drill the field.

The production profile of the Alex development project is redolent

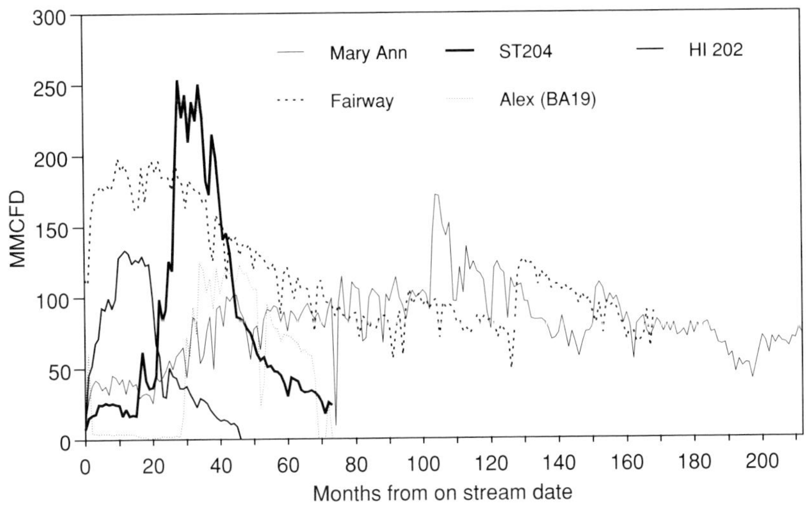

Sources: ASOGB, MMS

Figure 5.6: Production Profiles from Inception of Selected Deep Gas Offshore Fields

of a Norphlet field in its erratic behaviour. Production at Alex through one well began in October 1999, and a peak rate of 87 MMCFD was achieved only one month later, whereupon the well developed mechanical problems that led to its being plugged and abandoned. Production could only resume more than two years after this incident. In contrast, both the ST204 and HI202 fields had a comparatively trouble-free time, and quickly achieved spectacular output figures. The most prolific well in the ST204 field, for instance, produced 118 MMCFD of gas and 8 MBD of condensate at its peak. This production rate is three times as high as the maximal flow rates achieved at the Fairway field, which were seen at the time this field was developed as being 'high but consistent with reservoir properties'.[61] Such production rates have been achieved by fracturing as many gas zones per well as possible.

In cash flow terms, the production profiles of the ST204 or HI202 fields are highly attractive for any company, of course. El Paso, for instance, recouped the USD 16 million it cost to drill the first ST204 producing well in less than two months. However, this mode of exploitation is a byword for very quick production declines, but these receive far less publicity.[62]

Even if the MMS deep gas reserves estimates of 55 TCF were right on target, they would only be enough to satisfy US natural gas demand, at its 2003 level, for slightly over two years. Furthermore, there is no reason to assume that these estimates will indeed prove accurate. The pattern of development in the Norphlet formation once again provides a salutary example in this regard. As indicated before, the offshore Norphlet formation was seen at one point as having the potential to eclipse Prudhoe Bay in terms of natural gas potential. However, cumulative production in state and Federal waters from offshore Norphlet formation fields from 1981 and up to May 2004 inclusive is only 2.4 TCF and 0.94 TCF, respectively. Remaining reserves (assuming that 75 percent of gas in place is recovered), for their part, can be put at around 3.2 TCF and 1.6 TCF, respectively.[63] Such figures are respectable enough in themselves (and certainly better than nothing), but nevertheless they are a far cry from the Prudhoe-like '26 TCF plus' figures that were thrown around with abandon in the heady days of the early 1980s[64] (one should not forget, either, that Prudhoe Bay's 26 TCF of gas reserves were taken off the books in 1983, after the companies active in the field spent twenty fruitless years trying to devise an economic way to transport this gas to market).

5.6 Whither the US Natural Gas Market?

To sum up, the deep gas royalty relief initiative was intended to help offset the high cost and high risk associated with drilling deep wells, with a twofold objective in mind: the 'recovery of some otherwise uneconomic gas resources' and the 'accelerated recovery of some marginally economic gas resources'.[65] Whether such an intervention was really warranted by the economics of deep gas operations is questionable. After all, the attraction that small and large companies alike feel for the deep gas play predated royalty relief, and was in no way dependent upon it. Finally, in the light of the pedigree of operations in the offshore Norphlet trend, the attempt by MMS to present the GOM deep gas as an almost virgin play is disingenuous, to say the least. In all, royalty relief for deep gas appears to have been enacted for the benefit of an *established* province that was already quite attractive, as opposed to a marginal exploration play whose prospects would have been stunted had it not been carefully nurtured with tax breaks. To cap it all, the price thresholds below which deep gas royalty relief is available for companies have been set at preposterously high levels (Table 5.3).

There are also plenty of indications that oil companies have been content to use royalty relief chiefly to enhance the rate of return of profitable projects, rather than to undertake truly marginal ones. Indeed, so lackadaisical has the drilling response been to the deep gas

Table 5.3: Price Thresholds for Deep Gas Royalty Relief and Observed NYMEX Gas Prices, 2000–2005

		-- *Gas Price Thresholds (USD/MMBTU)* --				*Average NYMEX*
	Applicable	*Sale 178*	*Sales held*	*Sales held*	*For deep gas rule*	*Nearby Delivery*
Year	*inflation rate (%)*		*2001–3*	*2004–5*	*30 CFR 203*	*(USD/MMBTU)*
2000	-	3.50	5.00	4.33		
2001	2.20	3.58	5.11	4.06		
2002	1.20	3.62	5.17	3.36		
2003	1.80	3.69	5.27	5.49		
2004	2.62	3.79	5.41	9.34	9.34	6.18
2005	2.76	3.89	5.56	9.60	9.60	8.96

Source: MMS

fiscal incentives that, in 2003, a high MMS official seemed to question openly whether the US government was really getting value for money in exchange for the USD 150–220 million that the deep gas royalty relief initiative is expected to cost (in present value terms) over the next 16 years,[66] when he complained in an industry forum that 'the level of [deep gas] activity really has to increase to have [any] impact' on the decline in GOM shallow water gas production.[67] Unfortunately, this situation is unlikely to change significantly within the timeframe envisaged by the deep gas royalty relief initiative, in part because deep gas wells take so long to drill but also because many players have deferred from committing to drilling programmes until they see the results achieved by others. The ExxonMobil Blackbeard prospect provides a good example in this regard. The first well at this prospect was expected to take around a year to drill (it took more time because of the disruption caused by the 2005 hurricane season). During this time, besides tying up a rig which was not available for use elsewhere (at a time of an acute shortage of rigs capable of drilling deep wells), Blackbeard also inhibited drilling by other companies who, in the words of an officer at one of them, are 'like a bunch of birds on a wire watching the … well go down … [in order] to see how long it takes, how much does it cost and … [whether it] finds anything'.[68] In the event, the well had to be abandoned in August 2006 after it encountered higher than expected pressures.

Deep gas royalty relief certainly could not and did not prevent US natural gas price from spiking once more in 2003, 2004 and 2005 (not least because deep gas wells can easily take up to 300 days to be completed, so their drilling ties up parts of a rig fleet accustomed to completing even offshore wells in thirty days). The 2003 price spike led

Paul Horsnell to conclude that 'twice within two years, the market to achieve balance has had to spike prices to levels where immediate and severe demand destruction occurs. That is not the sign of a market that is either working well or playing a constructive role in the US economy.'[69] How severe has the demand erosion caused by the spikes been? Byrne put the figure for the late 2000 episode at around 4.5 BCFD.[70] He also warned that 'industrial users were slow to respond in 2001, but with volatility and uncertainty unavoidable in the future, *a repeat performance in 2003 will likely affect capital decisions for years to come*'[71] in gas-sensitive industries that compete internationally: ammonia, methanol, chemicals, aluminium, textiles, steel, paper. The repeat performance did in fact occur, prompting Byrne to posit that, given the inability of future supply to keep up with the pace of demand growth, 'the 30 TCF demand market that analysts and consultants project by 2010 (or even 2015) is out of the question,'[72] largely because US industry – which is the largest consumer of natural gas in the country at 31 percent of demand – 'will shift overseas to low-cost, gas-rich regions'.[73]

In the wake of the 2000–2003 price spikes, the notion of a 30 TCF per annum US gas market lost whatever credibility it might have enjoyed up to that point. As stated before, this idea had featured most prominently in an NPC policy document published in December 1999 ('one of the few government or privately-sponsored studies that offered any substantial basis for believing that it would be possible to significantly expand North American supplies of natural gas above 1999 levels').[74] This document 'provided the basis for many of the assumptions used by the … EIA in its subsequent annual forecasts of supply and demand in the US market', and it also 'played an important role in justifying decisions by power plant developers to build more than USD 100 billion in new gas-fired generating units'[75] over the 1999–2003 period (Figure 5.7). Not only did this building spree rival in magnitude the previous largest addition of new generation capacity (which took place during the 1970s), it also exceeded forecasts by a very substantial margin: the 1999 NPC study assumed that 144 gigawatts of new gas-fired capacity would be built by 2015, but in actual fact, new capacity installed to 2005 exceeded 200 gigawatts (out of a total newly installed capacity of 220 gigawatts).[76] This building spree was also unprecedented in terms of its reliance on a *single* fuel (the previous boom had witnessed the construction of coal, nuclear, gas and oil generation capacity, in contrast).[77] To cap it all, and again in marked divergence with historical trends, most of the new power generation turbines had no fuel switching capabilities.

Unfortunately, less than four years after publishing this very bullish

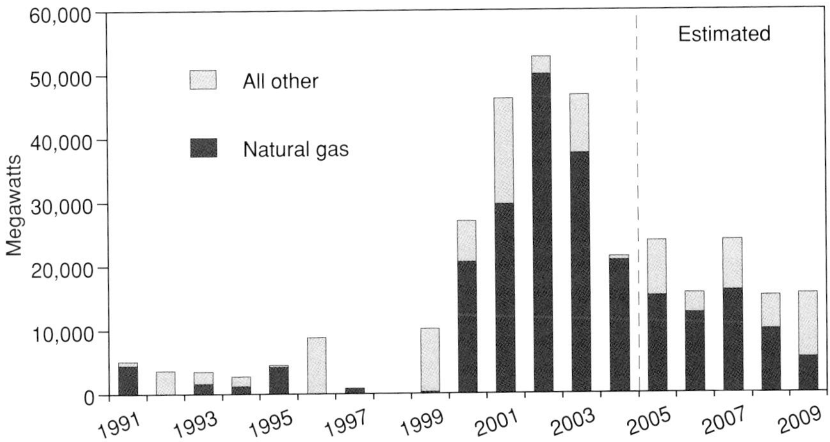

Source: DOE

Figure 5.7: Net Additions to US Electric Generation Capacity by Utilities and Independent Power Producers, by Fuel Type, 1991–2009

take on the prospects of the US gas market, the NPC came out with a new report[78] in which radical adjustments were made to the volume of natural gas expected to come from so-called traditional basins in North America (a definition which encompasses all basins lying to the south of the Arctic Circle, including some from which any gas is yet to be extracted). Given the implications of these downward revisions (and the dire economic prospects that they would entail for all the new gas-fired generation capacity added throughout the late 1990s), the NPC could not have undertaken them lightly. However, the decision was thrust upon it by a calamitous decrease in US output: by late 2002, the annual production figure recorded for the whole of North America was an alarming 6 BCFD below the Council's forecast for that year (and producibility for 2005 was seen as at least 20 percent lower than had been projected in 1999). In the specific case of GOM, the 1999 NPC study projected production in the whole province to climb strongly to 20 BCFD by 2005 and 22 BCFD by 2010. In contrast, the 2003 study projects a flat GOM production of only 14–15 BCFD.[79]

Looking further forward, the NPC has also had to reduce its estimates for 2015 production by more than 7.5 TCF, thereby highlighting a yawning 3.6 MMBOED hole (in BTU equivalent terms) relative to the expected US consumption for that year. As if this were not enough, the NPC report also pointed out that even at very high natural gas prices (USD 8/MMBTU), whatever incremental supply response could be elicited from traditional basins would be insufficient to bridge the gap against the forecast demand, particularly in the short to medium term. NPC postulated that, under certain conditions, US natural gas output *might* – after a longish spell – increase to a level comparable or slightly higher (1 BCFD maximum) to that recorded before production in the GOM shallow water began to decline rapidly. Unfortunately, even though the price assumptions underlying this scenario (USD 5/MMBTU in 2002 dollars) are likely to occur, the same cannot be said for some of the other conditions that are equally important if the scenario is to materialise, above all the lifting of the drilling moratoria on most of the Lower 48 Federal OCS and the reduction in permitting times in environmentally sensitive onshore basins.

Given the very low price elasticity of supply that now prevails in the tired traditional North American gas basins, the common view appears to be that equilibrium in the natural gas market will only be re-established through further, permanent, demand destruction in the industrial sector. Indeed, this is increasingly seen as the only way in which the majority of US consumers may be spared the rigours of a crippling natural gas price shock that could compare to the 1970s' oil shocks in severity. For obvious reasons, the prospect that vast swathes of American industry will be decimated on account of high natural gas prices is deeply unpalatable in the USA. However, there are disturbing indications that not even the sacrificial offering of US industry on this altar may be enough to disperse the gathering storm clouds in the natural gas market, chiefly because, as Weissman stresses, there is surprisingly little substance behind the widely held belief 'that natural gas prices above USD 6/MMBTU are not sustainable' because – the argument goes – such price levels inevitably bring about 'large reductions in industrial use – which in turn … rapidly bring back prices to more "normal" levels'.[80]

There is no question that the high prices recorded since the 2000 price spike have led to a significant downward shift in natural gas consumption by industrial concerns. However, given that most of the demand from large single point consumption facilities (i.e. aluminium smelters, ethylene, fertiliser and chemical plants) has already evaporated, the annual rate at which industrial demand will continue to be destroyed is bound to slow down significantly, and will certainly not approach the 3.5–4 BCFD that many analysts seem to think is possible.[81] The slowness with which the industry is coming to terms with this notion is providing fertile soil for procrastination and complacency regarding the solution of the natural gas problem, but has brought few tangible policy initiatives beyond the slightly expedited approval of a number of LNG import facilities.

A good example of the wishful thinking surrounding the issue of industrial demand destruction can be found by looking at the spin put on the unexpectedly large injection into storage that took place from April to July 2003 (a 344 BCF year-on-year increase, equivalent to 2.8 BCFD of consumption). Because this event followed a fairly severe price spike (Figure 5.8), it was presented as proof positive not only of the equilibrating power of industrial demand destruction but also of the fact that prices greater than USD 5/MMBTU are not sustainable even in the short run. Unfortunately, the rigorous dissection of electricity generation statistics for those months carried out by Weissman revealed that the increase in the volume of gas injected into storage was handsomely exceeded by the decrease in the volume used to generate electricity, which itself was the product of the abnormally low air conditioning loads in key urban areas in the eastern USA that accompanied a cooler-than-normal summer.[82] The volumes freed up by industrial demand destruction actually observed during this period were 3–4 times smaller than these figures (such is the importance of weather patterns over economic cycles in dictating US/North American gas demand).

Ominously, 'if temperatures … had been more like the summer of 2002, and/or the resurgence in the economy that began in August had begun 60 or 90 days earlier, we might have well seen USD 8 to 10/MMBTU natural gas prices …during the time of the year when natural gas prices historically are [supposed to be] at or near their lowest point'.[83] Under these changed conditions, consumption for electricity generation would probably have come in 200 BCF over the

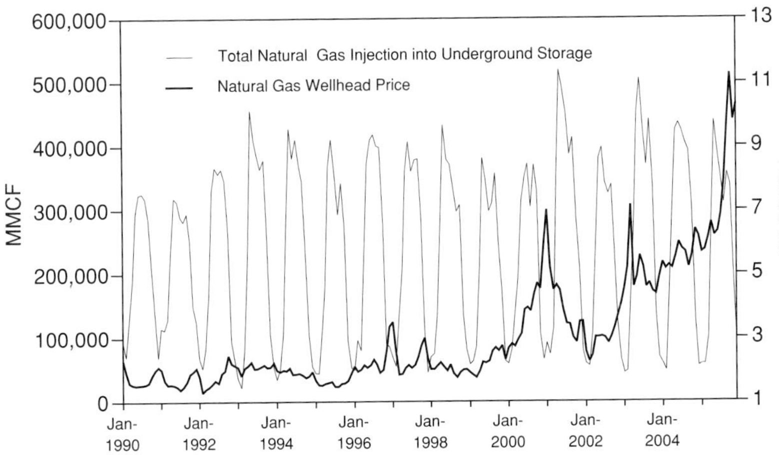

Source: DOE

Figure 5.8: Behaviour of Selected US Natural Gas Sector Indicators, 1990–2005

figures recorded for that summer, and this would have required local distribution companies to bid up spot prices in order to drive out of the market an equivalent volume of industrial demand, or else miss their storage targets for that winter. Indeed, had April to July gas use for electricity generation been similar to the figures recorded during 2002, the natural gas storage deficit at the end of July would have amounted to 1 TCF, a quite catastrophic figure.

It is clear, then, that the natural gas buffer that allowed the US gas market to emerge from the 2003 winter having undergone only a modest price upsurge was in fact the product of a fortuitous set of circumstances that policymakers should not automatically assume will be repeated. Although there was another short-term price spike in January 2004 (which especially affected New York and Toronto), the continent-wide natural gas market had a similar reprieve in 2004, when summer temperatures were again significantly cooler than the norm. Sooner or later, though, the fear was that the US Midwest would have one of the sweltering summers for which it is justly (in)famous and the effect on injection rates would translate into some gruesome price action in the North American gas market the subsequent winter[84] (with volatility likely to be exacerbated by the fact that the majority of the gas-fired generation fleet relies on non-firm gas transmission capacity).[85] As it was, the scenario of record high natural gas prices materialised during 2005, in the wake of the catastrophic 2005 hurricane season, which had a far larger and longer lasting effect on injection rates than any warmer than usual summer could have had.

Demand destruction in the industrial sector, then, is probably not going to be the silver bullet that will deliver US natural gas consumers from the bane of high prices. What is more, the disappearance of a relatively stable industrial load is bound to put consumers even more in thrall to the vagaries of the weather than usual, as the volume of gas injected and withdrawn from storage will come to depend almost exclusively on summer air conditioning, the violence of hurricane seasons, and winter heating loads. It is important to note that the secondary summer demand peak originated by the greater use of gas for power generation significantly shortens the summer injection season (as it only allows injection to take place during off-peak electricity demand hours). Gas use in power generation also subtracts flexibility from the whole transmission and distribution system, not only because power plant demand is highly variable on a monthly, daily and even hourly basis, but also because it creates 'an hourly demand profile that is even more pronounced than that of a traditional residential/commercial load profile'.[86] Furthermore, the overall fuel switching capability in the generation system has also decreased because only a minority of the new gas-fired plants can switch to oil burning, mainly due to air emissions permitting constraints. Obviously, as the NPC acknowledges, 'the decreasing ability of power generation to switch fuels for economic and reliability purposes places greater strains on gas supply, pipeline and gas storage infrastructure, and organised power pools in meeting the growth of power generation, particularly peak demand periods'.[87]

In the future, even relatively modest changes in temperature may translate into sudden increases in US gas consumption that could, in a matter of weeks, wipe out any accumulated storage surpluses, or else eat into volumes meant to be used in shaving seasonal heating peaks. A future repeat of the 2005 hurricane season (seen as a likely eventuality by many climatologists given hurricane strength cycles, quite apart from factors related to global warming) can achieve the same thing in a matter of days, of course. Furthermore, although seasonal disruptions in the natural gas market have been associated with winter up to now, summer months promise to be just as interesting in the future. This is because air conditioning load can vary dramatically according to the geographical location of a heat wave. There are some places in the USA where this sort of incremental load can be met entirely by coal-fired plant even if temperatures soar, whereas in other places even a small increase in mean temperature can have a major impact on gas demand, as relatively small and inefficient gas-fired plant is brought on stream. In this regard, it is important to point out that 2006 witnessed the first ever natural gas stock draws recorded during summer months.

To conclude, the USA will face a natural gas supply/demand imbalance throughout the remainder of the present decade. The gravity of this imbalance will be exacerbated by the fact that natural gas-fired generating units will have to meet virtually all of the incremental electricity needs of the American economy up to the year 2010, as the dash for gas of the late 1990s and the long lead times necessary to obtain planning permission to build coal-fired capacity have left the USA with no short-term alternative in terms of marginal electricity generation options (despite the natural gas supply problems, the NPC has readily acknowledged that 'permitting and siting a new coal facility will remain a formidable challenge').[88] Moreover, this construction boom has led to most regions in the USA having 'ample to surplus generating capacity … [with] ongoing generation capacity requirements' restricted to limited pockets, like New York City.[89] Obviously, this lack of capacity requirements is a significant disincentive for investment in further power plant construction. Taken together, the above factors suggest that the marginal generating gas costs at new-build gas-fired plants will determine the price of electricity in the USA in the short to medium term. Granted, over the 2002–2025 period, coal is expected to account for 33 percent of new US electricity generation capacity (87 gigawatts out of 264 gigawatts).[90] Unfortunately, 75+ gigawatts of this capacity may only be built over the 2016–2025 period (and DOE data indicate that out of the 63 gigawatts of coal-fired capacity that *could* be built between 2004 and 2010, only 16 percent has progressed as far as the development or construction stage, as shown in Figure 5.9).

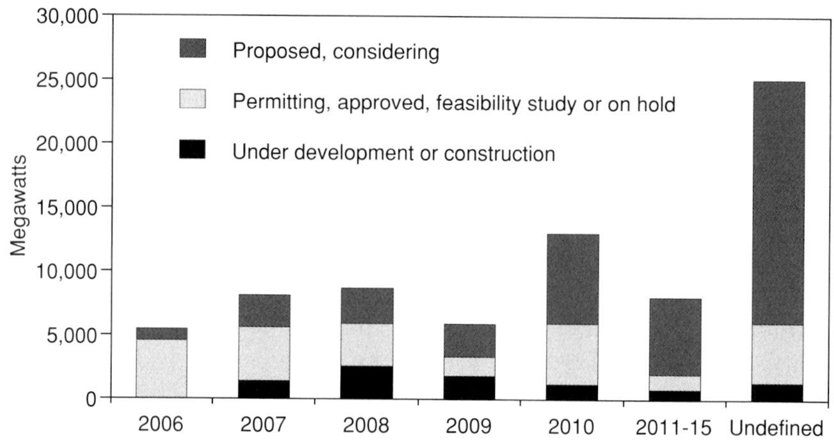

Source: Klara and Shuster 2005

Figure 5.9: Additional US Coal-fired Electricity Generation Capacity, by Project Status and Estimated Onstream Date, 2006–2015

Even as late as 2003, and despite the grievous production decline rates in the GOM shallow water province and a desultory drilling response to the deep gas royalty relief initiative on the part of the industry, there was still some residual hope that a 30 TCF per year market might materialise by the turn of the present decade if natural gas production from the deepwater province managed to hit some other-worldly output targets. By early 2004, though, few industry promoters spoke of the 30 TCF figure due to a profound change of circumstances and perceptions regarding US natural gas supply. So radical has this change been that the achievement of the unrealistically high deepwater output targets (supposed to make the 30 TCF market a reality) might not be sufficient for the North American market as a whole to avoid a major supply shock in coming years.

LNG imports are also bound to be insufficient to stave off this outcome. Consider the following: ConocoPhillips' LNG project, officially announced in July 2002, is expected to deliver 0.90 BCFD of gas into the US market. This may sound like a lot of gas, but it is in fact barely enough to compensate for the supply that the USA lost during the third quarter of 2003 alone. In fact, the gap between the 2010 output for currently active fields that the NPC forecast back in 1999, on the one hand, and the output now seen as likely to be realised from these same fields, on the other, may be so wide that plugging it in its entirety would require the gas from ten LNG trains[91] (a sobering thought indeed, when one considers that the ConocoPhillips project has a USD 5 billion price tag, even though it is based on cheap Qatari gas). Indeed, as Weismann points out, if the Alaskan pipeline is either not completed or delayed, the USA will somehow have to meet 87.7 percent of its incremental estimated demand on the basis of LNG imports.[92] Thus, the DOE puts US imports of LNG at no less than 15 percent of the total US supply of natural gas by 2015 (12 BCFD, from 1.75 BCFD in 2004), and 21.7 percent of total supply (17.5 BCFD) by 2025. It is worth bearing in mind that, in 2002, the DOE (which was actually less bullish on gas supply than the NPC) stated that LNG imports 'were not expected to become a major source of U.S. supply'[93] in the future.

In light of the above, and given that incremental LNG demand is also expected to increase rapidly in other parts of the world, it is difficult to disagree with the Canadian National Energy Board's conclusion in the sense that, up to 2010 (and possibly beyond that), 'it will be difficult to meet the expected demand for natural gas in North America from indigenous production and the available LNG import capacity', which will make it equally difficult to avoid 'periodic tight conditions, characterised by extreme price volatility and an ongoing need for adjustments on the consuming side of the market',[94] even if hurricanes do not contribute to make things worse. However, if the MMS forecasts of GOM gas production, in general, and deep gas production, in particular, are anything to go by, it would appear that US governmental circles are still in a state of denial about this, which will complicate the design and implementation of sensible policies that will allow the market to weather the storm until more frontier (Alaska, McKenzie Valley) and unconventional (chiefly coal bed methane and, much further out in the future, gas hydrates) gas resources come on stream, and LNG import capacity rises significantly.

NOTES

1 In other words, a well drilled anywhere within the Western or Central planning areas or the narrow sliver of the Eastern planning area that is

open to drilling.

2 To enjoy royalty relief deep gas wells have to be in production within five years of the promulgation of the final MMS royalty relief rule (April 2004), or within six years if the lessee is able to obtain a special one-year extension from MMS.

3 These were the 1999 NPC forecasts. By the time this study was published, Canada's National Energy Board had already pointed out that gas deliverability from the Western Canada Sedimentary basin was in marked decline, from 16.6 BCFD in 2001 to 16.3 BCFD in 2002. It expected deliverability by 2002 to be no more than 15.9 BCFD (National Energy Board 2003: 1).

4 The NPC's take on the perspectives of the US natural gas market was the most optimistic of all, although shortly after its publication, the Gas Research Institute came out with a report that put 2015 US gas demand at 33.5 TCF. In contrast, DOE sponsored a study of its own to look at the natural gas issue, prompted by fears that there had been an unexpected acceleration in the decline rates of GOM gas fields. This study (EIA 2000) considered a dozen different combinations of prices, technological progress, decline rates, lifting of drilling moratoria and so on, and concluded that the NPC's 30 TCF market (and, by extension the 2 USD/MMBTU that underpinned it) was feasible only under one of these combinations. Unfortunately, the conclusions contained in this report were politically inconvenient at a time of a dash for gas, and nobody paid a great deal of attention to them.

5 *Federal Register*, 69 (16): 3493.

6 Mexican natural gas imports averaged 912 MMCFD and 1.067 BCFD in 2003 and 2004, respectively, up from 592 MMCFD in 2002 and only 292 MMCFD in 2001. Imports are expected to continue growing at an annual rate of 4–5 percent until such time as the recently discovered Lankahuasa field is brought on stream, something that is unlikely to happen before 2008. Proved and probable gas reserves at Lankahuasa amount to only 80 MMBOE, so this field will in any case not be a silver bullet to solve the natural gas shortage in Mexico, let alone North America (see Lajous 2004).

7 At one time, Ladyfern was thought to contain over 1 trillion cubic feet of natural gas, and the field was expected to make up around a quarter of Canada's natural gas production for some time. The reserves estimates have been slashed by more than half, though. Furthermore, while Ladyfern production was 700 MCFD in 2001, it fell to 400 MCFD in 2002 and only 100 MCFD in 2003. As a result of this, and lower production figures for other parts of Canada, marketable gas production in 2003 experienced a 5 per cent year-on-year contraction. Moreover, this contraction occurred even though 1000 gas wells a month were being drilled, compared to only 300 wells a month in 1997.

8 National Energy Board 2004: 3.

9 Scenarios for oil sands development suggest that the natural gas requirements

associated to this activity will grow from 0.6 BCFD in 2003 to 1.2–1.6 BCFD by 2010 (*ibid.*: 12).

10 Out of a total possible undiscovered gas resources in GOM of 193 TCF; *Federal Register*, 69 (16): 3493.

11 Dyman and Cook 2001: 6.

12 *Federal Register*, 69 (16): 3493.

13 Wade, Plater and Kelley 1999: 3. Development activities were suspended in 1982 when it was discovered that Mobil contractors were surreptitiously discharging water contaminated with drilling muds. Mobil had to pay a USD 2 million fine.

14 Norphlet gas is best described as 'a hot, sour, high pressure, corrosive mixture of methane, hydrogen sulphide, carbon dioxide and free water' (*ibid.*: ix).

15 Rice *et al.* 1997: 224.

16 Interestingly, when Mobil filed for its original Mobile Bay drilling permit back in 1970, natural gas was selling for about 0.25 USD/MMCF.

17 MMS 2001a: 2.

18 *O&GJ*, 14 January 1985: 25.

19 In the 1981 State of Alabama lease sale, the first one held after Mobil's discovery, Exxon exposed a total of USD 303 million in 13 bids (equivalent to a quarter of total money exposed in the sale). Exxon submitted the winning bids on seven leases, for which it paid a total of USD 255 million. The company's bidding aggressiveness can be gauged from the fact that it submitted a bid of USD 137.3 million on one lease, which the next highest bidder only valued at USD 5.8 million. After Exxon, the biggest spenders in Norphlet leases and operations were Unocal, Chevron and Shell. As the discoverer of the play, Mobil achieved a strong production position for a smaller outlay in leases: Exxon paid 26 per cent per acre more than Mobil, and bought twice as much acreage.

20 The bill may rise further if ExxonMobil loses its appeal against a verdict that the company had wilfully underpaid royalties from 13 Norphlet wells to the state of Alabama. In December 2000, a jury disagreed with Exxon's arguments that the leasing terms allowed it to deduct processing costs before paying royalties and that no royalty payments were due on natural gas used as part of the production process. The jury awarded the state of Alabama USD 87.7 million in compensatory damages and USD 3.42 billion in punitive damages (the latter figure came from tripling the value of Exxon's annual production from the 13 wells). The case was retried after ExxonMobil objected to the exorbitant punitive damages award, whereupon, in November 2003, a jury reached the decision that ExxonMobil should be ordered to pay USD 63.6 million in compensatory damages and USD 11.8 billion in punitive damages. At the moment of writing, the Supreme Court of Alabama is still considering ExxonMobil's appeal against this verdict.

21 Wade, Plater and Kelley 1999: ix.

22 The threshold was 3.50 USD/MMBTU for leases assigned in Sale 178.

23 Baud *et al.* 2002: 44.

24 There are numerous additional provisions for sidetrack wells. All the details of the final deep gas royalty relief rule can be found in *Federal Register*, 69 (16): 3492–3514.

25 Melancon and Durr 2004: 2.

26 *Federal Register*, 69 (16): 3496.

27 *Ibid.*: 3504.

28 *AAPG Explorer*, October 2002: 14.

29 Given the imaging difficulties posed by deep formations, re-shooting the whole shallow water for 3-D seismic would appear like an irresistible proposition. However, the length of the seismic offsets required to image a target below 15,000 feet makes this proposition unviable, because the abundance of fixed structures and debris in the GOM shallow water makes it impossible to haul 30,000 feet long streamers. However, it is possible to haul such a streamer in a 2-D survey, so many companies are sharpening their interpretation skills in order to identify prospective targets which can then be subject to the full 3-D treatment (*Hart's E&P*, April 2003: 67–8).

30 Reeves, Kuuskra and Kuuskra 1998: 134. For instance, the Mary Ann field (located at a depth of 21,100 feet and with a reservoir temperature of 213° C) contains 9 percent of molecular hydrogen sulphide (Hunt 1996: 439). In combination with water and carbon dioxide, hydrogen sulphide gives rise to a physical process called hydrogen embrittlement, which causes most metals to become brittle and crack. Thus, high-carbon steels have to be used to drill for and transport natural gas with high hydrogen sulphide content. These special steels cost as much as ten times as much as the ordinary steel used in standard gas wells.

31 *AAPG Explorer*, October 2002: 39.

32 *Hart's E&P*, April 2001: 52.

33 *Petroleum Economist*, June 2002: 7.

34 The discovery well on ST 204 cost USD 26 million, with the subsequent two wells costing USD 16 million each. Drilling costs have decreased thanks to the use of PDC bits and synthetic base oil drilling mud (both of which allow for much increased rates of penetration), as well as expandable casing (which allows for smaller casing sizes at the onset of drilling and decreases the overall time it takes to drill a well).

35 *Federal Register*, 69 (16): 3504.

36 Early estimates for ultradeep well drilling costs put these at around USD 60 million (*PON*, October 6 2003: 1–5). However, Shell's Shark well, drilled to a depth of more than 25,000 feet, 'only' cost about USD 30 million to drill. The Shark well encountered producible sands, but proved dry. In contrast, the Chevron/Anadarko/Nexen Knotty Head well (in 3500 feet of water and a TVD of 34,189 feet) cost USD 140 million to drill. The well has been said to have found 500–600 MMB of light oil in a good quality reservoir (*PON*, 21 December 2005: 1).

37 The compressibility of gas means that a reservoir at 10,000 feet will hold around five times as much gas as an equal-sized reservoir at 2000 feet.

38 Rice *et al.* 1997: 227. Outside the Norphlet Formation, porosity tends to decrease with increasing depth, although the range of porosity values at different depths throughout the GOM region varies greatly (Dyman *et al.* 1997: 35).

39 The first deep gas well in Shell's Brazos block BA19 recorded an average output of 65 MMCFD (with a maximum rate of 87 MMCFD) of gas with 4.1 percent carbon dioxide content and 16 ppm of hydrogen sulphide. The well produced for 32 days before completion problems (damage to the well during shutdown) forced the company to plug and abandon it. The prospect was re-drilled in 2001. Total development costs came to USD 28.5 million. Spinnaker's High Island HI202 field achieved a maximum output rate of 160 MMCFD, and the same company's Resolute prospect in High Island block HI197 is expected to achieve a maximum production rate of 100 MMCFD.

40 Even some of the deepest dry gas fields in the Smackover-Norphlet formations have been found to contain small amounts of condensate (Hunt 1996: 437). MMS estimates that around 20 percent of the hydrocarbons expected to be produced as a result of deep gas royalty relief will be condensates; see *Federal Register*, 69 (16): 3492.

41 The sum of the maximum well test rate for the 45 deep gas completions in 2001–2 was 1.24 BCFD (for an average of 27.7 MMCFD per well). In the case of the 20 completions in the 15–16,000 feet depth interval, the sum of maximum test rates was 275.5 MMCFD (equivalent to 22 per cent of the total from all completions), for a 13.8 MMCFD average production per well. For the 12 completions in the 16–17,000 feet depth interval, the sum of maximum test rates was 275.5 MMCFD (equivalent to 31 percent of the total from all completions), and the average production per well was 32.2 MMCFD. One of these completions tested at a maximum rate of 80 MMCFD and three others at more than 50 MMCFD (whereas no completion below 16,000 feet exceeded a test rate of 25 MMCFD). Finally, in the case of the 13 completions beyond 17,000 feet, the sum of maximum test rates was 582.8 MMCFD (equivalent to 47 percent of the total from all completions), and the average production per well was 44.8 MMCFD. Two of these completions tested at a maximum rate greater than 100 MMCFD and three others at more than 50 MMCFD (see MMS 2003: 4).

42 *Petroleum Economist*, June 2002: 7. ST204 for a time was the most prolific gas field in GOM shallow waters. Development costs for the field were around USD 200 million (USD 130 million for drilling and USD 60 million for platforms and pipelines). Its total finding and development costs worked out at 0.50 USD/MMCF of gas equivalent.

43 *Hart's E&P*, April 2001: 50.

44 MMS revised its estimate based on a better understanding of 'deep potential derived from recent deep discoveries that are now producing (Anadarko's Hickory, El Paso's ST204, and Shell's Alex Discoveries)', on the one hand, and 'recently announced large discoveries', as well as 'new seismic data acquired and processed using the latest technology to improve imaging

at increased sub-surface depths … [and the outlining and mapping of] conceptual plays', on the other.

45 Lessees can obtain an SOO on leases under exploration for subsalt deep gas prospects, so long as they have acquired and interpreted full 3-D depth migrated geophysical data beneath the salt sheet and over the entire lease area, and have also submitted to MMS a reasonable schedule of work leading to the commencement of drilling. For an SOO on ultra-deep wells, the same requisites have to be met. Before requesting the suspension, the operator has to have conducted some additional data processing or interpretation of the geophysical information with the objective of identifying a potential ultra-deep hydrocarbon-bearing geologic structure or stratigraphic trap. The lessee also has to demonstrate that additional time is necessary to complete current processing or interpretation of existing geophysical data or information; to acquire, process, or interpret new geologic or geophysical data or information that would impact the decision to drill the same geologic structure or stratigraphic trap; or to drill into the formation.

46 Byrne 2003: 35.

47 *AAPG Explorer*, October 2002: 13.

48 *Ibid.*

49 Sale 190, held on March 2004, witnessed the highest signature bonus payments in three years (USD 369 million). Shallow water blocks accounted for USD 174 million in high bids.

50 Closest bids to LLOG's were Newfield Exploration (USD 21.8 million), a consortium formed by Houston Energy and William G. Helis Co. (USD 18.5 million), and another consortium formed by Houston Exploration Co. and Gryphon Exploration (USD 12.5 million). Out of the 13 bids submitted for the block (all by independents), seven were higher than USD 6 million (*PON*, 21 August 2003: 1–4).

51 *PIW*, 23 June 2003: 6.

52 Byrne 2003.

53 *Federal Register*, 69 (16): 3504.

54 *O&GJ*, 10 May 2004: 32.

55 Two significant deep gas discoveries were announced in 2003: JB Mountain and Mound Point. The former lies in Federal waters (South Marsh Island), the latter in Louisiana waters. Both were found by McMoRan Exploration.

56 *Federal Register*, 69 (16): 3507. Wood Mackenzie, for its part, envisions the industry drilling around 40 deep gas wildcats a year to 2010. If these wells succeed in finding 15 BCFGE per well, between 700 and 1800 BCFD of gas could be added to GOM shallow water production (*O&GJ*, 10 May 2004: 33).

57 If this were the case, these particular wells may end up producing as much as 1.2 TCF of gas over the next eight years. If, against all expectations and the long-term declining trend in discovery sizes, the half-life of these wells turns out to be closer to four years, then they may produce as much as 2.4 TCF of gas over 16 years.

58 *O&GJ*, 10 May 2004: 33.
59 Wade, Plater and Kelley 1999: ix. During 1996–7, for instance, at least 150 MMCFD of planned production was lost on account of failed wells and wells that are producing a fraction of their start-up rate, chiefly due to scaling problems.
60 *Ibid.*: 122.
61 *Ibid.*: 104.
62 As El Paso has found to its cost; during the third quarter of 2004, El Paso's natural gas output was 30 percent lower than for the third quarter of 2003, and less than half of the output reached two years earlier (*PON*, 22 December 2004: 2).
63 Reserves estimates from Wade, Plater and Kelley (1999: 163), minus cumulative production for the 1997–2004 period.
64 Norphlet resources are unlikely to grow significantly, even though 'both the USGS (state waters) and MMS (OCS) assessments assume that the largest pool in the play is still undiscovered'. As the NPC points out, 'good seismic coverage and data suggest the largest fields are already discovered. The play is well explored in Alabama where access is not an issue. Possibly a larger pool might yet be found offshore Florida where there is limited seismic. The ChevronTexaco Destin Dome discovery is in the Norphlet and has a thick gas column, but reservoirs are thin, of poor reservoir quality and structurally segmented' (NPC 2003, v. IV: II–52).
65 *Federal Register*, 69 (16): 3505.
66 *Ibid.*: 3492.
67 *PON*, 6 May 2003: 1.
68 *PON*, 26 May 2005: 5.
69 *JP Morgan Weekly Oil Data Report*, 26 February 2003: 1.
70 Byrne 2003: 36.
71 *Ibid.* Italics in original.
72 *Ibid.*: 31.
73 *Ibid.*: 38.
74 Weissman 2003, part I: 1.
75 *Ibid.*
76 NPC 2003, vol. III: V–17.
77 *Ibid.*, vol. II: 90.
78 *Ibid.*
79 *Ibid.*, vol. II: 225.
80 Weissman 2003, Part I: 5.
81 The NPC is expressly counting on industrial demand destruction to moderate consumption growth in the medium term (NPC 2003, vol. II: 9).
82 In 2003, up to late July inclusive, there had only been one day in which the highest temperature in Atlanta had been above 90°F, whereas in a normal year the expected value would have been 33 days.
83 Weissman 2003, Part I: 6.
84 As the NPC puts it, a sudden increase in the demand for storage in the order of 25 percent relative to a normal year would have a potential for 'even

greater price spikes and demand destruction than what was experienced in 2001 and 2003' (NPC 2003, v. II: 261).

85 As of year-end 2002, firm transmission capacity contracted by power generators amounted to 30 BCFD. Approximately 57 percent of gas-powered generation capacity relied on non-firm transmission arrangements for its supplies (*ibid.*, v. V: T16–17).

86 *Ibid.*, vol. II: 263.

87 *Ibid.*, vol. III: V–26.

88 NPC 2003, vol. II: 30.

89 *Ibid.*, vol. III: V–18.

90 Klara and Shuster 2004: 3. In the 2004 iteration of this report, these authors estimated that coal would fire 42 percent of new US electricity generation capacity by 2025 (112 gigawatts out of 267 gigawatts).

91 Simmons 2003: 17.

92 Weissman 2005, Part 1: 7. At these volumes, by 2025, the annual contribution of LNG imports to the US balance of trade deficit would be in the order of USD 110 billion (in 2005 money).

93 *Annual Energy Outlook* 2002: 82.

94 National Energy Board 2004: 5.

CHAPTER 6

HISTORICAL OVERVIEW OF THE DEEPWATER

Upstream activities in GOM have certainly come a long way since their shallow beginnings. The first genuine deepwater development project in GOM (and, indeed, anywhere in the world) came on stream as far back as 1979. However, up until the mid-1990s, deepwater production in the region expanded very slowly (Figure 6.1), even though a quarter of the leases assigned during the 1980–1990 period were located in deepwater.

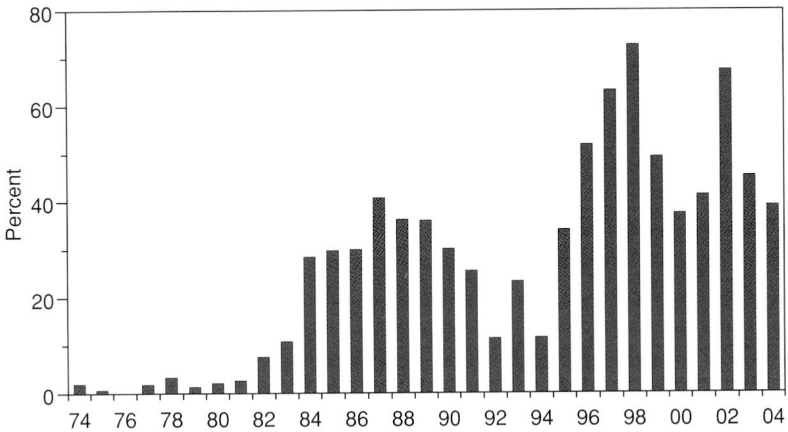

Source: MMS

Figure 6.1: Deep Water Leases as a Proportion of Total Leases Assigned in the GOM Federal OCS, 1974–2004.

The average depth of these leases (which, as Figure 6.2 indicates, have accounted for about 95 percent of cumulative deepwater output up to 2001) is 3600 feet. Their development was very protracted chiefly because offshore technology first had to mature in provinces where, as was not the case in GOM, oil companies could find prospects at more amenable depths (i.e. between 1500 and 2500 feet) that represented a natural progression from those which the industry had already conquered in GOM up to the mid-1980s. That is why, as Figure 6.3 shows, around half of the productive GOM deepwater blocks sold in the decade starting in 1983 had a cycle time of eight years or more between

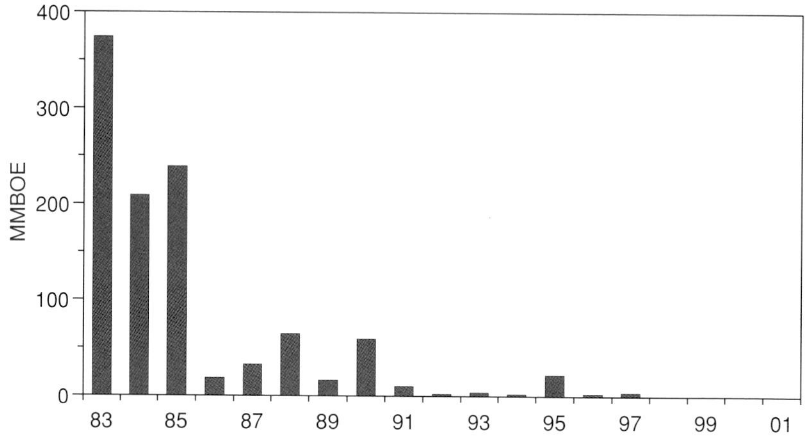

Source: MMS

Figure 6.2: Cumulative GOM Deepwater Production by Year of Lease Assignation, 1983–2001

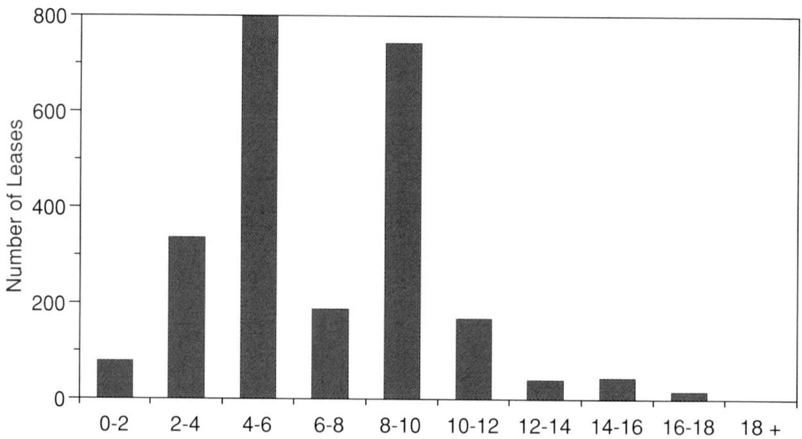

Source: MMS

Figure 6.3: Years Elapsed between Lease Assignation and First Oil for Successful Deepwater Projects in ihe GOM Federal OCS, 1983–1993

lease assignation and first oil. Once protracted teething problems were overcome in the province, however, incremental deepwater production began to dominate the path of GOM's overall oil output profile. The purpose of this chapter is to relate how this came about.

6.1 The Deep Frontier: Chronicle of a Conquest Foretold

In the early 1990s, the GOM deepwater province burst seemingly out of nowhere to become, in a short space of time, the hottest worldwide exploration play at the turn of the twentieth century. This suddenness is misleading, because deepwater E&P activities in GOM in fact had a very long gestation period whose origins can be traced back to a couple of scientific initiatives of the late 1950s to mid-1960s: Project Mohole and DSDP. The former was a US government sponsored effort to drill into the Earth's lower crust and upper mantle, to learn more about the composition and geologic history of the planet. Mohole was terminated in 1966, and even though it failed to secure its primary aim, it yielded many invaluable insights and innovations into the problems of drilling at extreme depths.[1] For its part, DSDP was led by a consortium of leading U.S. oceanographic institutions (called the Joint Oceanographic Institution for Deep Earth Sampling or JOIDES), with the objective of investigating the evolution of ocean basins by core drilling of ocean sediments and underlying oceanic crust (among other things, DSDP research empirically validated the theory of plate tectonics).

Five holes drilled in the initial stage of the DSDP programme in the Gulf of Mexico area, (in waters that are now under Mexican jurisdiction) encountered hydrocarbon shows, with one of them drilling through oil saturated caprock (Figure 6.4 and Table 6.1)[2] in Challenger Knoll, in the Sigsbee Escarpment. [3] At the time, the received wisdom in the oil industry was that the Sigsbee Knolls could not be salt domes (but they are), and that even if they were, they could not have a caprock (but they do), and that even if there was caprock present, there would be no petroleum below it (but there is).

The early 1970s witnessed an increase in the attention devoted to the prospectivity of deepwater areas (the 1974 World Petroleum Congress was specifically dedicated to this topic, for instance). The timing behind this surge of interest owed nothing to coincidence, and everything to the OPEC revolution. As Horsnell says, 'when oil companies had access to the easy acreage, there was really no great incentive to make quantum leaps in technology. [They] could of course have invested in a research programme to allow you to drill in ten thousand feet of water, but there was no earthly reason to do so'.[4] With the easy acreage gone, though, companies had no choice but to prospect for oil in progressively deeper waters, as Figure 6.5 shows, at times seemingly regardless of cost. For instance, Esso's deepwater drilling off Thailand in 1976 set the company back USD 100,000 per day in rig costs (the equivalent of nearly USD 260,000 in 2001 money), at a time when

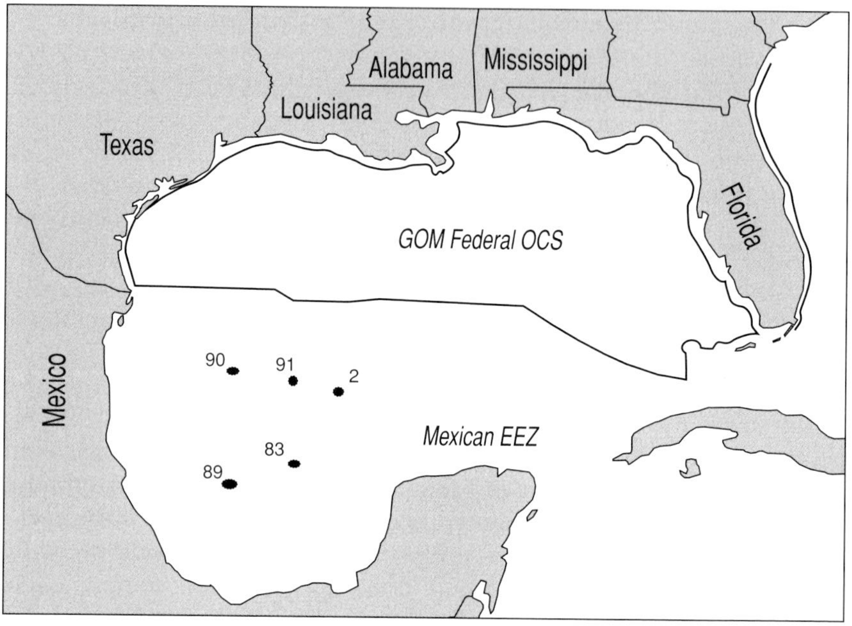

Source: Martin and Foote 1981b

Figure 6.4: Location of GOM DSDP Holes that Encountered Hydrocarbon Shows

Table 6.1: Details of Hydrocarbon Shows Encountered in DSDP Holes in the Gulf of Mexico

DSDP Drilling Site Number	*Location*	*Water Depth (feet)*	*Interval (feet)*	*Period*	*Nature of Show*
2	Challenger Knoll, Sigsbee Basin	11,900	340–480	Jurassic	Oil and gas saturated caprock
88	Salt plug at base of Campeche Slope	8,400	180–520	Pliocene–Pleistocene	Methane and ethane, biogenic
89	On rise near base of slope at south end of Sigsbee Basin	10,200	730–1,250	Miocene–Pliocene	Methane with traces of ethane, biogenic
90	Western Sigsbee abyssal plain	12,400	430–2,560	Miocene–Pleistocene	Methane in upper cores of interval, trace of ethane
91	Central Sigsbee absyssal plain	12,500	530–2,800	Miocene–Pleistiocene	Methane

Source: Hedberg, Moody and Hedberg 1979: 295

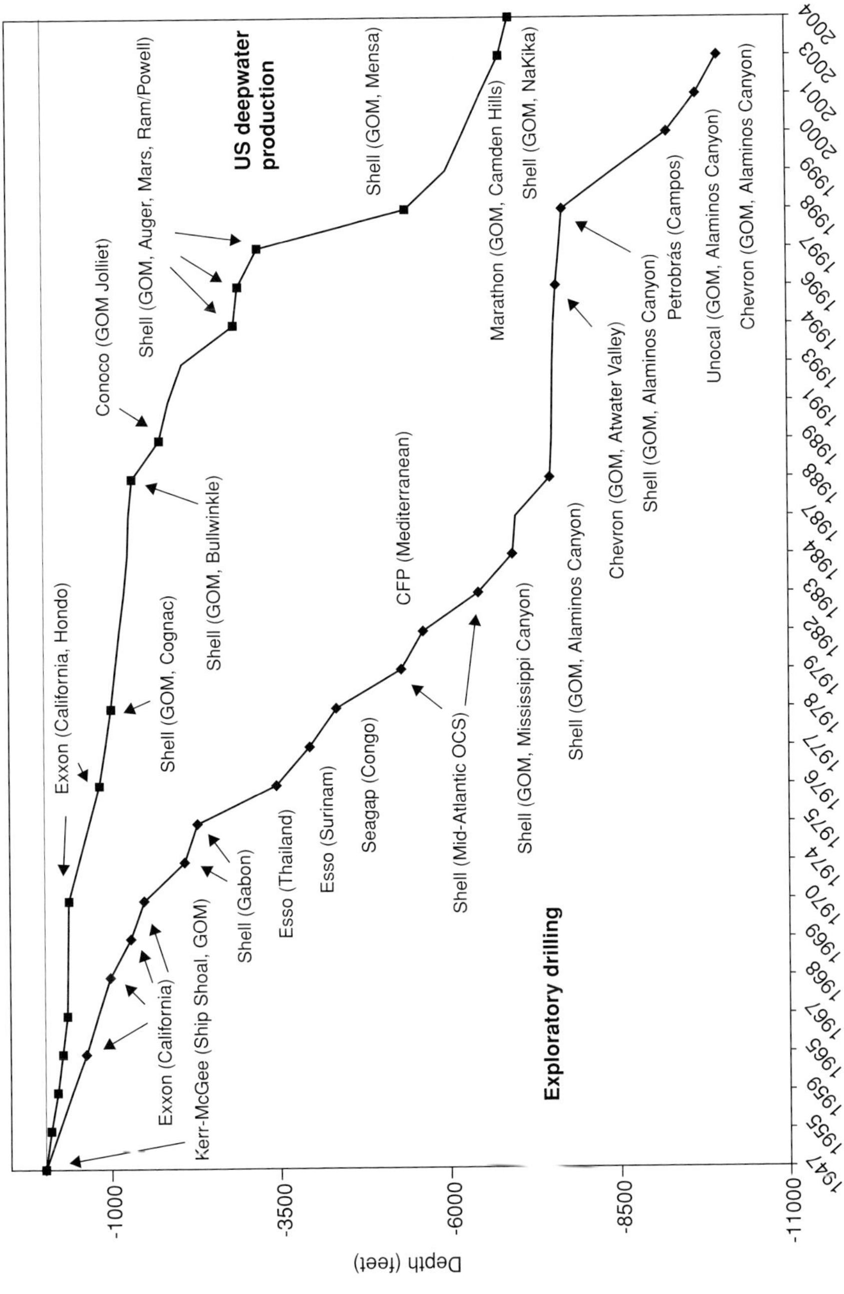

Figure 6.5: Progression of Offshore Exploratory and Production Drilling (Worldwide and USA), 1947–2004

the average cost of drilling and completing an oil well in the USA was only USD 150,000.

GOM was the place where these companies' deepwater endeavours first bore fruit. Given the pedigree of offshore activities in the region, this was as logical as it was fortunate, in that at least it allowed oil companies to take the plunge (literally) into hitherto uncharted areas within the reassuring confines (political and otherwise) of their own backyard. In its initial stage, from 1974 to 1989 (when Conoco installed the Jolliet tension leg platform), the development of the GOM deepwater province was carried out by taking extant technology and making it bigger, thicker, heavier, taller and, inevitably, more expensive. This approach, seemingly close in spirit to the golden rules of Soviet engineering, was really what the times called for, as the technological state of the art left precious little scope for subtlety and finesse.

The first genuine deepwater development project in the world (i.e. the first one to pierce the symbolic 1000 feet depth threshold) can be credited to Shell, which led a consortium that developed a 200 MMBOE find located in 1090 feet of water in four blocks located in the Mississippi Canyon area. The project, christened Cognac, set a number of records at the time of its completion: the three piece structure was the largest ever installed, at the greatest water depth (1025 feet), with the largest number of wells (62), and the heaviest steel platform (59,000 tons). At a total cost of USD 464 million (equivalent to USD 8.68/BOE in money of 2001), it was not the most expensive offshore development to that date, but it was not far off the mark either, especially on per barrel of output basis. Including leasing and exploration costs in the calculations takes total upfront expenditure by the Cognac consortium to a daunting USD 13.96/BOE (again in money of 2001).

The brute force approach embodied by Cognac was not widely imitated, most probably on account of the up-front costs. In 1981, Unocal installed a platform in a prospect lying beneath 955 feet of water, at a fraction of Cognac's outlay (USD 90 million). One of the reasons for the low cost of this development project (christened Cerveza in a none too subtle dig at Shell's perceived extravagance) was that the jacket could be built in one piece instead of three, thanks to the existence of a barge with a length of 650 feet and a launching capacity of 42,000 tons (crucially, this was a cost-cutting option that had not been available to the Cognac consortium).[5] Cerveza spawned a sister project called Cerveza Ligera but its light jacket approach proved to be just as much of a technological dead end for deepwater development as Cognac had been. For its part, Shell continued with its tried-and-tested maximalist approach in the Bullwinkle project, which was basically Cognac writ on an even grander scale: the platform's one piece jacket, lying in 1353 feet of water, was at the time the world's tallest, and although it had two fewer well slots (60) it weighed around 30 percent more (77,000 tons). This platform cost 20 percent less to build than Cognac, chiefly because it was built and launched in one piece (thanks to the existence of giant barges). The other department where Bullwinkle came in a distant second to Cognac was leasing costs: Shell paid USD 34.5 million for the three blocks where the field was found, a paltry sum in comparison to the colossal amount that the blocks harbouring Cognac cost the consortium that developed it.

6.2 Bullwinkle and the Transformation of Deepwater Economics

Even as the Bullwinkle jacket was being installed in 1989, Shell engineers were predicting that the likes of it would never be built again, and time has proved them right. Developing the Mars field with a fixed platform, for instance, would have required nearly four times more steel than was employed at Bullwinkle (290 million tons), at a cost of at least USD 3 billion. That is not to say that developing Bullwinkle by means of a fixed platform was a mistake, though. By taking the conservative approach of extrapolating extant technology to its very limits (rather than tinkering with more promising but ultimately unproven methods), Shell was able to bring this particular field onstream far more quickly and cost effectively than would otherwise have been the case. Furthermore, its investment in both Cognac and Bullwinkle has been repaid handsomely: cumulative oil production for these fields as of 2001 stood at approximately 170 MMB and 102 MMB, respectively. Moreover, Bullwinkle (whose location at the edge of the deepwater was particularly favourable) was turned into one of the first major processing hubs built around subsea production in 1997 (when it was expanded to handle 200 MBD of oil and 320 MMCFD of gas). Hub operations have proven extraordinarily profitable for Shell,[6] so one has to presume that Unocal's jibes must have resounded in Shell's ears all the way to the bank.

At a time when Bullwinkle was still on the Shell drawing boards, other companies were looking at ways of solving the problem of commercial deepwater production. In its Lena prospect (located in 1017 feet of water in Mississippi Canyon block MC281), Exxon pioneered the use of a guyed tower, which some studies indicated could be the optimal solution for water depths between 1000 and 2000 feet. In the end, Lena also turned out to be a pricey one-off, but in many ways this project is emblematic of the sometimes far from obvious hurdles that have had to be overcome in order to make the deepwater amenable to development. Suffice it to say that an apparently minor detail like launching the Lena tower on one of its ends (the taken-for-granted procedure in the industry up to that point) would have required beefing up the structure to the tune of 6 percent of its weight (i.e. the equivalent of 3000 tons of steel), merely to enable it to sustain the cantilevered loads that would have been generated during the tilt.[7]

Notwithstanding the visibility of landmark projects like Bullwinkle and Lena, the GOM deepwater severely tested the patience and staying power of the oil industry. Up to the late 1980s, 74 fields had been discovered at depths greater than 600 feet, but a small mean field size meant that leases on thirty of those fields had expired without production, while a further two fields had produced for less than two years (cumulative production at these fields was a paltry 0.5 MMB of oil and 5.4 BCF of natural gas). Out of the remaining fields, only twenty (with an average field size of 75 MMBOE) were proven, with 14 of them in production. A further 22 fields (of a significantly smaller average size) were active but unproved, with development decisions pending.

At that point, the fields that would eventually launch the deepwater boom had already been discovered (Ram/Powell in 1985, Auger in 1987, Mars in 1989). Shell – the company that would do the most to get the boom going – was still publicly playing down the commercial prospects of the GOM deepwaters, ostensibly because of the low flow rates that it had encountered at its various discoveries.[8] However, Shell's lamentations were merely a ruse to hide its deepwater hand from other companies: as Figure 6.6 shows, Shell was very busy at this point accumulating deepwater acreage, and was streets ahead of its competition in the process of deepwater prospect identification. Indeed, it had stolen a march on the rest of the industry during the late 1970s, when it had undertaken extensive seismic shoots all over the Atlantic ocean, from the vessel *Lady Glorita* during the return legs of a number of exploration voyages (seismic lines were shot, without authorisation, in areas that would later be incorporated into the EEZs of a number of countries). It was the data collected in these trips that first alerted Shell to the presence of enormous deepwater structures offshore West Africa, for instance but, at that point offshore technology was still much too primitive for the company to be able to do anything about its privileged information.

The next milestone in GOM deepwater development came in 1987, when Conoco began to install a tension leg platform (TLP) at the

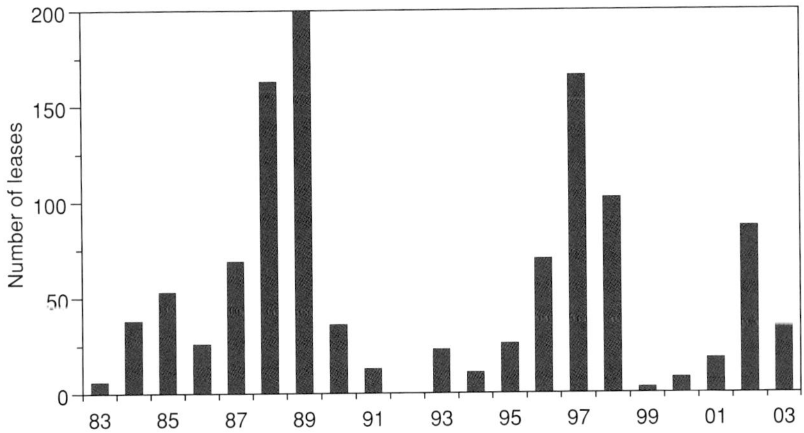

Source: MMS

Figure 6.6: Shell Annual Deepwater Lease Acquisitions in the GOM Federal OCS, 1983–2003

Jolliet prospect (discovered in 1981), located in Green Canyon block GC184, in 1760 feet of water. TLPs, whose use had been pioneered by Conoco in 1981 in the North Sea's Hutton field, were originally seen as the technology of choice for small fields that had a brief production lifespan. A large part of their attractiveness supposedly lay in the fact that much of their structure could be removed from one location and taken to another.[9] In practice, this has proven a largely empty promise: nearly all TLPs currently in use in the world have been built from scratch.[10] However, they have exceeded expectations in almost every other relevant department and, as a result, they have become virtually synonymous with deepwater activities in the more glamorous GOM fields (their capital costs are too high for smaller fields, and they are unsuitable for ultradeepwaters). TLPs have accounted for the lion's share of the cumulative GOM deepwater output from the installation of Jolliet onwards.

More or less at the same time as Jolliet was halfway through its installation process, a truly momentous event was taking place at the Bullwinkle field, where Shell was gearing up to start production in earnest. Originally, the company thought that Bullwinkle wells would produce about 4 MBD apiece (i.e. roughly double the rate of Cognac wells, and very much at the outer boundary of the best sustainable rates ever achieved at shallow water wells). However, production testing revealed that the wells could sustain much higher flow rates without any attendant loss of pressure (in fact, the highest single well production rate ever achieved in Bullwinkle was 8.425 MBD). At that point, those in the know about GOM deepwater production (most of whom were in Shell's employ) reached the conclusion that compaction and diagenesis of deepwater reservoir sands had been minimal because of relatively recent and rapid sedimentation, which meant that well productivity in the sandstone reservoirs of GOM deepwater fields could approximate – perhaps even exceed – those achievable in Mexico's famously prolific offshore carbonate reservoirs.

At a stroke, the economics of deepwater production were radically transformed by this discovery. For instance, Shell originally thought that the development of Auger would require drilling thirty wells, which would allow the company to produce 45 MBD (assuming a production rate of 2500 BD per day per well, with 18 wells operating at any one time). In fact, Auger ended up needing only 14–17 high capacity wells (i.e. able to sustain production rates greater than 10,000 BD each), and total production capacity at peak more than doubled the value originally estimated, which obviously allowed for its costs to be recouped much faster.[11] Unsurprisingly, development decisions on the most attractive deepwater strikes followed Shell's pleasant surprise at Bullwinkle in relatively quick succession: Auger in 1989, Amberjack in 1990, Pompano in 1992, Mars in 1993.

As can be appreciated in Figure 6.7, output from these various fields

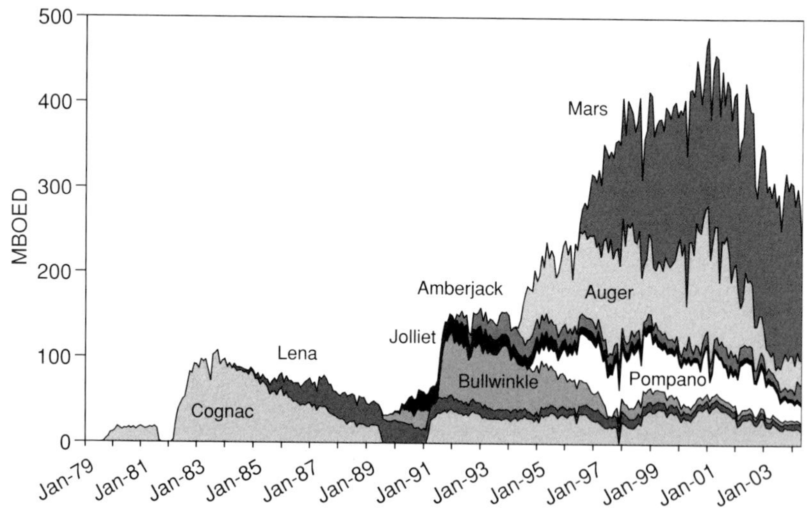

Source: MMS

Figure 6.7: Production Profiles of Early Deepwater Development Projects in the GOM Federal OCS, 1979–2004

represented a major turning point in the evolution of the production profile of the GOM region. And although there is no denying the excellence (even world-class stature) of some of these strikes, Horsnell is surely right when he points out that it is

> a measure of the extent to which opportunities for developing new oil provinces in the world have narrowed, that such a surge in interest in deepwater oil occurred in the late 1990s. Over that period oil prices were more often weak than strong, and companies continued to stress that they would achieve growth in profitability by constant reduction of costs. These are hardly the optimal conditions to launch oneself in frontier projects, involving far greater than usual technical cost, risk and managerial attention.[12]

Hardly optimal the conditions might have been, but given an overall lack of prospects worldwide, oil companies were forced to try their luck in the deep, or resign themselves to eventual extinction.

6.3 Conclusions

The initial phase in the development of the GOM deepwater province can be said to have come to an end with Shell's announcement of its development plans for Mars. According to industry lore, Shell geologists identified a couple of interesting seismic reflections in two blocks located in Mississippi Canyon (MC763 and MC807), and the team putting together the company's acquisition strategy in an upcoming lease sale decided, almost as an afterthought, to put a bid on these blocks. Although the prospect was seen as very high risk, Shell nonetheless decided to drill it, selecting it over other – supposedly less risky – blocks. Shell convinced BP to come on board in order to spread the development risk (which was still substantive by any measure), and then proceeded to find a field that, until the advent of Thunder Horse, could boast of being the largest US oil strike since Prudhoe Bay.

The Mars development announcement was a watershed event because it convinced the oil fraternity at large that, far from being dead as an oil play, 'in terms of potential, raw barrel numbers, the fiscal terms of the play and the available technology, including infrastructure available, the Gulf [was once again] the place to be'.[13] Deepwater fever gripped the industry, and companies that had remained on the sidelines during the MMS auctions held in the 1980s began a belated effort to put together a good lease inventory for drilling. The result of their enthusiasm is apparent in Figures 6.8 and 6.9, which plot the winning bids submitted for individual deepwater blocks before and after 1993.

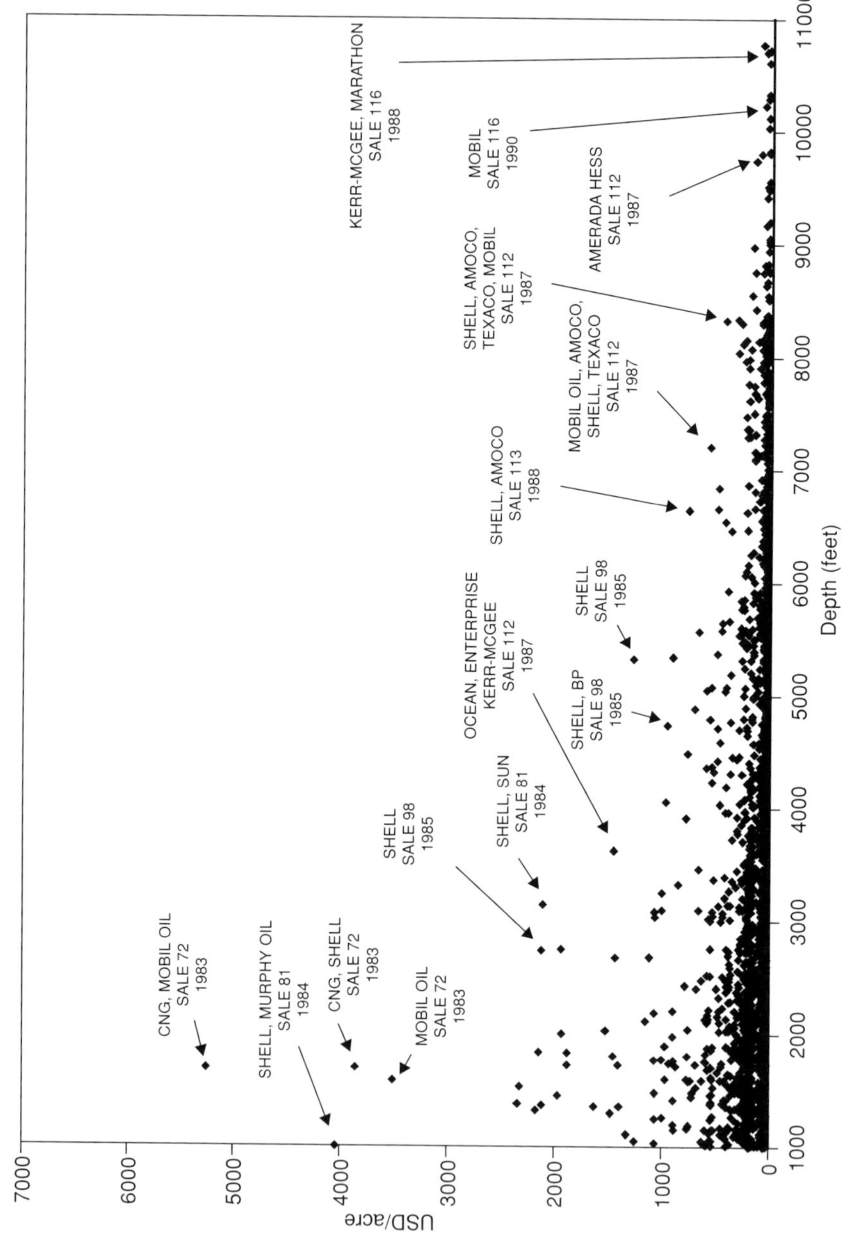

Source: MMS

Figure 6.8: Distribution of Winning Bids for GOM Deepwater Acreage, 1983–1993

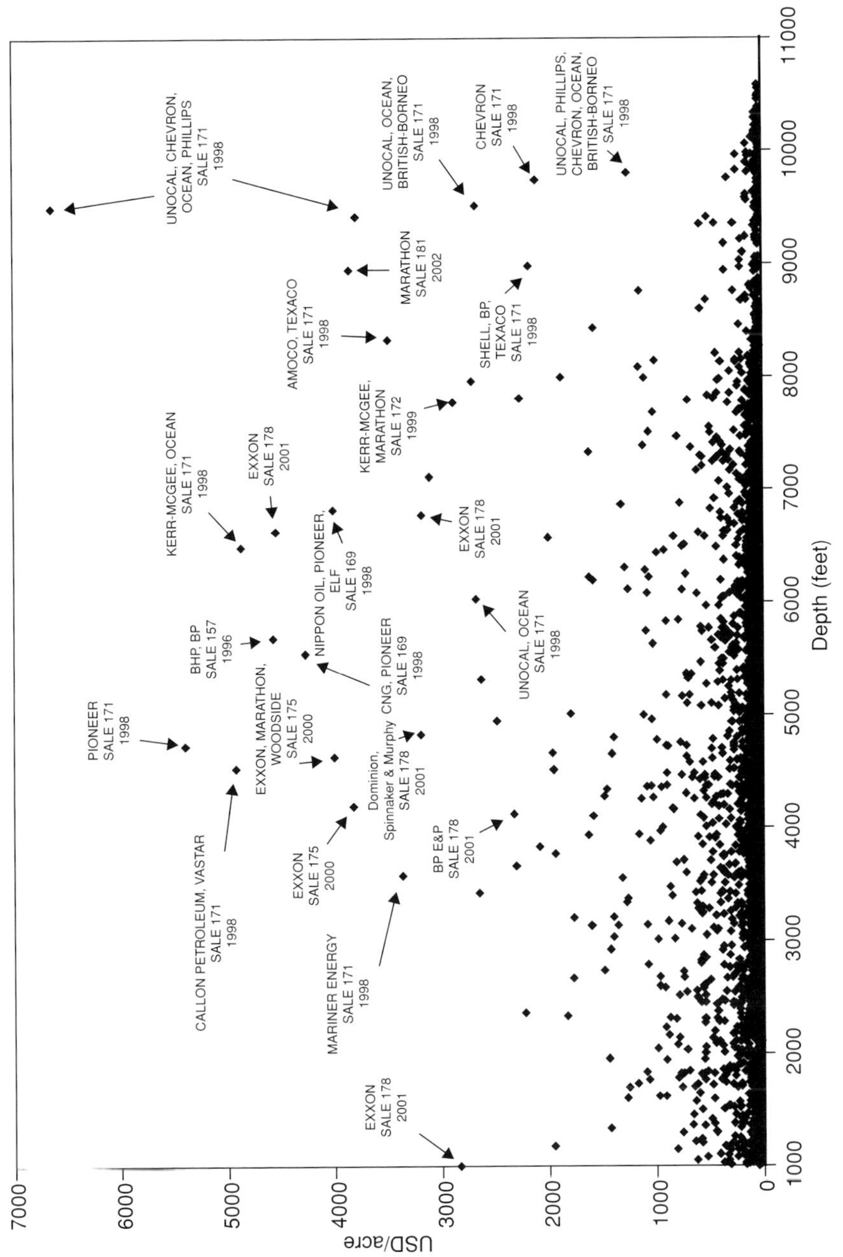

Source: MMS

Figure 6.9: Distribution of Winning Bids for GOM Deepwater Acreage, 1994–2004

As can be appreciated, the highest overall bids registered after 1993 were submitted in lease sales held toward the end of this period. By this time, offshore technology had caught up with GOM deepwater conditions and, thanks to the efforts of engineering and service contractors, its use had become commonplace outside the rarefied circle of the largest oil majors.

Although oil companies have grabbed the headlines throughout the whole opening up of the deepwater province, the key protagonists in this process have in fact been the service companies. A significant percentage (around 75 percent) of the oil industry's upstream capital expenditure is sourced externally, and whereas it used to be the case that companies (especially majors) bought supplies, equipment and services on a straight fee basis from contractors, today the relationships between the buyers and the providers of oilfield services are full-scale 'outsourcing alliances that go beyond ordinary transactional arrangements and involve the sharing of risk and reward'.[14] It is only the propensity exhibited by oil companies to 'bask in the glow of the technological strides made' that has obscured the fact that 'the real innovators in this story have perhaps been the service companies'. [15] So, while the geological concepts that suggested oil existed in these new environments by and large came from the E&P companies, a significant share of the know-how and technology to get the oil out came from the service companies.

Engineering and services contractors have made a vital contribution to the overall expansion of the industry's operational envelope and capabilities, partly by developing technology able to cope with progressively more demanding operating conditions and smaller fields, all across the whole spectrum (from surveying to imaging to riser and platform design, and so forth), and partly by taking on the role of systems integrators (a function comparable in many ways to the one that prime contractors play in other high technology industries, notably aerospace). Nowadays, service companies orchestrate 'alliances and contractual relationships involving suppliers, service providers, and even other operating companies ... to reduce overall system costs and cycle times and to ensure access – sometimes pre-emptively – to crucial technology and inputs'. [16] These networks, which 'are most relevant in technologically complex frontier regions ... where exploration and development are expensive and risky',[17] have underpinned the efforts of non-majors in the deepwater province, and have played no small part in the frontier success stories of even the very largest oil firms.[18]

As the oil industry moves into ever harsher operating environments and the mean size of strikes shrinks, the importance of service companies can and will only increase. After all, as Horsnell highlights, 'given the large-scale run down of the internal capabilities of most major oil companies, and the apparent, and almost certainly short-sighted, downgrading of the importance of engineering and research within company budgets ... the bold predictions that the industry makes for its own future capabilities will [have to] be primarily delivered by the contractors and not the oil companies'.[19] There is an awful lot riding on these bold predictions, in corporate as much as in strategic and geopolitical terms. Fortunately, the capability of service companies – as well as operators – to move into ever deeper waters is an issue in cost terms, but not in terms of drilling technology *per se*. Moreover, as production technology continues to evolve, faster completions and well interventions will lead to reductions in development costs, while advances in subsea technology will enable the industry to unlock stranded reserves and enhance recovery at existing fields.

All that is good news of course. However, whether the industry's predictions for deepwater output materialise or not will depend primarily on how large the as yet undiscovered deepwater resource base turns out to be. Most observers see the deepwater creaming curve as being consistent with what one would expect for a basin that is very much in an immature exploration phase, with many fields (including some large ones) awaiting discovery. This interpretation is often underscored by superimposing the deepwater creaming curve on the shallow water creaming curve (Figure 6.10). Needless to say, much solace is drawn from the clear contrast between both curves, the implication being that if deepwater reserve additions follow a similar pattern to the one seen in the shallow water sub-province, 'then many years of deepwater success should lie ahead'.[20]

Such an assertion begs the question of exactly how many years of success lie ahead, because it is not valid to pose a like for like comparison between the shallow and deepwater provinces. To start with, deepwater fields are depleted very quickly, which means that even if cumulative production in the shallow and deepwater provinces does turn out to be similar (a far from certain eventuality), the aggregate production profile in both provinces will be very different, with the deepwater having a markedly shorter lifespan. Furthermore, most of the tail-end of the shallow water creaming curve is made up by reserves from fields whose size would pose seemingly insurmountable economic obstacles to development were they to be found in the deepwater.

If one casts the shallow water curve aside, then, the profile of the GOM deepwater creaming curve flattens significantly when reserves are plotted against wells drilled. This profile, moreover, is very different from the one from curves of genuinely frontier deepwater areas (like

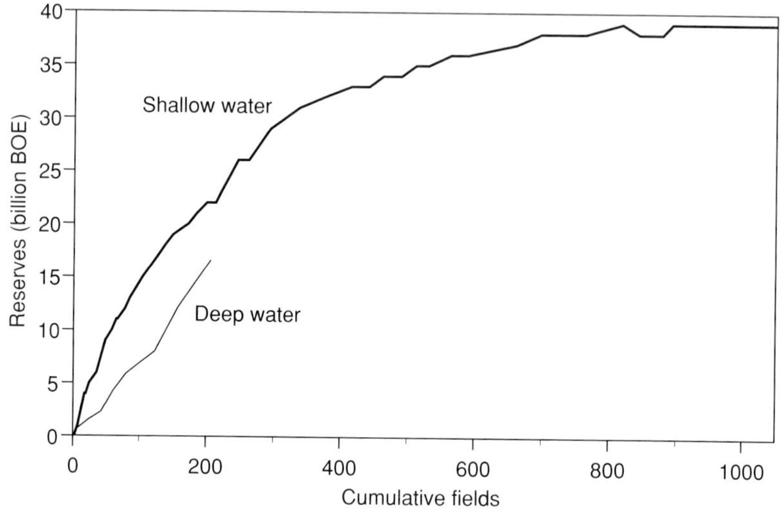

Source: MMS

Figure 6.10: Creaming Curves for GOM Shallow Water and Deepwater Sub-provinces, as of 2002

Angola, shown in Figure 6.11), which has prompted some observers to posit that, perhaps, the GOM deepwater might in fact be quite close to reaching maturity.[21] Such a conclusion apparently goes against the grain of MMS estimates of ultimate deepwater reserves, which the agency puts at 71,000 MMBOE,[22] out of which only 20 percent has been found (i.e. 56,400 MMBOE remain to be discovered).[23] Indeed, MMS ultimate deepwater reserve estimates compare very favourably with its own estimates of shallow water ultimate reserves, which the agency puts at 65,000 MMBOE (out of which 49,800 MMBOE have already been discovered).

Although the MMS figures appear superficially reassuring, digging a little more suggests that one of the methodological cornerstones of this assessment may be seriously flawed; namely, the USGS estimates of the likely thickness and extension of Palaeogene (Palaeocene, Oligocene, Eocene) strata throughout the deepwater area. The overwhelming majority of deepwater discoveries to date have been made in Pliocene and Miocene strata, with very few – and for the most part, modest – strikes thus far (Chinook, Great White) in deep Palaeogene formations (the so-called Wilcox play). None the less, the USGS has extrapolated the thicknesses and extension of Pliocene and Miocene strata to the Palaeogene, and this has led it to conclude that enormous volumes of hydrocarbons are yet to be found in Palaeogene rocks. Crucially, though, Palaeogene

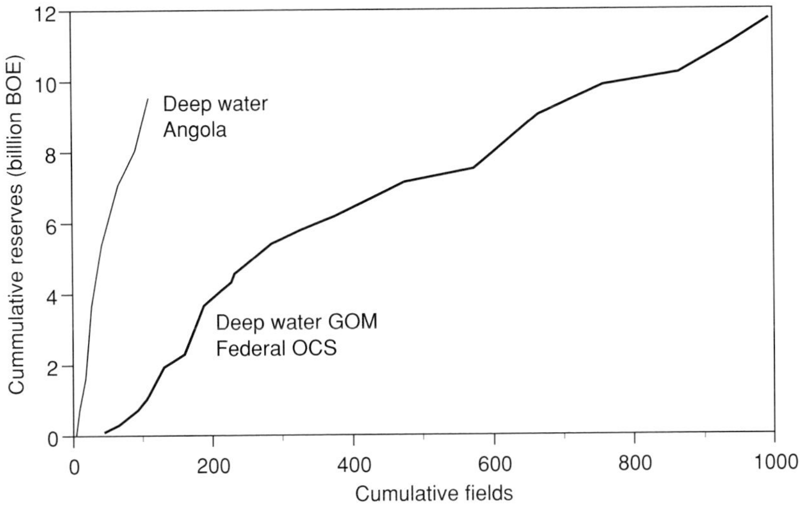

Source: MMS

Figure 6.11: Creaming Curves for Selected Deepwater Provinces, up to 2003

strata of the requisite thickness, quality and extension seem to lie mainly in the Mexican sector of the GOM deepwater, whereas strata in the American sector appear to be thinner, much less extensive and often broken up by salt intrusions (in fact, in the Eastern Planning area, the Palaeogene is absent).[24] Indeed, pessimism regarding Palaeogene prospects, supported by robust empirical evidence and rigorous geological analysis (not to mention a long string of expensive, unsuccessful wells), led the geologist in charge of OPEC's world oil supply analysis at the time of writing to posit that a likely figure for GOM deepwater reserves yet to be found could be 4000 MMBOE,[25] which is a far cry indeed from the nearly 57,000 MMBOE estimated by MMS.

Given the above, reservations about the bullishness of the USGS (and, by extension, that of MMS and DOE) regarding the Palaeogene do not appear out of place. Healthy scepticism regarding the prospects of the Palaeogene (in particular) and the GOM deepwater (in general) is clearly widespread within service companies, as witnessed by their reluctance to invest in expanding their capacity up until the end of 2004 inclusive, despite the manner in which the runaway demand for their talents (reflected, for instance, in historically high rig fleet utilisation rates) has overstretched them over the 2002–2004 period.

We shall retake the critically important question of the expected lifespan of the GOM deepwater in a subsequent chapter. First, though, we will address the issues of its deepwater resource endowment, the production profile of its deepwater fields, the evolution through time of deepwater oil and gas output, and the impact on future deepwater production and reserve addition trends derived from the foreseeable allocation of exploration capital between oil and gas activities in North America.

NOTES

1 See Pratt, Priest and Castaneda 1997: 120–36.
2 See *The Leading Edge*, July 1999: 722. Martin and Foote 1981a: 172.
3 Martin and Foote 1981b: 55. The hole was drilled to a depth of 472 feet. It encountered an immature oil of post-Cretaceous age. As a result of this surprising finding, subsequent DSDP holes in GOM were drilled in places where the chances of accidentally encountering an oil or gas reservoir were seen as minimal.
4 Horsnell 2000: 78–9.
5 Roesset 1999: 106.
6 The Bullwinkle hub currently handles subsea production from, among others, the Rocky, Troika, Angus, Manatee and Aspen fields.
7 Danaczko, Pichini and Rowe 1985.
8 Wallace, Duberg and Kirkley 2003: 24.
9 Pratt, Priest and Castaneda 1997: 274–5.
10 However, the Hutton TLP was redeployed in 2002. In contrast, up to 2005, the FPSO *Petrojarl1* had been redeployed 17 times between the UK and Norwegian sectors of the North Sea.
11 Roesset 1999: 111.
12 Horsnell 2000: 78–9.
13 Stouffer and Knight 2002: 2. The phrase came from an official in charge of deepwater activities at Unocal.
14 Ernst and Steinhubl 1997: 153.
15 *Ibid.*
16 *Ibid.*
17 *Ibid.*
18 As can be readily appreciated by examining in detail the various technological components that must come together to bring to fruition deepwater developments like the ExxonMobil-led Hoover/Diana, BP/Oxy's Horn Mountain, BP/Shell's NaKika, Kerr-McGee's Red Hawk and the BP-led Mad Dog and Holstein (see the special supplements that *Hart's E&P* dedicated to these projects in March 2002, January 2003, April 2004, December 2004 and April 2005, respectively).
19 Horsnell 1999: 62.
20 Eskew and Jones 2001: 80–1.
21 See Sandrea 2004.

22 Crawford *et al.* 2003.

23 These figures were calculated by assigning each producing field and reservoir to one of 92 GOM hydrocarbon plays (on the basis of Bascle *et al.* 2001), and then estimating the number of undiscovered accumulations in each play, assuming a lognormal size distribution for all accumulations within a play, modelling frontier or conceptual plays on similar but more mature plays, and applying cumulative growth factors to all fields (it has been widely observed that oil and gas fields tend to 'grow' throughout their lifetimes, in proportions that vary according to their type and size. Reasons for this growth vary widely, but include areal extensions of existing reservoirs, discoveries of new reservoirs, and improvements in production procedures. Field growth also reflects the understandable conservative bias of any estimate prepared early in the life of a field. A detailed discussion of reserve appreciation and cumulative growth factors may be found in Lore *et al.* 2001: 49 ff).

24 Even volcanic clastics have been found in some particularly poor quality Oligocene formations in the American sector.

25 Sandrea 2004.: 20–1.

CHAPTER 7

DEEPWATER OIL AND GAS PRODUCTION

The resource endowment of the deepwater sub-province has proved to be almost diametrically opposed to that of the shallow water. In other words, it is an oil- rather than a gas-prone region. The main reason for this is that most of the GOM deepwater rocks are comparatively cool, on account of three factors. Firstly, the source rock for most of the deepwater area is an Upper Jurassic kerogen that generates substantial quantities of natural gas only when subject to very high temperatures (as it is, gas-to-oil ratios in all US Gulf Coast fields – both on- and offshore – only show a sharp increase at reservoir temperatures greater than 305°F). Secondly, the temperature gradient at the top of the sediment column starts at a fairly low level (39°F, or 4°C).[1] Thirdly, subterranean thermal gradients in the deepwater area are modest (with underground temperature increasing at a rate of only 1.0–1.25°F per 100 feet of depth).[2] Thus, despite the enormous pressures encountered at 25,000 feet below sea level, typical reservoir temperatures at this sort of depth over the majority of the deepwater area are only 200–300°F (which is considerably less than the 415°F registered at a depth of 21,000 feet in the Mary Ann gas field, for instance[3]).

Deepwater natural gas discoveries have been concentrated in relatively restricted areas: the upper slope portion of all lease areas (i.e. the shallowest portion of the deepwater), on the one hand, and the East Breaks and Mississippi Canyon administrative divisions, on the other.[4] This explains in large part the smaller average size of deepwater gas discoveries, as reservoirs in the slope tend to be thinner and less continuous than elsewhere. Beyond a depth of 1200 feet, the majority (70 percent) of deepwater discoveries have been oil, with a further 10–12 percent involving fields that harbour both oil and gas.

In terms of magnitudes, a clear division is also apparent between deepwater oil and gas finds. Whereas 60 percent of the 55 deepwater oil fields discovered up to 2000 fell within the large or giant categories, 87 percent of the 31 deepwater gas discoveries fell within the small or medium categories. Indeed, not one of the thirty largest deepwater discoveries up to that year had been a gas field, and only around 10 percent of the fifty largest discoveries had been gas fields. Likewise, whereas 31 shallow water fields have turned out to harbour at least

1 TCF of recoverable reserves (most of it non-associated gas), to date only one deepwater field (Thunder Horse) has been found to have this much gas. Furthermore, most of the reserves at Thunder Horse are constituted by natural gas liquids, rather than dry gas. This is quite a common characteristic in deepwater fields, and it is responsible for the significant increase in the GOM output of natural gas liquids and condensates that has accompanied the unfolding of the deepwater boom (Figure 7.1).

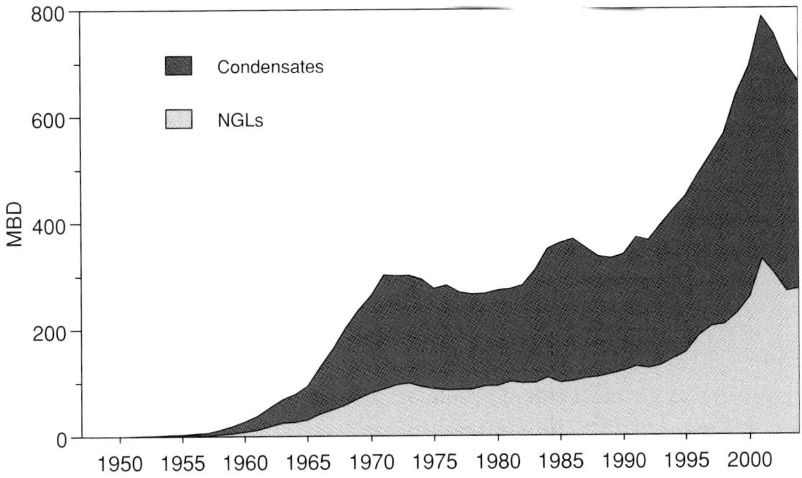

Source: MMS

Figure 7.1: GOM Federal OCS Output of Condensates and Natural Gas Liquids, 1947–2004

7.1 Deepwater Field Profiles

The production profile of GOM deepwater gas fields is very similar to that of small shallow water fields (as well as the new crop of fields in the deep gas sub-province), namely, a very quick rise to maximum output and an equally quick decline leading to abandonment. The key difference between shallow water and deepwater gas fields involves the larger share that is accounted for by liquids in many deepwater gas projects.

Figures 7.2 and 7.3 plot the production profile of two representative deepwater gas projects, the Shell-operated Mensa (located in Mississippi Canyon blocks MC686, MC687, MC730 and MC731) and the ExxonMobil-operated Mica (located in Mississippi Canyon blocks

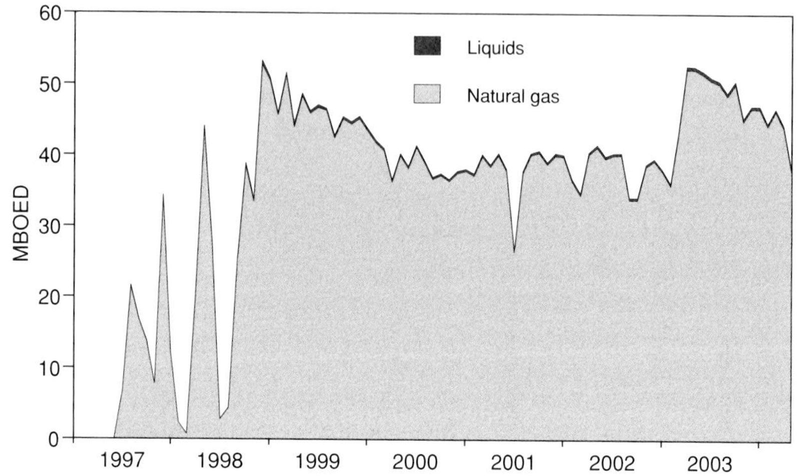

Source: MMS

Figure 7.2: Production Profile of Mensa Natural Gas Development, 1997–2004

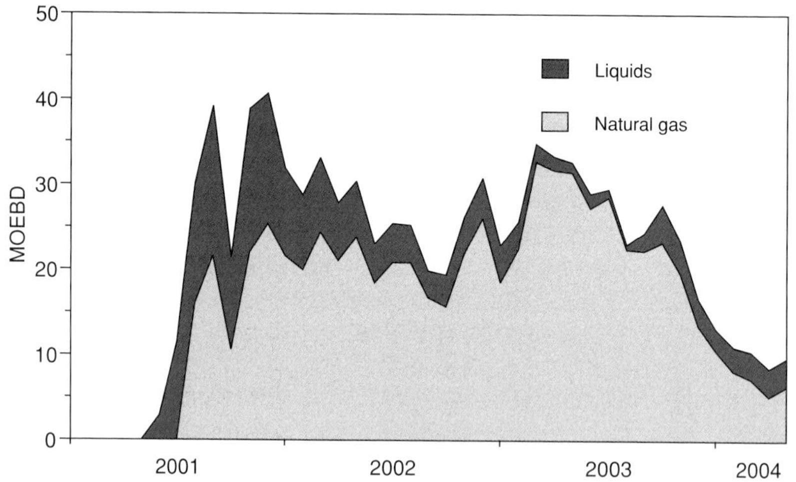

Source: MMS

Figure 7.3: Production Profile of Mica Natural Gas Development, 2001–2004

MC167 and MC211). Output of liquids by the former is negligible, but quite significant in the case of the latter.

The production profiles of GOM deepwater oilfields display less uniformity, as they depend both on field size and field vintage. In comparison to traditional shallow water oilfields in their class size, the first deepwater fields to come on stream displayed both a quicker ramping up to peak output and a more pronounced decline thereafter. Figure 7.4 plots the evolution of output at the Cognac development. Oil and gas production from temporary facilities at the field began during September 1979 and was briefly suspended to allow production from permanent production facilities to begin in March 1982. A second, much longer, production hiatus is associated with a major field redevelopment program initiated in July 1989, and which required a platform shut-in until March 1991.

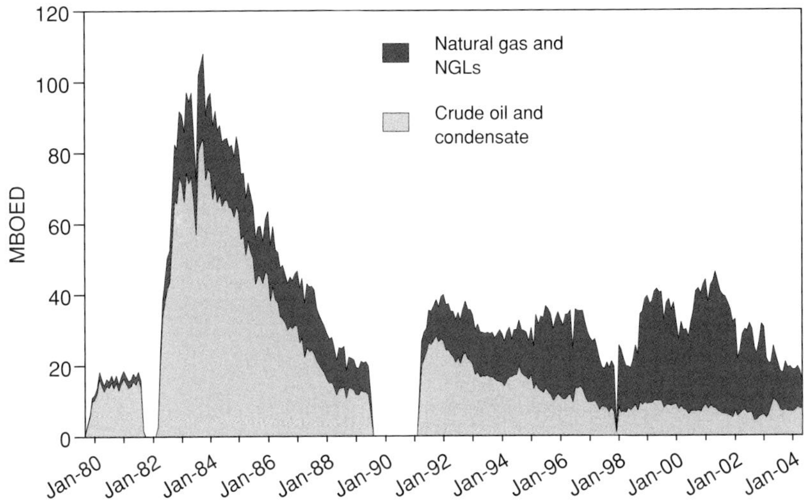

Source: MMS

Figure 7.4: Production Profile of the Cognac Development, 1979–2004

The production profile of the Bullwinkle development is plotted in Figure 7.5. As in Cognac, production initially was carried out from temporary facilities, with production through permanent facilities starting two years later. In a further parallel to Cognac, production activities at the Bullwinkle platform were temporarily suspended approximately six years after their commencement, albeit in this case the purpose of the shut-down was to give the platform a new lease of life by transforming it into a deepwater production and processing hub.

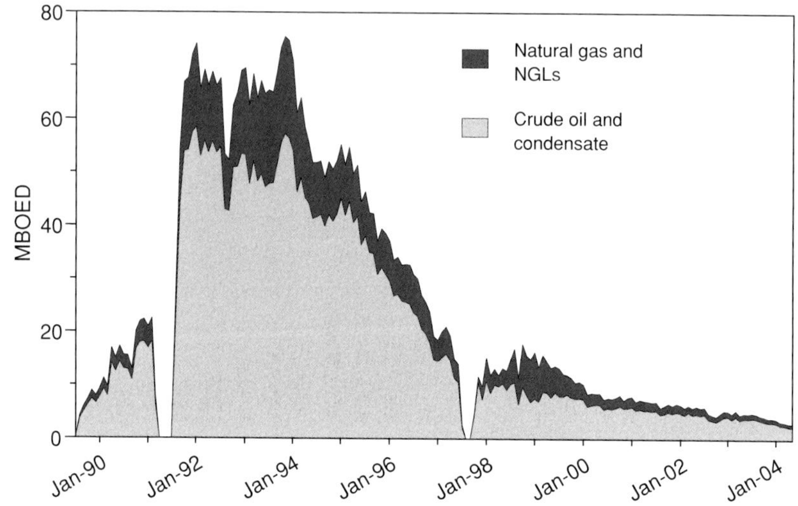

Source: MMS

Figure 7.5: Production Profile of the Bullwinkle Development, 1989–2004

The large deepwater fields of the onset of the boom period (i.e. Mars) have very different reservoir economics and production profiles than earlier fields like Cognac and Bullwinkle, due partly to technological progress, and partly to the less incremental nature of their development. The time to peak of fields such as Mars (Figure 7.6) is briefer, and they will probably have a less protracted economic lifespan, although post-peak production was initially expected to go on for a relatively long period of time. The production profile of the largest among all the newer vintage deepwater developments, Thunder Horse/Thunder Horse North, is expected to be similar (Figure 7.7).

In contrast, as can be seen in Figures 7.8 to 7.12 (which show the production profiles of the Brutus, Hoover/Diana, Mad Dog, Petronius and Ursa development projects), the production profile of smaller deepwater fields is compressed, sometimes (i.e. Petronius) extremely so. This sort of field may take barely over a year to peak, at relatively high production rates, whereupon their production curve goes into a short decline phase characterised by a very steep slope. Furthermore, whereas production at larger deepwater developments has tended on the whole to exceed initial expectations, the opposite appears to be the rule for the smaller projects to have come on stream.

Figure 7.13, which plots the average depth of producing fields weighted by annual production levels, makes it easy to gauge the impact

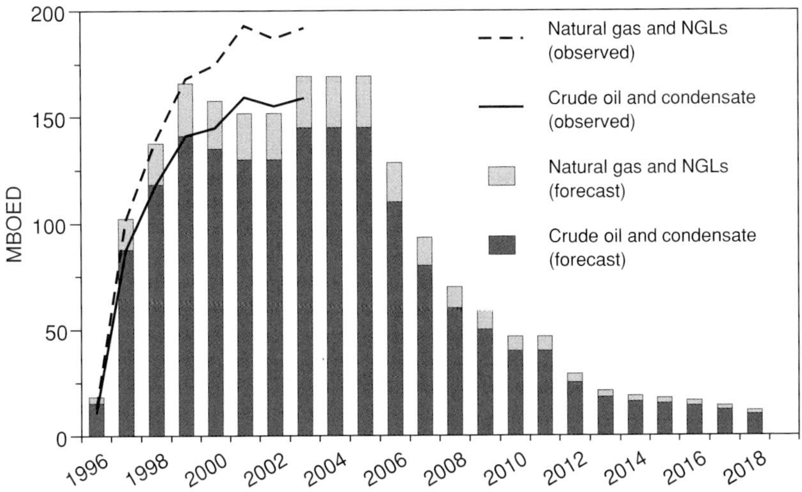

Sources: Riddle, Snyder and George 2001, MMS

Figure 7.6: Production Profile of the Mars Development, 1996–2019

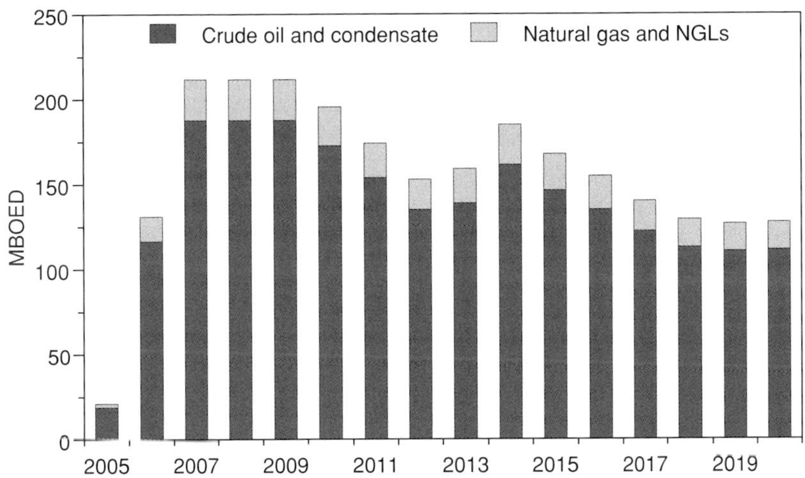

* Production is now not expected to start until 2008

Source: BP

Figure 7.7: Production Profile of the Thunder Horse/Thunder Horse North Development, 2005*–2020

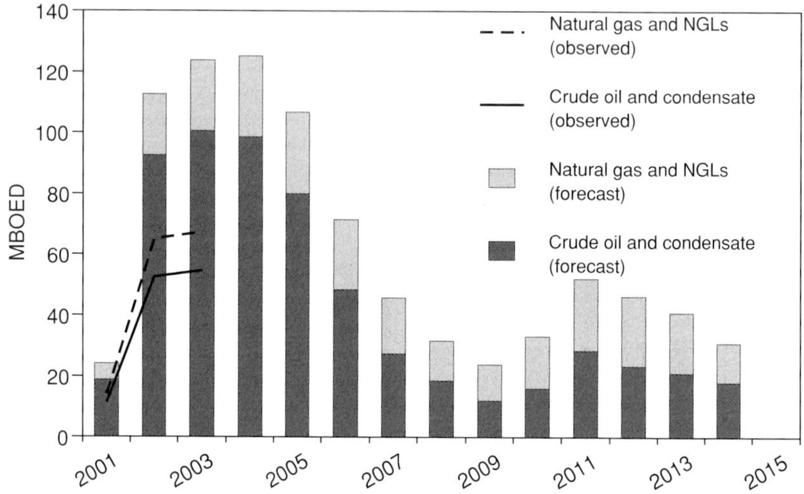

Sources: Riddle, Snyder and George 2001, MMS

Figure 7.8: Production Profile of the Brutus Development, 2001–2015

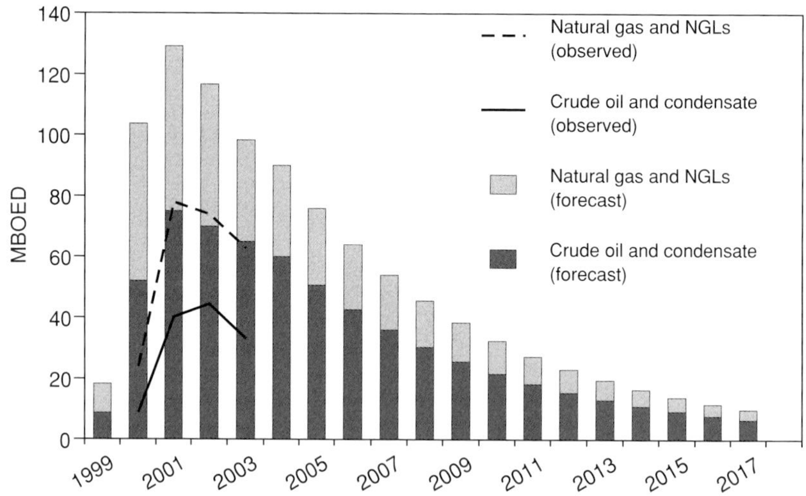

Sources: Riddle, Snyder and George 2001, MMS

Figure 7.9: Production Profile of the Diana/Hoover Development, 1999–2018

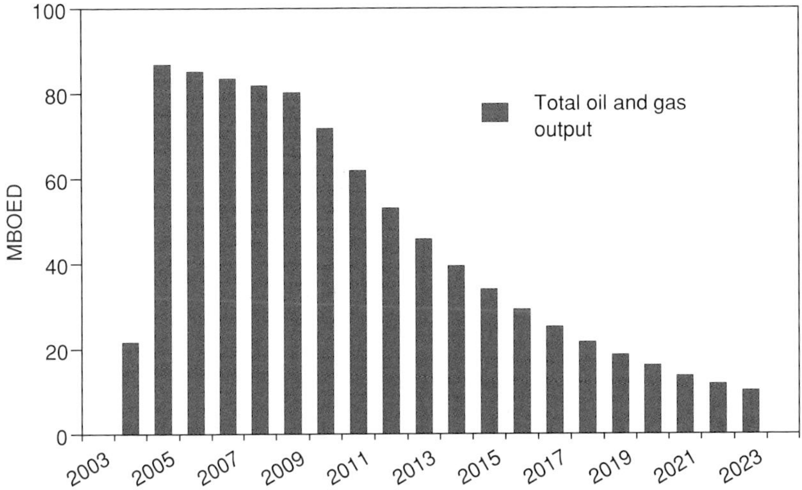

Source: Matharu and Lucas 2002

Figure 7.10: Forecast Production Profile of the Mad Dog Development, 2003–2024

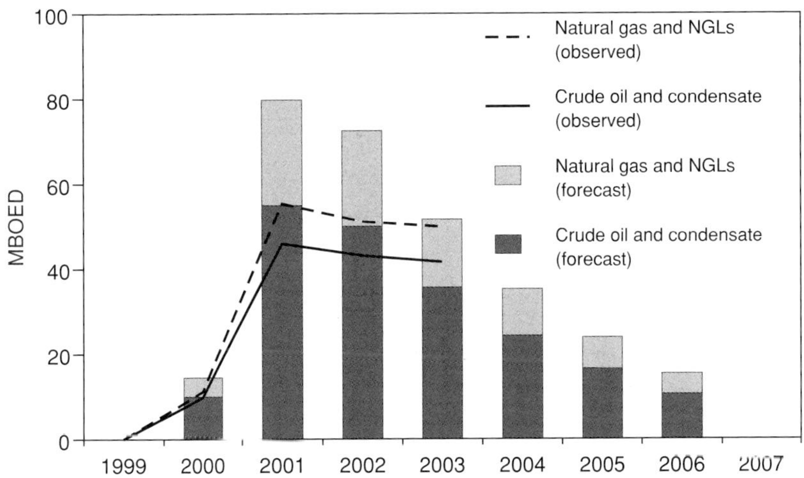

Sources: Riddle, Snyder and George 2001, MMS

Figure 7.11: Production Profile of the Petronius Development, 1999–2007

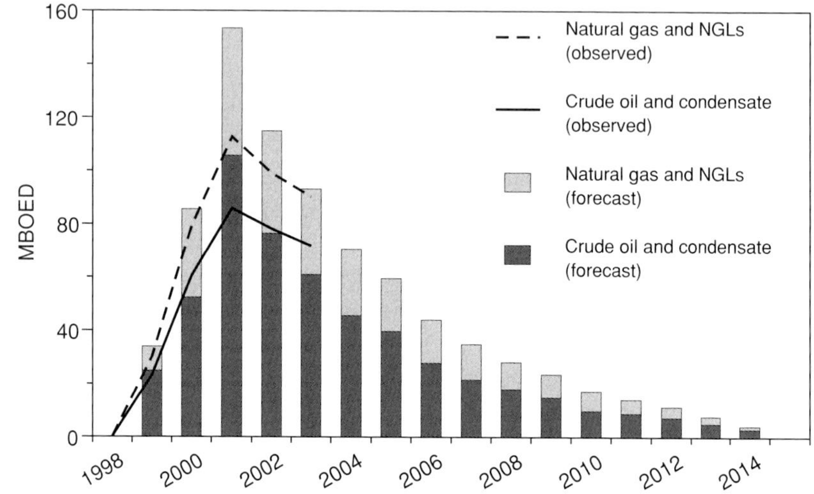

Sources: Riddle, Snyder and George 2001, MMS

Figure 7.12: Production Profile of the Ursa Development, 1998–2015

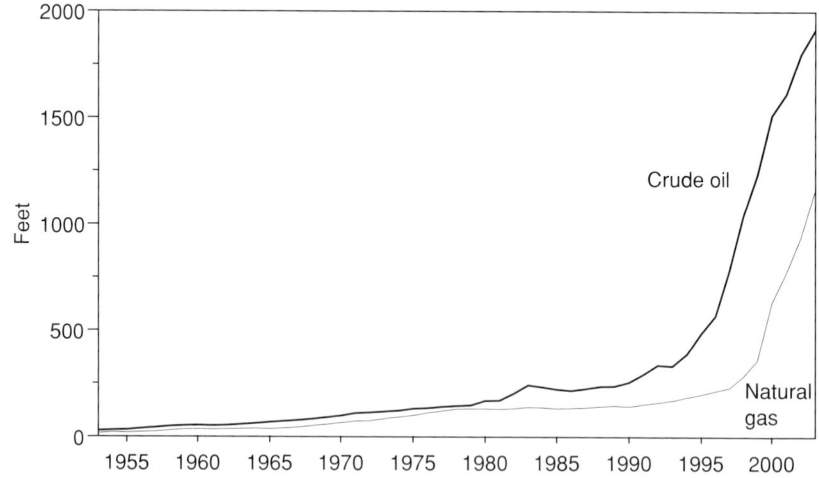

Source: MMS

Figure 7.13: Weighted Average Water Depth of GOM Production, 1953–2003

that deepwater fields have had not only as far as overall GOM output goes but also in terms of the very character of offshore oil activities in the region. As can be appreciated, in 1970 the weighted operating water depth average for crude oil production was just 100 feet. By 1990 it had only reached 250 feet, but then it began to increase rapidly as soon as the first deepwater fields came on stream during the mid-1990s, reaching 1000 feet in 1998 and 1500 feet (i.e. a sevenfold increase on the figure posted ten years before) in early 2000. Indeed, as has been pointed out already, according to the commonly accepted definition of the deepwater threshold (1000 feet of water or more), deepwater crude oil production is now the norm rather than the exception in GOM.

In contrast to the situation in oil, the volume weighted average depth for natural gas production activities in GOM only pierced the 1000 feet threshold in 2003. At first glance this might seem reassuring, in that a large part of GOM gas production, which supplies a quarter of the gargantuan US demand, is still taking place under relatively undemanding conditions. This notion, though, is misguided. As we explain below, in light of the onset of accelerated production decline in shallow water fields, it is actually a cause of great concern that the production-weighted average has not increased more than it has, given the number of years that the deepwater province has been open to large-scale development.

7.2 The Deepwater Gas Problem

Throughout the late 1990s, deepwater natural gas production estimates tended to be very bullish. The conventional wisdom saw annual GOM deepwater output increasing nearly fourfold (to 4.5 TCF) between 2000 and 2010. However, the key reason why deepwater gas supply was forecast to expand with such vigour was that it had been earmarked to cover slightly more than half of an equally vigorous increase in overall US gas demand for that period. In other words, if deepwater output failed to reach these lofty targets, the 30 TCF per annum gas market scenario, so beloved of the oil industry and policymakers alike, would not materialise.

Given the importance of future deepwater output figures in the overall gas supply scenario, surprisingly scant attention was devoted to examining the conditions that would have to be met for these targets to be reached. Foremost among these were the impressive rates at which reserves (over and above the 28–35 TCF discovered up to 2000) would have to be added and brought on stream. As Nehring indicated back in 2001, unless more large and giant gas discoveries were made in the deepwater, the sub-province would never, in fact, become 'the predominant component of increasing domestic gas production ... [as] projected'.[5] However, Nehring's pleas for caution were drowned amid the din caused by the fact that the discovery rate for new GOM deepwater fields accelerated greatly between 1995 and 2002 (with more than 60 percent of all confirmed deepwater finds occurring over this period, thanks to better drilling techniques and a greater availability of rigs). Rather unfortunately, though, most of these discoveries tended to be small fields, generally harbouring less than 150 BCF of recoverable reserves. Thus, in absolute terms, the volumes of gas found during that period were not particularly impressive: for instance, during 2000, about 1.8–2.4 TCF of gas was discovered and barely 300 BCF of this volume was constituted by non-associated gas.

This pattern of resource discovery had (and has) serious implications for the future of GOM deepwater gas output, for one reason: whereas reserves in large GOM fields display a marked tendency to grow over time, small fields have displayed the contrary tendency to disappoint or, at best, to remain unchanged. Thus, if future deepwater discoveries were to continue to fall mainly in the 30−300 BCF of reserves bracket, the oil industry would be hard pressed even to double annual deepwater gas production from the 2000 figure of 1.4 TCF, let alone to get it to reach 4.5 TCF. Indeed, Nehring concluded that, unless more large and giant gas discoveries were made in the deepwater (against the geological odds), the share of output from this province in total US domestic gas production would never even remotely approach the figures that the 1999 NPC report had optimistically projected.

According to Nehring, a more likely scenario would involve an annual reserve addition rate of 4.2 TCF, which would have seen deepwater output increasing rapidly to a peak of around 2.5 TCF (6 BCFD) by 2004, and then decreasing slowly (at an annual rate of 6 or 7 percent) through to 2005, with the decline accelerating thereafter as the crop of fields discovered up until 2000 in the GOM upper slopes was developed to the full. Concurrently, shallow water output would have continued to decline, so this deepwater production profile would have translated into a slow net increase in overall GOM annual gas production through 2007 (up to a maximum of 6.2 TCF).

Nehring prepared this output scenario on the basis of a defensible outlook for future deepwater production, meant to make maximum use of the available knowledge on deepwater discoveries to estimate the timing and the quantity of future discoveries and their production profiles, as well as the projected production additions from both known and future discoveries. As Figure 7.14 shows, Nehring's figures were strikingly at variance to those generated by means of the 'abstract, top-down approach' embraced by what he called 'gas establishment forecasters', who – in Nehring's opinion – complacently took 'overall estimates of resource potential' as their point of departure, and then reached a future production number on the basis of simplistic 'abstract analytical constructs such as reserves added per well drilled'.[6]

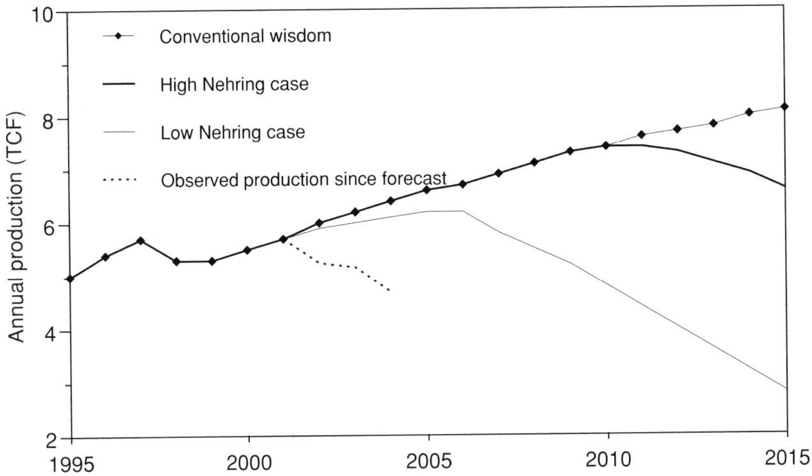

Source: Nehring 2001a

Figure 7.14: Natural Gas Output Forecasts for the GOM Region to 2015

Given the degree to which Nehring's predictions differed from those of the industry at large, it was always tempting to disqualify them as the ramblings of an inveterate pessimist, not least because this made it unnecessary to think about what could happen if a chunk as large as 3–4 TCF of the forecast US gas supply for 2010 failed to materialise as planned. After all, and as has already been pointed out, the lead times involved in planning, obtaining permits for and building certain types of large infrastructure projects are such that, in 2002, it was probably already too late for the USA to embark upon an alternative strategy to meet its incremental electricity demand with generation capacity not fired by gas, at least as far as the 2004–2008 timeframe was concerned. Looking forward in time, this raised the alarming prospect that the growth prospects of the US economy at certain points during the present decade could conceivably be hampered because of electricity supply constraints.

Nehring's natural gas output predictions took into account the likely behaviour of discovered natural gas fields as well as the likely magnitude of natural gas reserve additions, of course. However, they also made allowances for the fact that, due to geological constraints, a largish proportion (if not the majority) of future deepwater natural gas output would be produced in association with oil, thereby making the production path for natural gas highly dependent on both the magnitude of deepwater oil reserve additions and the NGL content of the associated gas.

In order to gauge the impact of future oil activities on gas production, Nehring took as his point of departure the 163 deepwater fields whose existence had been confirmed at the time he prepared his estimate, and then projected future reserve growth estimates (ranging from 5 to 100 percent) for each one of these fields, according to empirically established patterns related to both their type and size. On the basis of these projections, Nehring estimated the total ultimate recovery from these fields at 18,640 MMBOE (out of which 11,200 MMBOE, 1851 MMBOE and 5589 MMBOE would be crude oil, natural gas liquids and natural gas, respectively). Nehring then complemented this figure with low and high estimations for future hydrocarbons reserves additions up to 2015, likely size of finds, gas–oil ratios, likely development lead times and production profiles. The estimated range of recoverable resources associated with Nehring's low and high cases went from 26.6 billion BOE to 34.0 billion BOE, respectively (with new discoveries of gas resources ranging from 45.7 TCF to 56.2 TCF, respectively). Nehring saw annual gas reserve additions for the period 2000–2015 at 4.2 TCF for the low case and 6.6 TCF for the high case.

Nehring's high reserve addition figure was significantly lower than the 7.4 annual reserve addition rate (6.9 TCF until 2010 and 8.5 TCF thereafter) underlying the unbridled output expansion scenarios that policymakers clearly took for granted in the late 1990s. Unsurprisingly, his deepwater gas production forecasts (2.5 TCF in the low case and 3.1 TCF in the high case, with the peak occurring in 2007–2008) also came in way short of the 4.5 TCF that conventional wisdom required for total US gas supply to reach the 30 TCF mark.[7]

Nehring's views on future GOM deepwater oil discoveries and production were actually quite bullish. His reserve addition estimations, for instance, implied an increase of nearly 100 percent on discovered GOM hydrocarbon resources through to 2010. Likewise, his estimates of ultimate recovery from existing fields *exceeded* contemporary industry forecasts for these fields by nearly 30 percent.[8] Last but not least, his estimates for shallow water decline rates for gas underestimated the actual rates observed since.

One would have thought that, on the strength of the above factors, Nehring's bearish predictions on the natural gas output front would have been all that harder to dismiss, so making it easier for 'domestic gas supply planners' to resist the temptation to 'rely on Gulf of Mexico for an indefinite future'.[9] Unfortunately, these predictions went very much against the prevailing mood in both industry and policymaking circles and, to use a biblical turn of phrase, Nehring's cautionary conclusions 'sank as lead in the mighty waters'. However, the passage of time seems to have vindicated him in full: the NPC's supply estimates have been revealed as being based purely on aspirations, rather than hard data. If anything, Nehring can be said to have erred on the side of optimism. After all, his high production scenario implied an annual reserve addition average of 6.6 TCF, a figure significantly in excess of the 4.3 TCF annual average recorded over the 1983–1998 period, and greater even than the 5.6 TCF in new discoveries and revisions recorded during 2000 (by all accounts, a banner year for GOM exploration). Moreover, the peak production year for GOM as a whole seems to have come four years earlier than Nehring expected.

Despite the significant increases in deepwater gas production achieved in recent years, it has become increasingly obvious that the lack of giant non-associated gas discoveries in this province will never allow production to reach the levels necessary to counteract increasing declines in non-associated gas production, while simultaneously meeting a rising US demand. Given the degree to which the deepwater GOM gas-producing potential was overstated, alternative sources of supply (Canadian gas, Alaskan gas, coalbed methane and LNG) will clearly have to be developed if US gas demand (and US electricity generation) are not to be curbed in coming years by greatly increased prices. Unfortunately, current estimates regarding potential supplies from these alternative sources appear to be just as inflated – if not in volumes certainly in timing – as those from the deepwater were in their time. Perhaps by way of admission to former exaggerations, the aforementioned 2003 NPC study still sees natural gas output by 2015 (70 BCFD) as being a calamitous 30 percent lower than the figure it had forecast back in 1999.

7.3 Deepwater Oil Prospects: Local Surpluses?

The situation for future deepwater crude oil production in the short to medium term is quite different to that of natural gas, chiefly because the availability of deepwater GOM crudes is set to continue rising at least during the next five years or so. Having said that, future deepwater crude output levels, although respectable, will certainly not be enough to wean the US economy from its fix of imported oil, as some of the more delusional members of the 'American oil from American soil' brigade dared to dream at one point.

Figure 7.15 shows one of the more outlandish forecasts in this regard. As can be appreciated, this forecast suggested that the deepwater output contribution could be of a magnitude to allow US crude oil production to equal its 1970 peak of 9.8 MMBD by 2020 and then continue increasing up to 10.7 MMBD by 2030. Given the decline rates in both the Lower 48 and Alaska, this would have required that deepwater GOM output exceed Saudi Arabian output levels for quite a stretch of time.[10] Thus far, though, the welcome contribution of deepwater oil to US oil supply has proven insufficient to arrest the ongoing decline in this country's total oil production or, indeed, even to put a dent in the rate of growth in US crude imports (which averaged around 10 MMBD in 2004).

Up to 2002 or so, MMS expected deepwater oil and lease condensate production to peak around 2009–2011 at a level of *circa* 2.3 MMBD (a figure which implies annual production increases of 150–200 MBD). In oil industry circles, though, this was always seen as an optimistic figure, and the likely peak was put at 1.9 MMBD (see Figure 7.16). In its most recent long-term forecast, MMS seems to have come round to this point of view. Of course, regardless of which prediction comes

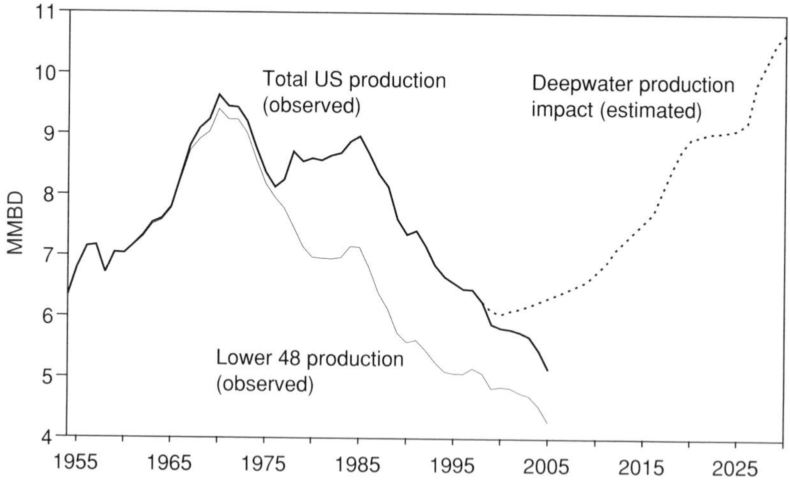

Source: Anderson and Boulanger, 2002

Figure 7.15: US Crude Oil Production Profile, 1954–2030

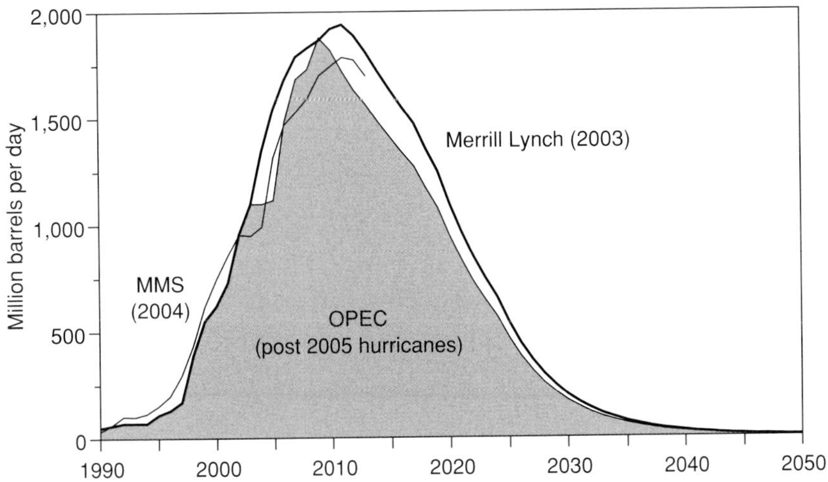

Sources: Merril Lynch, MMS

Figure 7.16: Crude Oil Production Profile for the Deepwater GOM, 1990–2050

closer to reality, a significant part of deepwater production increases will be negated by the decline in shallow water output, so the likely annual net gain in total GOM oil output until the deepwater peak will be, at the very best, around 100 MBD.

It is indisputable that, 'the deepwater of the Gulf of Mexico can rightly claim to be America's new frontier and … a world class hydrocarbon province'.[11] Nevertheless, the blind faith that many policymakers and not a few analysts seem to place in the bountifulness of the deepwater GOM is rather surprising. After all, until relatively recently, the behaviour of future production projections pointed consistently towards an *underperformance* of the deepwater relative to expectations (and quite *conservative* expectations, at that). This underperformance was somewhat obscured by the fruition of giant projects like Auger and Mars, which on the whole tended to exceed expectations (at least at an early stage in their lifetimes).

In order to appreciate these points, one need only look at the low and high projections of total GOM crude output that MMS has published on an annual basis since 1997 (the annual projections – which represent the average December output for each projected year – are collated in Table 7.1, together with the end-of-year production levels actually recorded).[12] Up until 2000, the path of actual output tended to be closer to the low projection profile (hardly surprising, since the high projections assumed that new technology would entirely offset decline rates in shallow-water fields), albeit still undershooting the target.[13] Output

Table 7.1: Five-year GOM Crude Oil and Condensate Output Projections (MMS) Versus Actual Annual Production, 1997–2008

		1997	*1998*	*1999*	*2000*	*2001*	*2002*	*2003*	*2004*	*2005*	*2006*	*2007*	*2008*
Low Projection MBD	1997	1,230	1,240	1,545	1,660	–	–	–	–	–	–	–	–
	1998	–	1,226	1,493	1,592	1,606	1,666	–	–	–	–	–	–
	1999	–	–	1,488	1,514	1,537	1,434	1,381	–	–	–	–	–
	2000	–	–	–	1,428	1,441	1,442	1,331	1,315	–	–	–	–
	2001	–	–	–	–	1,376	1,498	1,440	1,437	1,526	–	–	–
	2002	–	–	–	–	–	1,585	1,498	1,430	1,666	2,002	–	–
	2003	–	–	–	–	–	–	1,530	1,639	1,756	1,797	1,580	–
	2004*	–	–	–	–	–	–	–	1,562	1,868	2,006	2,050	2,098
ACTUAL		1,128	1,218	1,356	1,433	1,562	1,566	1,537	1,518**	–	–	–	–
High Projection MDB	1997	1,300	1,407	1,730	1,932	–	–	–	–	–	–	–	–
	1998	–	1,347	1,667	1,816	1,874	1,976	–	–	–	–	–	–
	1999	–	–	1,731	1,825	1,910	1,846	1,836	–	–	–	–	–
	2000	–	–	–	1,660	1,733	1,786	1,707	1,733	–	–	–	–
	2001	–	–	–	–	1,547	1,753	1,749	1,805	1,967	–	–	–
	2002	–	–	–	–	–	1,789	1,743	1,719	2,041	2,478	–	–
	2003	–	–	–	–	–	–	1,706	1,876	2,052	2,139	1,926	–

* Mean projection
** January–June

Source: MMS

in 1997, 1998 and 1999 was 20 MBD, 8 MBD, and 132 MBD lower than the latest low projections for those years, respectively. The year 2000 marked the first time that observed production exceeded the low forecast, although by a very small margin (5 MBD). Since then, there has been no observable tendency for this exceeding of expectations to recur regularly (if anything, the opposite appears to be true). It should also be pointed out that, again until recently, the trendline for these projections was heading downwards. To cite an example: for the year 2000, the low projection was 232 MBD lower than that made three years previously, while the high projection was 272 MBD lower. The year 2003 marks only the first time in which this trend was reversed (and in the year 2004, MMS abandoned the practice of producing low and high projections).

Consider also Figure 7.17, where we have superimposed the growth profiles of the shallow and deepwater booms (the first going from 1964 to 1973, the second from 1990 to 2003). This graph shows, among other things, that deepwater volume growth is not completely unprecedented in terms of its impact on total GOM output. Indeed, the early years of the deepwater boom appear rather tame in comparison to the early years of the shallow water boom. Although the two series begin at the same level, by year 4 the path of deepwater production is lagging far

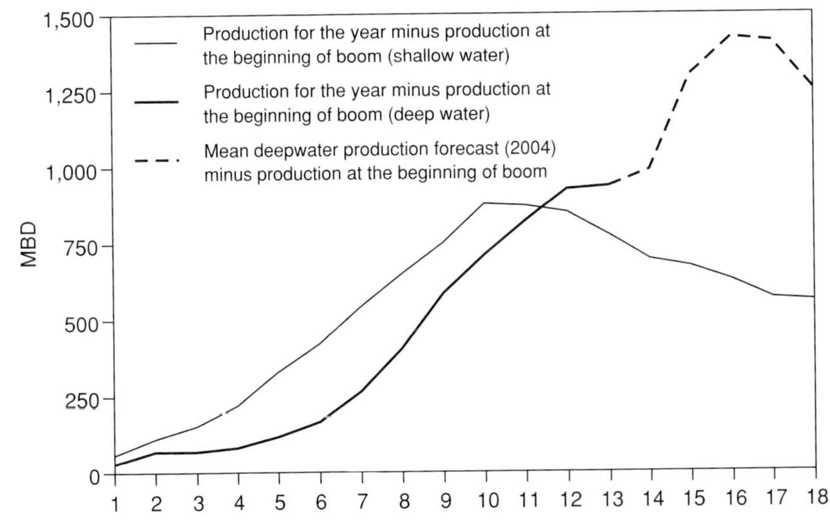

Source: MMS

Figure 7.17: Comparison of the Crude Oil Production Booms in the GOM Shallow Water (1964–1973) and Deepwater (1990–2008) Sub-provinces

behind that of shallow water production. It is not until year 7 that growth rates become comparable again, and full catch-up in terms of absolute output only occurs between years 11 and 12, as shallow water output peaks and then stagnates.

Notwithstanding this significant mismatch between expectations, on the one hand, and reality, on the other, the notion that incremental deepwater production would be so large that it would swamp the downstream processing capacity in the USGC area (particularly in terms of desulphurisation) was common currency among industry watchers, all the more so since the arrangement of pipeline infrastructure onshore in Louisiana limited the primary market of deepwater crudes to refineries located along the Louisiana Gulf Coast. Given that, up until very recently, investment in desulphurisation (or any other investment in refining activities) was still seen as a waste of shareholders' capital, during the late 1990s this perception of impending glut led a number of GOM operators to devise complicated schemes to deal with this contingency, for instance, importing condensates from Algeria to dilute their deepwater volumes. Providers of pipeline capital, for their part, saw the answer as a large capacity line linking Louisiana with the Houston area refineries (the closest this scheme came to fruition was a joint Enbridge/LOOP project called Alligator, which would have linked LOOP/St. James with Texas City).

Many analysts saw in the desperate market conditions of 1997–1998 a proof of their warnings regarding the inadequacy of the GOM refining system to digest large incremental flows of medium sour deepwater crudes. Indeed, equity producers were forced to price deepwater crudes so keenly (in order to ensure the custom of logistically advantaged Louisiana refiners) that they found themselves competing head on with imported waterborne crudes that were much lower in quality (like Mexican Maya crude, say). This situation improved markedly from mid-1998 onwards, because the supply restrictions that affected the market for heavy sour grades during this period blunted the competition between deepwater crudes and imported heavy sour streams, as the former were able to move painlessly into spaces vacated or not filled by the latter. Nevertheless, even in the context of the tight supply panorama that has prevailed in the international oil market since late 1999, the general consensus until recently still was that the continued rise in the production of deepwater crudes would end up by generating local surpluses in the Louisiana Gulf coast, unless pipeline infrastructure were built to transport these incremental volumes to the Texas Gulf Coast and beyond.[14] However, such an eventuality now appears remote in the extreme.

Between 1998 and 2004, USGC deep conversion capacity expanded to such a degree (see Figure 7.18) that it managed to outstrip the supply of heavy sour crude to the region (initially, because of Mexican and Venezuelan compliance with supply restriction accords, and then because of recurrent political and operational problems in Venezuela). The volume-weighted average quality parameters of GOM Federal OCS production offshore Louisiana, it is true, have deteriorated markedly after the addition of lower quality deepwater blends to the production mix. However, during that same time, the average quality of the crude charge processed at Louisiana GC refineries has decreased in tune (Figure 7.19). Moreover, this has coincided with an enormous expansion of upgrading capacity in the region. Therefore, on an incremental basis, the quality deterioration of the imported crude oil charged to Louisiana GC refineries has far outstripped that of the GOM Federal OCS landed in this state.[15]

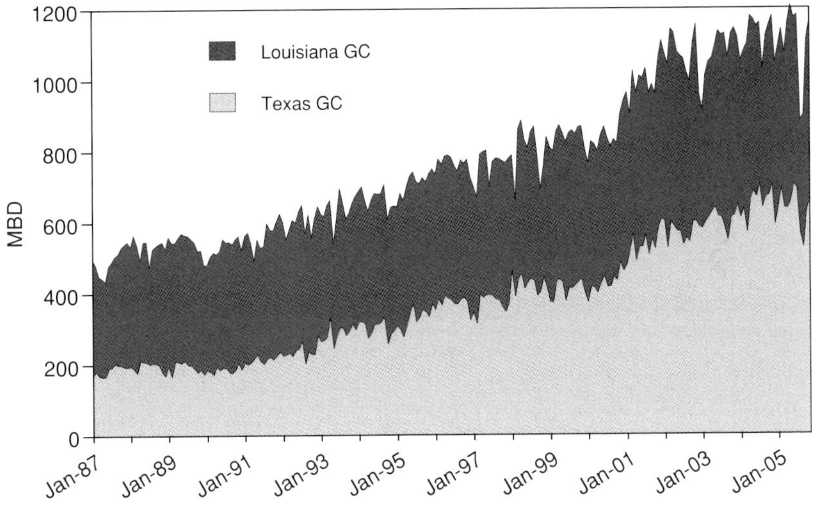

Source: DOE

Figure 7.18: Monthly Crude Charge to Coking Plants, by USGC Refining District, 1987–2005

Quite apart from the above, the quality of crudes from many of the deepwater fields to come on stream after 1999 turned out to be better than the quality of crudes from earlier fields (such as Mars). As a result of this, the addition of output from projects such as Ursa and Brutus to the Mars blend has translated into an improvement in the key quality parameters of this most important of deepwater blends (Figure 7.20). At the yield level, there has been a marked reduction

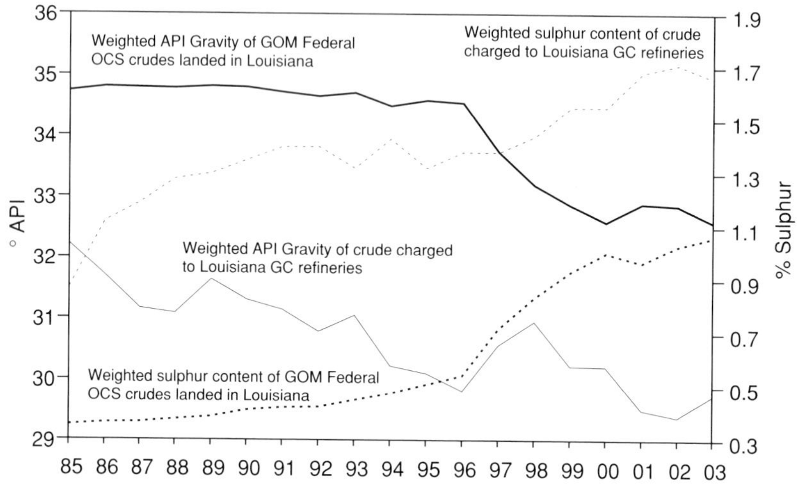

Sources: DOE, LDNR, MMS, Poseidon Pipeline LLC, Shell Offshore

Figure 7.19: Evolution of Key Quality Parameters for Offshore Crude Production and Refinery Crude Charge in GOM, 1985–2003

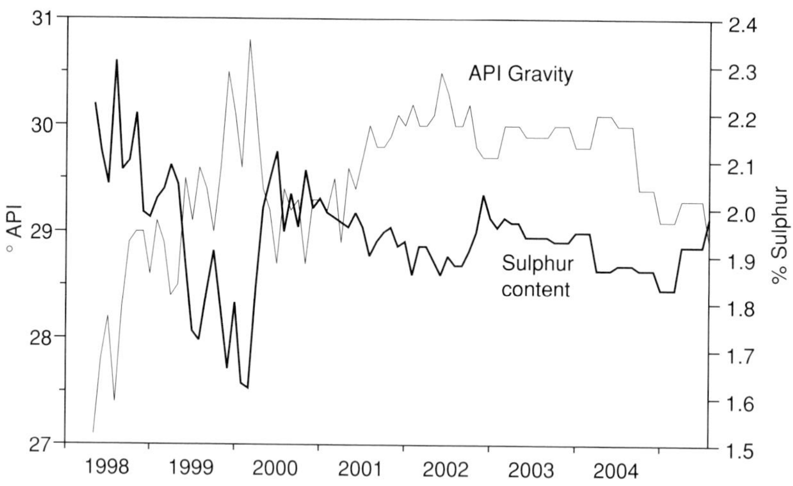

Source: Shell Offshore

Figure 7.20: Mars Blend, Behaviour of Key Quality Parameters, 1998–2005

in the residuum yield of the blend, while the LPG cut (33°–155° F) has more than doubled (Figure 7.21). Even greater improvements in API gravity and sulphur content are discernible in the quality of the Poseidon common stream (Figure 7.22).

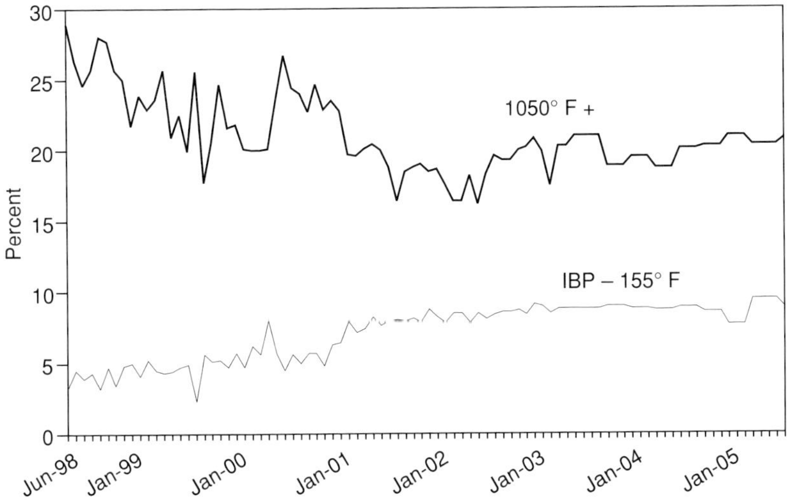

Source: Shell Offshore

Figure 7.21: Mars Blend, Monthly Variations in Volume Yields, 1998–2005

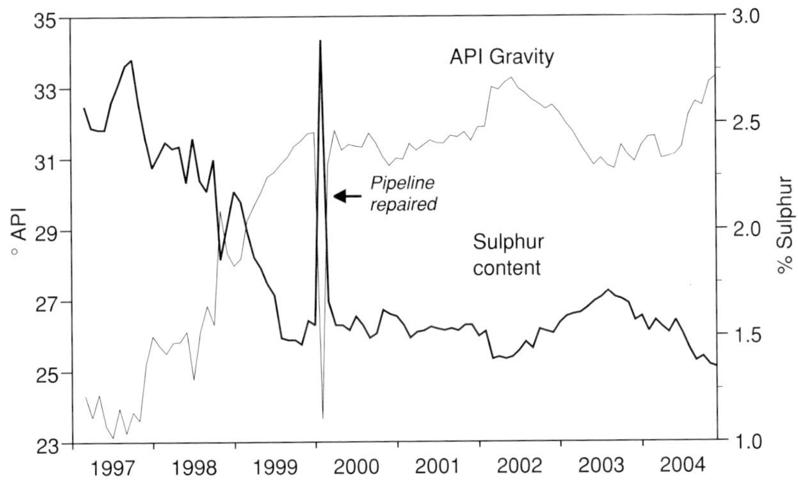

Source: Poseidon Pipeline LLC

Figure 7.22: Poseidon Blend, Behaviour of Key Quality Parameters, 1997–2004

In a nutshell, the foregoing means that an increasingly capable and expanding USGC refining system has had to cope with a lower deepwater contribution to its sulphur pool and a greater deepwater contribution towards the gravity of its total charge than had originally been expected. Furthermore, in the short to medium term, downstream processing capacity is bound to get even more breathing room. On the one hand, the single biggest new contribution to deepwater volume in coming years is slated to come from the Thunder Horse field, whose crude oil has rather benign characteristics (32.3° API, 0.75% sulphur content). On the other hand, the lower quality crudes from the Atlantis, Mad Dog and Holstein developments (which will be blended into a common stream with 30° API and 2% sulphur content) will be shipped through the Cameron Highway Pipeline System to Texas, rather than Louisiana. The size, number and sophistication of Texas refineries are such that the incremental volumes of Southern Green Canyon Blend will have a barely perceptible impact on the quality of the crude charge of the Texas GC refining system (and on the behaviour of the light versus heavy price differentials in the region).

7.4 Future Deepwater Oil Growth: How Fast?

The concerns regarding the capability of the US downstream to deal with incremental deepwater crude oil flows have also receded to the extent that deepwater volume growth slackened after 2000, a development that was rationalised in many circles as reflecting the relative behaviour of gas and oil prices. With the general crisis in natural gas deliverability in the USA (made dramatically evident by a catastrophic price spike early in 2000), gas prices began to rise even faster than oil prices. Given the severity of the deliverability problem, companies were more confident of a long period of high gas prices than they were about a similar bonanza for oil prices (even though they still paid lip service to the notion of a 30 TCF gas market which in principle was incompatible with very high prices). As a result, the relative attractiveness of gas development and exploration improved markedly, and gas projects jumped up to the very front of the queue in terms of their call on the investment capital available to the industry at large.

Consider the following figures (plotting weekly data for 1996 to 2004 taken from the Smith Bits rig statistics), which show the extent to which drilling activity in the Gulf has followed the trends in overall US activity. Figure 7.23 shows total US development drilling, broken down according to the nature of their primary drilling target (oil or gas). As can be appreciated, at the end of 1997, development drilling began to fall, when world oil prices entered a period of decline in the wake of OPEC's Djakarta meeting, held in November of that year. It began to rise again in March 1999, coinciding with the start of the recovery in

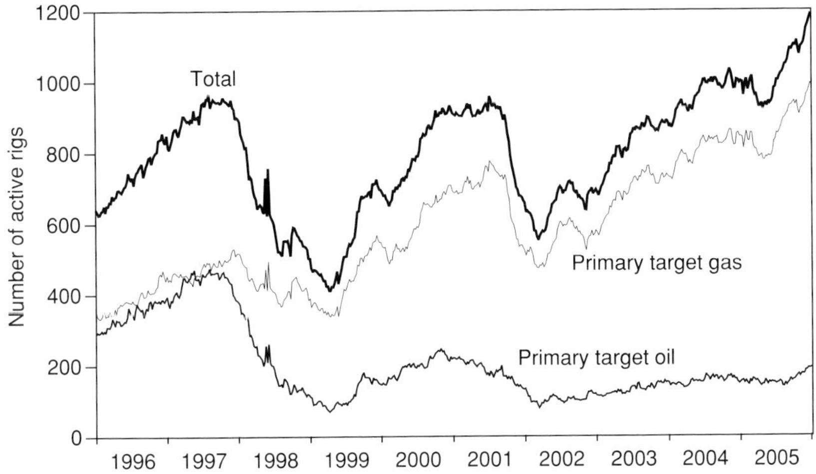

Source: Smith Bits

Figure 7.23: Total US Development Drilling Activity, by Type, 1996–2005

world oil prices, and by October 2000 the total had almost regained its 1997 peak. However, the increase in oil development activity was both very gradual and modest. As 2001 drew to a close, activity once again took a nosedive (because the market reached the conclusion that a major confrontation was brewing between OPEC, on the one hand, and Russia, on the other). By March 2002, these fears were forgotten, and overall activity started to increase once again, and in late 2003 it reached levels comparable to those recorded two years previously. Oil development activity never recovered from the 1998 crash: most of the upsurge in activity took place in gas development. To a certain extent, this is understandable, as the acceleration in depletion rates forced the industry to run at breakneck pace merely to stay in place (US gas well completions grew from 10,000 wells in 1999 to over 22,000 wells by 2001, and yet daily supply stayed as flat as it had been during the 1992–1999 period). Having said that, it is still a cause of surprise (and concern) to see that, despite extremely high nominal oil prices prevalent in mid-2004, oil development levels at that point were down to a third of what they had been in late 1997.

The situation was slightly different for exploratory drilling, in that from mid-1999 onwards there was a far greater overall increase in activity (Figure 7.24). The exploration effort during 2001 was especially intense, and it has remained at levels comparable to those recorded on the eve of the 1998 price crisis. However, virtually all of the increase

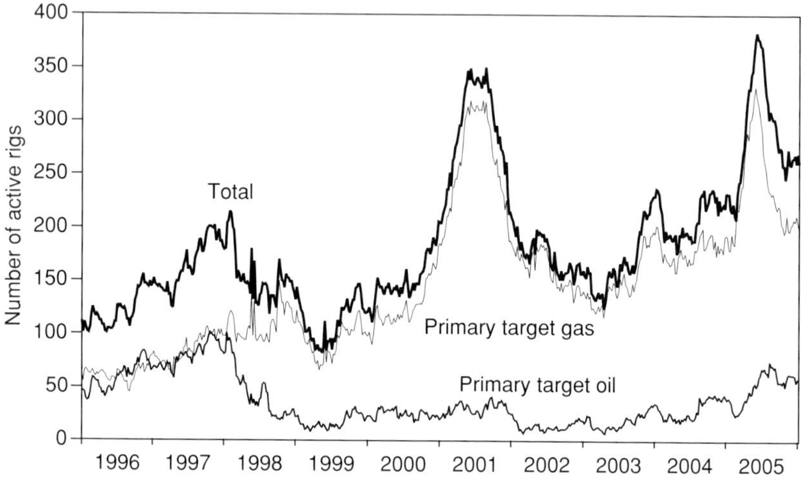

Source: Smith Bits

Figure 7.24: Total US Exploratory Drilling Activity, by Type, 1996–2005

in activity has taken place in gas exploration, with the profile of oil exploration remaining flat (at best) from the end of 1998 onwards.

Figures 7.25 and 7.26 show comparable statistical series for the GOM Federal OCS alone. In development drilling, activity has recovered from its 1999 lows, but again the increase has been heavily skewed towards gas development (indeed, the depression affecting oil development drilling was worse during 2004 than in 1998–9). In the case of exploratory drilling, oil exploration again remains at very low levels. Whereas at certain points in 1997 exploratory activity was heavily biased towards oil, by 2000 the focus had shifted almost entirely to gas. Thus, while the collapse in oil company cash flows after 1997 led to a sharp slowdown in overall GOM drilling activity, their recovery (with a vengeance) after 1999 has not led to an equivalent recovery in oil activity.

The allocation of exploration capital between oil and gas activities for the industry has already had an effect on the market for offshore leases, as witnessed by the improvement in shallow water acreage prices. So is it possible that the mobilisation of capital by many independent companies in the direction of the deep gas province and away from the deepwater will cramp future output in the latter province? The answer to this question is negative, chiefly because the companies responsible for the bulk of deepwater investment are the majors and the larger independents (so the investment shift by smaller players may well lead to an even greater concentration in the deepwater province). However,

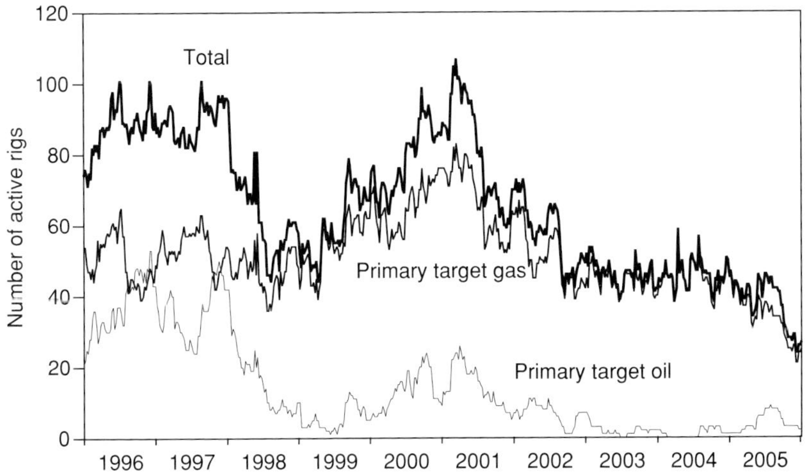

Source: Smith Bits

Figure 7.25: GOM Federal OCS Development Drilling Activity, by Type, 1996–2005

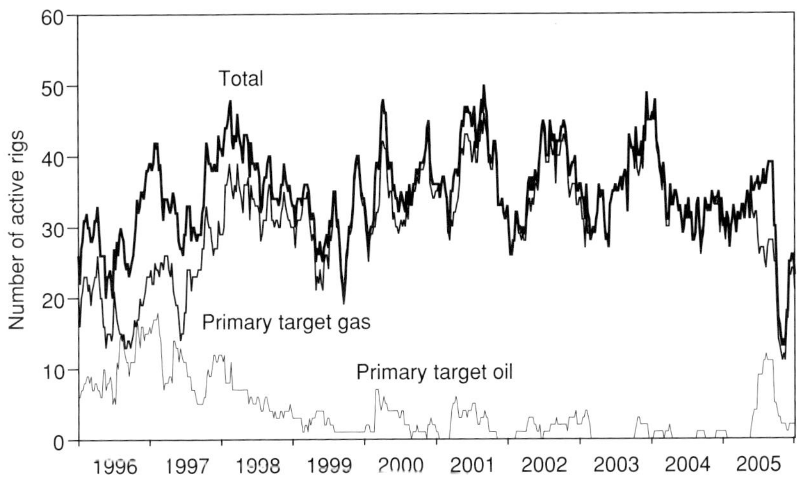

Source: Smith Bits

Figure 7.26: GOM Federal OCS Exploratory Drilling Activity, by Type, 1996–2005

this should not be taken to mean that GOM deepwater oil output will once again expand at pre-2001 rates, chiefly because the rig activity statistics quoted above point towards the possible existence of a deeper seated malaise affecting the GOM offshore.

The 2003 MMS average production estimates for the period 2004 to 2008 showed a welcome return to form for the deepwater sub-province, as a number of large finds (including Thunder Horse) were expected to be brought on stream. Shortly thereafter, though, the trend again reversed direction and underwent a change in sign (indicating a very significant year-on-year production decline of 190–215 MBD). The MMS personnel in charge of preparing these estimates on the basis of operator surveys professed to be unbothered about this reversal in the overall trend, attributing it to 'conservative operator estimates and the many uncertainties in making five-year projections', particularly since 'leases that [will] begin production by the year 2007 are not identified in the survey [but] will also increase oil and gas production beyond these projections'.[16] According to MMS, 'previous reports showed similar declines in the latter years of the forecast, but subsequent reports shifted this peak'. In fact, though, as Figures 7.27 and 7.28 show, none of the previous MMS future deepwater production rate reports for oil had ever forecast such a pronounced decline at the tail end of a projection period. Furthermore, what made the 2003 projection particularly striking was its contrast with the very bullish projection published only a year earlier (the difference between the two average forecasts for 2006 production comes to nearly 300 MBD).

Ironically enough (given the well publicised 'silver bullet' role envisaged for deepwater gas production in the context of the US gas market as a whole), the MMS projections for deepwater natural gas production were always considerably less bullish than those for crude oil (Figures 7.29 and 7.30). The underperformance of natural gas production relative to expectations has consequently been less noticeable (in other words, actual gas production has approximated the low MMS projection much more closely than that for crude oil). In contrast to the crude oil projections, all the MMS gas production projections show declines in the latter years of the forecast, with subsequent reports shifting the apparent peaks (albeit not by a great deal). Thus, the MMS medium-term forecast for GOM-wide gas production sees it as declining to just over 11 BCFD by 2007. However, even though the MMS extrapolation of deepwater trends based on the current crop of projects sees natural gas output in the sub-province declining to 1.37 BCFD by 2013, the agency sees a significant rebound in total GOM gas output, starting in 2008. By 2013, MMS sees total GOM gas output reaching the 13.5

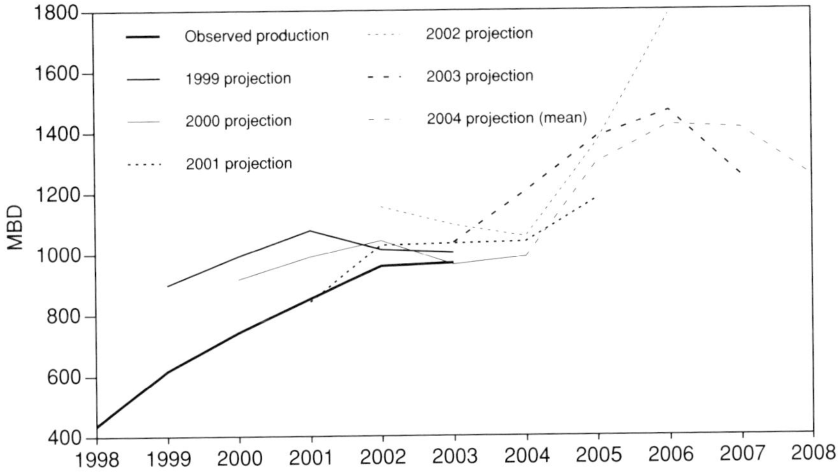

Source: MMS

Figure 7.27: MMS Five-year Deepwater Crude Oil Output Projections (High Scenario) versus Actual Deepwater Crude and Condensate Production, 1998–2008

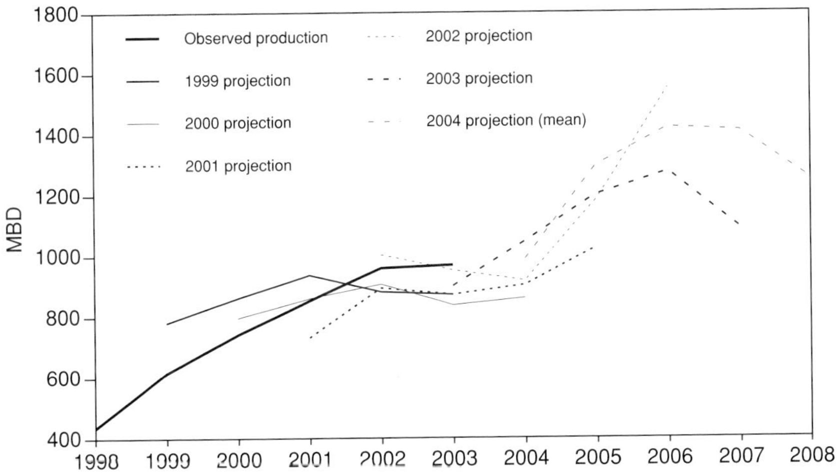

Source: MMS

Figure 7.28: MMS Five-year Deepwater Crude Oil Output Projections (Low Scenario) versus Actual Deepwater Crude and Condensate Production, 1998–2008

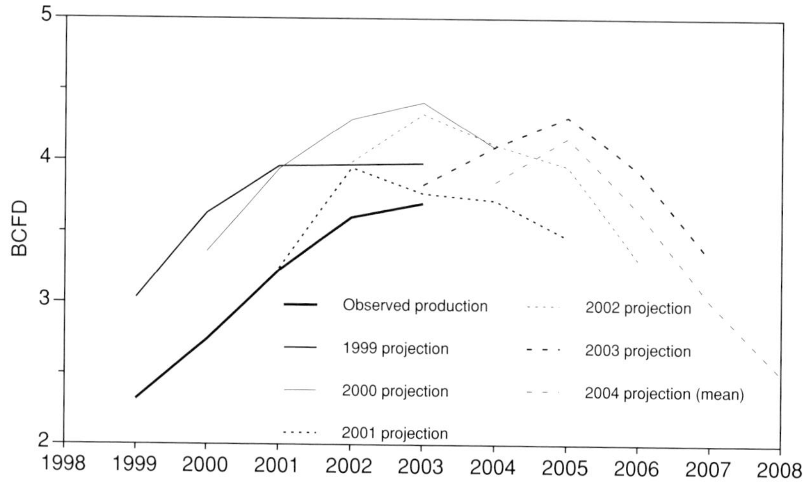

Source: MMS

Figure 7.29: MMS Five-year Deepwater Natural Gas Output Projections (High Scenario) versus Actual Deepwater Natural Gas and NGL Production, 1998–2008

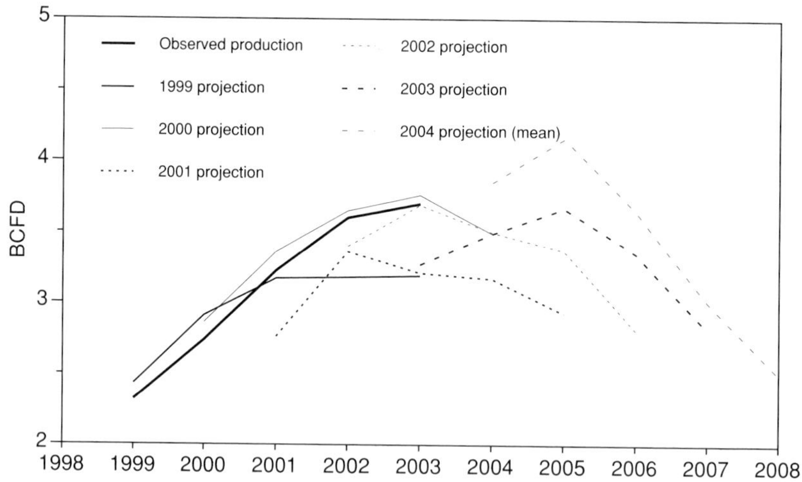

Source: MMS

Figure 7.30: MMS Five-year Deepwater Natural Gas Output Projections (Low Scenario) versus Actual Deepwater Natural Gas and NGL Production, 1998–2008

BCFD mark. However, 49 percent of this figure is highly speculative in nature, as it is expected to come from extrapolated (and to some, hopelessly optimistic) deep gas growth, industry announced discoveries in the deepwater and, finally, yet to be found reserves.[17]

Remarkably, at the time of their publication, the oil industry collectively did not ask (perhaps because it did not dare) whether the 2003 changes to the MMS crude oil production projections could potentially be the harbinger of gathering clouds on the deepwater production horizon. And neither was any significance attached to the fact that the traditional date for MMS to publish its 2004 production projections came and went, and no estimates were forthcoming. In the light of the havoc that hurricane Ivan wreaked on GOM offshore production towards the end of 2004, perhaps it was just as well that MMS did not go to the trouble of preparing a mid-year forecast for the years 2004–2008 (in the event, it released a ten-year production projection report covering the 2004–2013 period in early November 2004). After all, the protracted sequels that have followed Ivan would almost certainly have rendered worthless any near-term assessment of production. However, the longer dated MMS projections will not necessarily weather the storm (figuratively speaking) any better, not so much because the 2005 hurricane season surpassed that of 2004 in destructiveness, but rather for the simple reason that Ivan brought to the fore disturbing evidence that all may not necessarily be well in the deepwater sub-province, in terms of the health of some deepwater fields.

Hurricane Ivan pushed through the GOM without laying upstream facilities to waste in any highly visible way (aside from seven shallow water production platforms toppled or destroyed).[18] However, 25 days after the precautionary shutdown of many deepwater production facilities lying in the storm's likely path, over 450 MBD of oil production remained off-line, and production losses had already mounted alarmingly (reaching 15 MMB by the first days of October, and helping to bring total US crude oil production during September down to a 54-year low). By mid-October, given the tentative dates for bringing production capacity back on line at some fields, it became apparent that cumulative production losses would possibly reach the 40 MMB mark.[19] By year-end 2005, production losses due to Ivan had reached 50 MMB of oil and 32 MMBOE of gas, although both of these figures were dwarfed by the nearly 100 MMB of oil and 95 MMBOE of gas production lost due to the combined effects of hurricanes Rita, Katrina and Wilma (Figure 7.31).

Companies on the whole have tried to play down some of the more disquieting reservoir engineering implications that could be associated with the prolonged absence of so much oil and gas following these

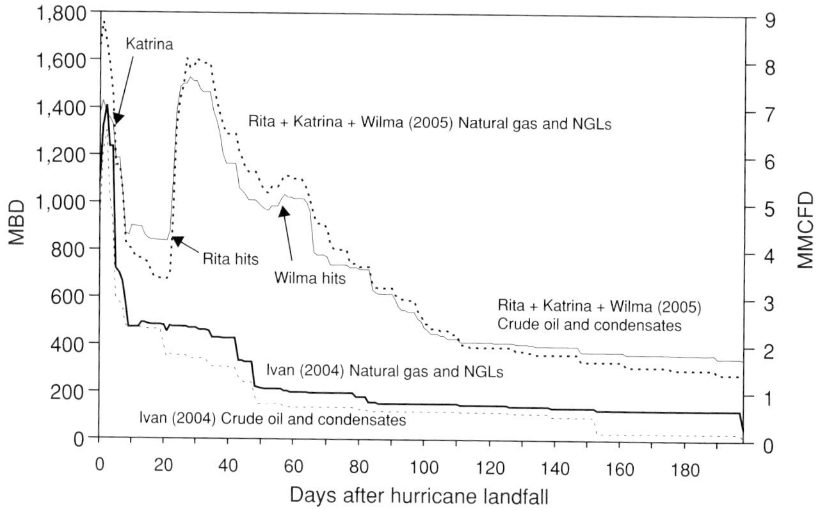

Source: MMS

Figure 7.31: GOM Federal OCS Crude Oil and Natural Gas Production Losses due to Selected Hurricanes, 2004–2005

natural catastrophes, attributing the production losses to damaged infrastructure as opposed to damaged fields. However, their caginess regarding some problem cases (Marlin, Popeye, Brutus) made their explanations ring somewhat hollow. As has already been mentioned above, Brutus (with the highest output among fields affected by Ivan) had already failed to live up to the expectations of analysts in terms of its productivity. Furthermore, even before Ivan struck, persistent rumours were circulating about possible 'fingering' problems in its reservoirs (i.e. its wells were starting to produce sand and water, and hydrocarbons were becoming compartmentalised and out of the reach of the wells).[20] In the deepwater, this is nothing short of a calamity, as wells there cannot be re-drilled or sidetracked cheaply in order to go after small, bypassed, and compartmentalised pools of oil and gas.

Brutus' sluggish behaviour after the passage of Ivan by no means constitutes an isolated phenomenon in the deepwater sub-province. As a matter of fact, mid-life crises in the form of niggling to severe production problems have manifested themselves in a number of deepwater fields, and the 2004–2005 hurricanes have either magnified many of them, or else turned them into terminal events. Take Typhoon, for instance. Hurricane Rita snapped the moorings of the TLP, causing the platform to capsize and rendering it beyond salvage and repair. Chevron and its partners thereupon gave up on the field, and sold it to a small operator

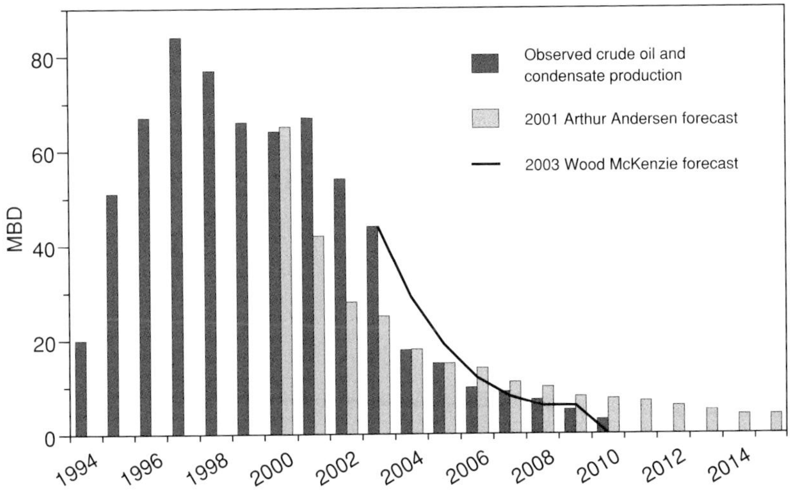

Sources: Ryddle, Snyder and George 2001, MMS, Wood McKenzie

Figure 7.32: Crude Oil Production Profile of Auger Development 1994–2015

(Energy Resource Technology) which intends to reactivate it. Whether Energy Resource Technology will succeed in re-developing the field using a re-usable, mobile floating production unit, is far from clear. However, although Rita appears to be the culprit of this outcome, the fact is that by the time this hurricane had formed, Typhoon output was already missing its design targets by a significant margin.[21] A less extreme illustration of the same phenomenon can be seen in Figure 7.32, which compares two production forecasts for the Auger field prepared in the years 2001 and 2003, respectively. The graph shows that Auger production exceeded expectations during the early part of the field's life, but the rate of decline since peak production has likewise exceeded expectations by a handsome margin (and production has now ceased altogether at two of the four blocks where the field is located). Although ultimate recovery under both forecasts is similar, the field now stands to be abandoned much earlier than planned. Indeed, reserves in Auger and many other deepwater fields have not expanded in time in a manner akin to that experienced in many shallow water fields. This is important because a very significant percentage of ultimate recovery in the shallow water has come in the form of hydrocarbons extracted over and above initial reserves estimates. Thus, the failure of deepwater reserves to grow is something that has profound implications for the longevity and magnitude of future output in the province. Therefore, it is to this issue that we now turn.

NOTES

1 McBride, Weimer and Rowan 1998.
2 See Smith 2002: 109. The abnormally low geothermal gradient encountered over much of the USGC depocentre can be ascribed to the very rapid process of sedimentation and associated overpressuring of formations (Selley 1985: 348).
3 Hunt 1996: 439.
4 According to Nehring, these are the only areas where either optimal rates of biogenic gas generation or thermogenic gas generation are likely to occur.
5 Nehring 2001a: 38.
6 *Ibid.*: 39
7 *Ibid.*
8 *Ibid.*
9 Nehring 2001b: 14.
10 The graph comes from Anderson and Boulanger 2002: 1.
11 Baud *et al.*, 2002: ix.
12 The low projection incorporates decline rates on shallow water production, whereas the high projection assumes that shallow water production remains level. Projections for the deepwater component of the total are based on operator plans as reported to the MMS.
13 Melancon, Bongiovanni and Baud 2003: 2.
14 Eskew and Jones 2001.
15 Only a small proportion of Federal GOM OCS crude production is landed in Texas.
16 Melancon, Bongiovanni and Baud 2003: 14.
17 Melancon *et al.* 2004: 17.
18 These platforms used to produce 3100 BD of oil and 9 MMCFD of gas.
19 In addition to this, there were some well-publicised problems with the quality of certain common steam offshore crude blends, notably LLS, after the passage of the hurricane (see *PON*, 29 October 2004: 5).
20 Potentially serious compartmentalisation problems have also been encountered at Anadarko's Marco Polo field (*PON*, 1 November 2004: 7), for instance.
21 Typhoon may prove to have one of the shortest life cycles of any GOM field. Chevron acquired the Typhoon lease in 1995, drilled the first well in 1998, began producing the field in 2001 and sold it jointly with BHP and Noble Energy for an unspecified sum in 2006 (together with the Boris and Little Burn oil fields). In a gesture that is at the same time a statement of intent and a corporate crossing of fingers on a large scale, the company has renamed the Typhoon field Phoenix.

CHAPTER 8

HOW LONG CAN THE DEEPWATER BOOM LAST?

There is no doubt that the deepwater GOM is a world-class hydrocarbons province, which has spawned some truly impressive technological achievements and also enjoys the added benefit of being located in a politically safe and predictable environment. As a result, GOM deepwater activities have, for more than a decade now, fired up the imagination of oil companies frustrated by the dearth of decent exploration prospects worldwide. For this very reason, forecasts about the province's long-term productive potential have betrayed a bullish bias. This is hardly surprising, given that for the most part they have been underpinned by an abstract, top-down approach to future discoveries that in the past was consistently used to draw rosy estimates of future US production, up until the very moment that Hubbert's predictions were proved right, and the forecasting party was spoilt for all and sundry.

Up until 2002 or so, such was the frenzy in GOM activity that there seemed to be little reason to worry about such details or, indeed, to believe that production in the province would not continue to climb at the rates seen since the onset of the deepwater boom in 1996. Between 2002 and the end of 2004, for instance, the number of projects in production went up by more than 50 percent, to 111. Over this period, annual GOM deepwater output increases continued to exceed 100 MBD of oil and 400 MMCFPD of gas, just as they had done every year since 1997. But even as deepwater output was close to reaching 1 MMBD of oil (and exceeding 3.6 BCFD of natural gas), the notion that production might peak at around 2.7 MMBD of oil by 2010 began to appear as less and less a sure thing even before the onset of the catastrophic 2005 hurricane season.[1]

This newfound uncertainty concerning the future prospects of the deepwater GOM had much to do with the fact that over the 2002–2004 period, the average number of operating rigs in the sub-province came down by 29 percent and the number of wells drilled dropped by 37 percent. Also, as French *et al.* highlight, 2004 data show a 'levelling off of interest for the greater than … 2,625 feet water-depth interval'.[2] A number of rationalisations to explain these trends were put forward by MMS: potential inconsistencies in the way in which wells are classified as either exploratory or development, a possible increase in the use of exploratory wells for production purposes, and longer cycle times as activities have moved to deeper waters (mainly due to operators having to devote much more time to planning and design activities). These explanations sound plausible, but even if they were true, they could not account for the fact that GOM oil rig counts declined or did not increase when oil prices reached stratospheric levels. This suggests that drilling activity did not increase either because there were simply no rigs available anywhere (even though statistics suggested otherwise), or because there were no crews available or, most soberingly, because there were simply not enough prospects worth drilling. The latter possibility, of course, is consistent with the hypothesis that the growth prospects of GOM deepwater output in the short to medium term (i.e. until 2009 or so) are being seriously overestimated, chiefly on account of flawed perceptions regarding the prospectivity of the Palaeogene. At the very least, premature as it might be to say that the GOM deepwater is close to being 'played out', rig counts from 2003 onward do seem to show that quite a few companies are indeed deploying their E&P capital elsewhere.

As related before in this study, reports concerning the imminent demise of GOM as a hydrocarbons play have proved premature on at least two previous occasions, and the time to write its definitive obituary even now is a long way off, not least because the year 2004 proved to be an especially fruitful one in terms of technological milestones.[3] However, even before the *annus horribilis* that was 2005, it had become abundantly clear that deepwater exploration and development activity was undergoing something of a lull, a lull that coincided in time with the near disappearance of global spare production capacity in oil, on the one hand, and a steadily worsening gas supply crunch in North America, on the other hand. Therefore, the related questions of how long the GOM deepwater boom can be expected to go on for and at what level the eventual production peak is likely to occur, have become imbued with an even greater significance than ever before.

The objective of this chapter is to analyse these questions from four distinct – but closely related – angles. The first one of these is the likelihood that the output contribution of ultradeepwater frontier areas to GOM production may increase significantly in the short to middle term, thereby delaying the peak point in the province by a few years. The second one is the role that FPSOs may or may not play in opening up these frontier areas to development activities. The third one is the production impact associated with significant improvements in resource recovery rates from fields exploited by means of subsea facilities. Last (but by no means least) is the role that aggressive fiscal incentives could play in ensuring that, even if industry perceptions about the attractiveness in geological terms of GOM deepwater deteriorate, investment capital would continue to flow into the sub-province, thereby sustaining – and possibly even increasing – production.

8.1 The Promise of the Ultradeepwater

For a long time, conventional wisdom saw GOM deepwater production peaking at a level as high as 2.7 MMB of oil around 2010. Reaching such a figure was always going to require many things, but foremost among them surely was the continuation of the annual reserve discovery rates achieved at the end of the 1990s. Unfortunately, reserve additions in some of the better established deepwater areas began to taper off sharply after 2000, which made it apparent that the required discovery rates would not be reached unless large commercial discoveries began to be made in what are, even as this is being written, ultradeep frontier areas: Atwater Valley, Keathley Canyon, Walker Ridge, Alaminos Canyon, the Sigsbee Escarpment and Amery Terrace. This is the reason why the focus for exploration activities has been steadily moving towards deeper and deeper waters (Figure 8.1).

Thus far, there has been no shortage of companies willing to take the plunge in even the most remote GOM ultradeepwater areas,[4] despite the fact that the reserves and discovery tallies from these areas are still

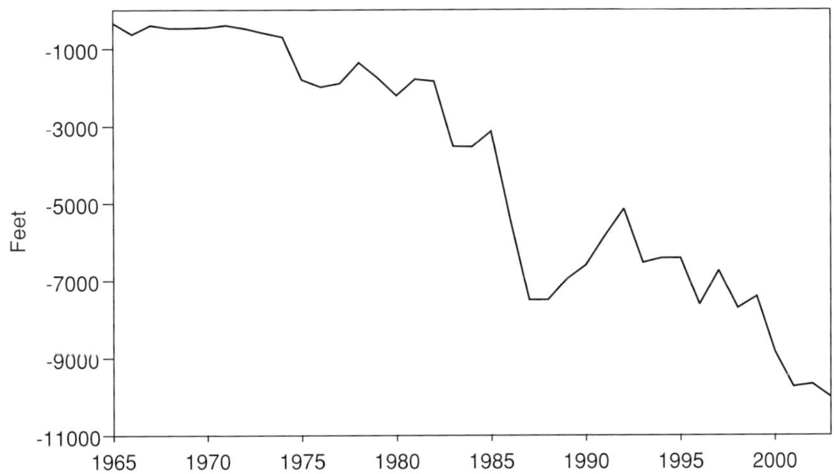

Source: MMS

Figure 8.1: Maximum Annual Water Depth for Exploration Wells Drilled in the GOM Federal OCS, 1965–2003

unimpressive (Figure 8.2). Significantly, during the period 2001–2004, 11 discoveries were made in water depths greater than 7000 feet, but at present there are no development wells in water depths exceeding 7500 feet. Suspicions have been cast on some of these discoveries (in particular Trident), but these ultradeepwater wells, drilled in pursuit of tantalizingly fuzzy seismic reflections, have seemingly opened up whole new geologic vistas.

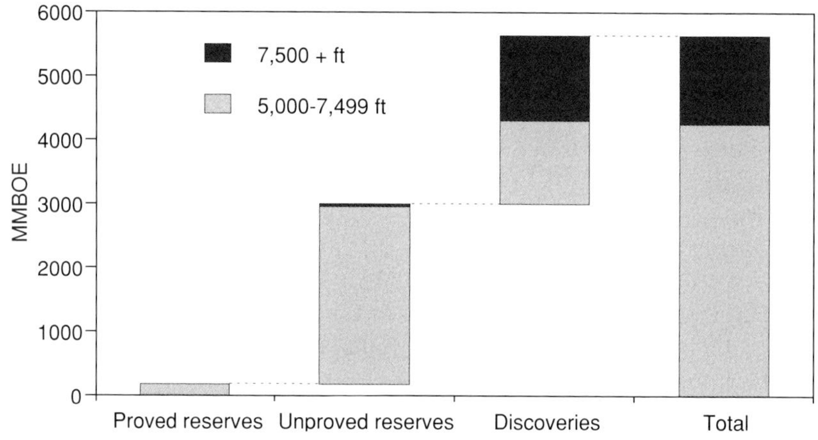

Source: MMS

Figure 8.2: Ultradeepwater Reserves (to 2000) and Unproved Reserves and Discoveries (to 2003) in the GOM Federal OCS

The most striking features in the ultradeepwater landscape are gigantic anticlinal structures called compressional box folds. These structures encompass 6000 feet or more of stratigraphic closure, and the larger among them dwarf in extension even the biggest North Sea fields (a feature that makes them easily capable of harbouring oil pools in the 1000–2000 MMBOE size range). More of these foldbelt anticlines may lie underneath the Sigsbee salt canopy, and some geologists believe that these sub-canopy structures could comprise the same type of fore-reef carbonate debris that harboured the famously prolific Mexican Golden Lane fields.[5]

Thus far, very few wells have been drilled either within these structures or their immediate vicinity. The meagre results achieved have been out of all proportion to the outlays involved, and this suggests that any assessment of ultradeepwater prospects calls for caution, first and foremost. In essence, the key question is whether the unsuccessful wells drilled in these areas reflect local adverse conditions (i.e. faulty seals, inadequate migration paths) or whether their failure is symptomatic of more intractable problems, notably a lack of hydrocarbon charge resulting from insufficient thermal maturity of the source rocks, and the poor quality and insufficient thickness and extension of Palaeogene sediments. At the moment, it is still too early to tell one way or the other, but it is certainly not necessary to agree with the Running-Out-of-Oil Malthusians to harbour serious concerns about some of the features encountered thus far in GOM ultradeepwater areas.

Take, for instance, the problem posed by salt in Keathley Canyon. Not only can salt sheets there routinely exceed a thickness of 20,000 feet in places, but they also lie under 4000 feet of water plus a sizable layer (approximately 2500 feet) of non-prospective sediment (to put these figures into perspective, one need only think that the discovery well of the Thunder Horse field involved a drill-string descending through 6000 feet of water and 2000 feet of salt to a record depth of 25,770 feet). One of the few wells drilled in the canyon so far – by BP in block KC255 in 1991 – targeted a mini-basin structure (broadly similar to the structures that harbour the Bullwinkle and Auger fields) located in 5800 feet of water and lying between massive allochthonous salt bodies (which cover the rest of the block). The well was dry, although the fact that it encountered abundant sand was seen as encouraging, because it silenced those who doubted whether good sands could be found in waters as deep as these in the first place.[6]

The drilling target for this particular well was chosen because it was seen as the best supra-salt prospect in this block and its immediate surroundings. Thus, the absence of a trap for this structure means that any further exploration efforts in the area surrounding the well would involve somebody staking their exploration capital on drilling in 4000+ feet of water, through approximately 23,000 feet of salt and non-prospective sediments, in the hope not only that there will be some prospective sediments below all this, but also that the migration of oil into those sediments will not have been impeded by the salt sheets that lie directly underneath them. One can hazard the guess that ExxonMobil, at least, will not be going down this particular path anytime soon, in the light of Lee Raymond's comment that 'the best thing ExxonMobil could have done after it drilled its first well in the Gulf was to never drill another again'.[7] Whether other companies will prove to be made of sterner stuff (or, in Mr. Raymond's perspective, whether they are more spendthrift with their shareholders' capital) is something that remains to be seen.[8] But, whatever happens, there is certainly no reason to assume that this sort of exploration venture will necessarily be crowned with success (let alone major success).

There has been more drilling activity in Atwater Valley than in Keathley Canyon, leading to seemingly interesting finds (K2, Champlain, Neptune, Shenzi) in the fold belt structures beneath the Sigsbee Salt canopy. On closer inspection, though, these have turned out to be largely disappointing. For instance, Champlain (in Atwater Valley block AT63) was initially thought to have encountered a 200-foot pay zone, and to harbour several hundred MB in reserves. In early 2005, after the reservoir was downgraded to the 50–70 MMBOE range, Unocal virtually gave away its 30 percent interest in the six exploration and appraisal wells drilled at Champlain in exchange for a small overriding royalty were the field ever to produce petroleum commercially. The previous year, the other major partners in the project (Chevron, ENI, BHP) had turned over their shareholdings to Norsk Hydro.[9]

Finds elsewhere in Atwater have tended to be on a non-commercial scale, although there have been a few successes (Merganser, Sturgis). Back in 2002, Kerr-McGee, the discoverer of the Merganser field (quite respectable in size, at 200–400 BCF of reserves), posited that anything between four and six fields of this same magnitude would have to be brought together before an Atwater development project could be considered viable.[10] By 2004, enough gas pools had been discovered in the area to justify the emplacement of a production hub with very long subsea tiebacks, and Kerr-McGee assumed the leadership of a collective development effort (since christened Independence) that may see the start of production by early 2007.[11] During these two years, Kerr-McGee examined a number of floating production options for these remote areas, including the use of FPSOs. Although the total absence of infrastructure in Atwater constitutes a significant hurdle for future development, this company concluded that the development alternative offering the most potential was the use of progressively lighter spars, even though this would necessitate the laying down of export pipelines. This view on export infrastructure, by the way, is diametrically at odds with that held in many industry circles and the banking and investment community at large.

Failures in Atwater Canyon have not been wholly disheartening, and most observers seem to believe that the poor outcomes at some sites reflect purely local conditions (although some have warned that they might be symptomatic of more diffuse factors that could potentially impinge on the prospectivity of whole swathes of the GOM abyssal plain). But conventional wisdom sees Walker Ridge, not Atwater Canyon, as the most promising of the GOM frontier areas. It is by no means hospitable, though: the shallowest blocks lie in 5000 feet of water and in the localities where leasing has been at its most intense (in the Mississippi

Fan foldbelt play), drilling targets often lie under mammoth salt sheets. On the positive side, though, these salt sheets are younger and less deformed than is the case elsewhere in the deepwater GOM, which makes seismic data less spectacularly difficult to interpret than in, say, Keathley Canyon.[12] In the midst of its salt sheets, Walker Ridge also seems to contain a number of mini-basin plays that are similar to oil-producing structures found in the Green Canyon and Garden Banks areas. Large thrusted folds found in the outermost reaches of this area have also been tested (Chinook, Loyal, Dana Point), although there have been few significant strikes to date. In addition, some deep wells drilled down to Lower Tertiary strata have achieved a measure of success (Cascade, Jack, St. Malo).[13] But, on the whole, the reason why the oil industry at large is keen on Walker Ridge is the similarities of many areas within it to a number of locations in the northeastern part of the Mississippi Fan foldbelt (within the Green Canyon administrative division) where some large finds (i.e. Mad Dog, Atlantis) have been made.

The hopes and ambitions of some companies have also been stoked up by the analogies that some have seen between the turtle structure harbouring the Thunder Horse field in Mississippi Canyon, and certain features within the Walker Ridge foldbelt.[14] Thus, after the 1999 spate of discoveries in Green Canyon, exploration activity in the Walker Ridge area picked up notably. Up to 2000, only three boreholes had been sunk there – not one of them in the Mississippi Fan foldbelt[15] – but this desultory drilling rate picked up sharply after Chevron, BHP, Kerr-McGee, Unocal/Spirit Energy and Marathon among others launched vigorous drilling programmes in this area. In contrast, Shell, ExxonMobil and BP maintained until recently a comparatively cool stance towards exploration in Walker Ridge, despite their extensive acreage holdings there. Whereas the former two seem to be sticking to their guns, BP has embarked on an ambitious exploratory drive focused on lower Tertiary prospects, of which Das Bump (drilled in late 2004) is supposed to be the most promising.

The Alaminos Canyon area shares with other frontier ultradeepwater areas many of the geological characteristics already highlighted in the paragraphs above, notably, ubiquitous salt. Its eastern and western margins are covered by relatively shallow salt lobe canopies that have not fused together. Running between these canopies lies a transition zone with large areas where no salt is present (the Alaminos Canyon itself is a deep depression whose perimeter is constituted by the distal edges of the lobate canopies).[16]

The Alaminos Canyon area also presents a few other drawbacks that are more peculiar to it. Amongst these one can mention that large parts of it lie within a zone where huge westward drifting eddies (250 kilometres in diameter and 2000 metres deep), drawn from the Eastern Gulf Loop Current, come into contact with the continental slope. The interaction between eddies and slope causes the former to decay and eventually dissipate (that is why this zone is designated as an 'eddy graveyard'). Due to the sheer size of the eddies, strong currents (with speeds of one metre per second or greater near the surface) can reach the bottom near the outer shelf edge, and the tempo of the decay process means that the currents can persist for weeks in these areas, thereby posing a significant engineering challenge (not to mention a potential safety hazard) for drilling operations.[17]

Drilling in the Alaminos Canyon area has also been slightly more complicated than elsewhere in GOM for political reasons. The locations that have been the focus of intense oil industry interest lie in close proximity both to the US–Mexican maritime border and to the western gap area covered by the drilling moratorium that was part of the delimitation of the Mexican and US EEZs. The drilling of the first exploratory well in this zone (1996) drew energetic protests from the Mexican government, on the grounds that any eventual oil extraction from this well could result in the drainage of oil from the Mexican side of the border (the bulk of the Perdido foldbelt lies under Mexican waters). To a large extent, though, this political impediment has vanished. The subsequent drilling of a handful more wells in the Alaminos Canyon border zone has not drawn any official protests, even though these wells lie closer to the border and their development prospects are somewhat brighter. The lack of interest by the Mexican government in these and other wells can probably be attributed to the triumph of Vicente Fox in the presidential elections of 2000, and the change in orientation (particularly with regards to the desirability of foreign investment) that Fox has tried to implement in Mexico, without any major success at the time of writing.

On the positive side, Alaminos Canyon does not appear beset with prospectivity problems. A good number of finds have been made in its northernmost reaches (Diana, South Diana, Hoover, Key West, Marshall, Madison, Rockefeller, Krakov). While for the most part modest in size, these fields have benefited for development purposes from the relative proximity of infrastructure associated with the East Breaks deepwater fields, as well as the siting of a production hub centred on the Hoover/Diana fields. In the future, it is entirely likely that more smallish fields similar to these will be found in the northern Alaminos Canyon area. As has been pointed out, though, the focus of the oil majors' expectations lies much further south, towards the US–Mexican maritime border. There, the interaction between the forces of deposition and salt sheet rheology has produced large intraslope formations (either remnants of submarine canyons blocked by salt uplift or depressions caused by subsidence upon salt withdrawal). These formations have inhibited the further downslope movement of sediment and, as a result, sands have accumulated there in truly extraordinary thicknesses.

As of late 2005, a number of prospects have been drilled in the ultradeepwater Alaminos Canyon (Figure 8.3): Baha (operated by Shell), Trident (Unocal), Toledo (Chevron), Great White (Shell), Tobago (Unocal), Tiger (Chevron). A couple more (Hadrian, Silvertip) are scheduled to begin drilling imminently. The first three are successive holders of the drilling depth world record (7718 feet, 9687 feet and 10,011 feet, respectively). However, nothing much by way of tangible production appears to be in the immediate horizon for any prospect in the southernmost reaches of Alaminos Canyon. The exploratory wells at Baha, Trident and Great White were all said at one point to have struck oil in respectable quantities: 300 MMBOE, 150–200 MMBOE and up to 1000 MMBOE, respectively.[18] However, at the time of writing, the Baha leases have been relinquished, and industry insiders have expressed serious doubts regarding the discovery status of Trident. Only the Great White discovery is slated for a full appraisal programme.

The obstacles in the way of Alaminos Canyon prospects making a significant contribution to GOM deepwater output (at least between 2008 and 2010) stand out in a particularly stark – and instructive

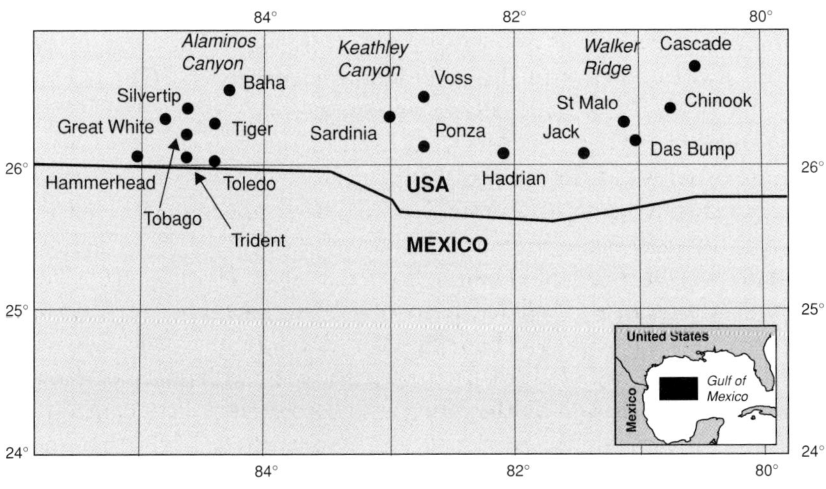

Figure 8.3: Location of Selected Ultradeepwater Wells in the Border Region of the GOM Federal OCS

– relief when one considers the economics of the Baha prospect. With a closure area of over 50,000 acres (Figure 8.4), an extension comparable to that covered by Washington D.C., Baha was seen as the 'largest untested anticlinal structure in the Gulf of Mexico visible with today's technology'.[19] Seismic sections of the prospect showed over 14,000 feet of stratigraphic section (and four different pay zones, one of which was hypothesised to consist of fore-reef carbonate debris) within the closure of the anticline. Thick salt sheets did not obscure the structure, and the reservoir rocks were very old, hence allaying any thermal maturity concerns (unfortunately, due to their age, the rocks are also very brittle and difficult to drill through). Furthermore, as well as being the largest, Baha was also the *shallowest* structure identified in the American sector of the Perdido foldbelt, which is a sobering fact indeed when one considers that Shell's original drilling target was located at no less than 22,000 feet below the mudline![20]

Baha's status as a discovery, just like Trident's, always generated suspicion (justifiably so, as subsequent events have shown). On the face of it, though, Baha seemed to have quite a few things going

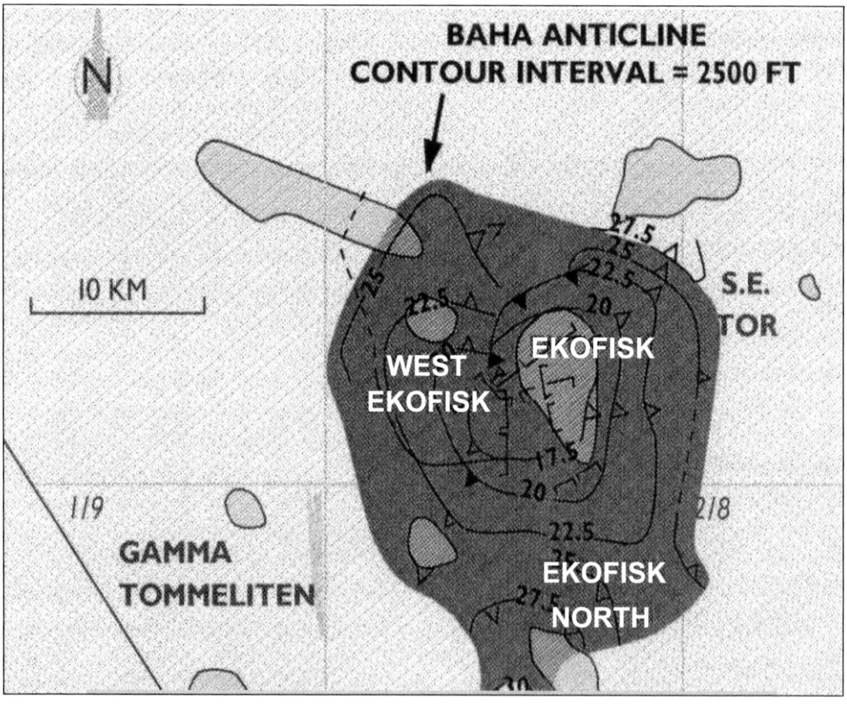

Source: Applied Earth Sciences Institute 1994: 11

Figure 8.4: Superimposition of Baha Structure on Ekofisk Basin

for it. However, it also had one major thing against it; namely, cost: USD 105 million (USD 120 million by some accounts) for the first well. Moreover, a fully-fledged appraisal and development programme, including transportation infrastructure, would probably not have left much change from a USD 5 billion kitty.[21] As a result of these costs, Shell was unable to convince its partners in the Baha wells to put up the cash for further appraisal drilling and, having failed to find other interested parties (as well as ruling out the option of going it alone), the company finally relinquished its lease on the Baha blocks in 2002.

The decision by Shell's partners to pull the plug on Baha shows that costs in the remote Alaminos Canyon are still so high that, even in a high oil price environment, a potential 300 million barrel prospect may still be a doubtful proposition in terms of generating enough revenue to pay its way. But if this is the case, how do many analysts, government insiders and even oil industry insiders reconcile their frankly bullish output outlook for GOM ultradeepwater areas with the very uncooperative finding, development and lifting cost figures in frontier areas like Alaminos Canyon? The answer to this question is that they have put their faith in FPSOs, a production technology that, to be fair, has proved remarkably effective in many petroleum provinces around the world, by sparing companies from having to discover enough reserves to justify the cost of building fixed infrastructure to ship hydrocarbons from remote areas (such as the furthest corners of the GOM ultradeepwater). However, there is by no means universal agreement that FPSOs will offer a viable proposition to unlock the riches of the ultradeepwater. For instance, Kerr-McGee's conclusions regarding development options for some of its Atwater prospects seem to suggest that this company believes that FPSOs will never amount to much in GOM, for reasons unrelated to their technical merits. It is to this issue that we now turn.

8.2 Will FPSOs Do the Trick?

An FPSO is a floating production facility (generally in the shape of a ship) that uses either a turret or a bow mooring system to maintain a geostationary position while producing hydrocarbons via subsea risers connecting wells in the sea floor with its onboard processing facilities. An FPSO will store crude in tanks located within the hull, and will periodically offload oil to either shuttle tankers or ocean-going barges for transport to shore or to market.

FPSOs entered the oil scene back in the mid-1970s. In 1975, Hamilton converted a semi-submersible rig Deepsea Pioneer into the world's first FPS, and deployed it to serve the Argyll field (and later, the Duncan field as well), located in 260 feet of water in the British sector of the North Sea.[22] Around the same time, Arco Indonesia Inc. installed a 66,000 DWT production vessel (a converted ship) in the Ardjuna field, offshore the island of Java. Then, in 1977, Shell installed the 35,000 DWT *Bahía Gaditana* at the Amposta field, located offshore the delta of the river Ebro (in Spain), in 190 feet of water. Petrobrás followed suit shortly thereafter (1979), installing an FPSO at its Garoupa field, in waters 550 feet deep.

Up until 1985, FPSOs were mainly seen as infrastructural exotica. After 1986, the confluence of rapid technological evolution and the spread of oil industry operations to more remote marine environments allowed this production method to gain an ever greater degree of acceptance: between 1986 and 1994, FPSOs were commissioned at a rate of two per year. From the mid-1990s onwards, their ranks swelled even more, with FPSOs being installed in almost every major offshore producing region of the world at an average rate of eight to nine per year. From 1999 to 2003, 13 FPSOs were installed worldwide, and it is anticipated that a further eight will be installed over the 2004–2008 timeframe. Up to the end of 2003 inclusive, 129 FPSOs had been deployed worldwide, which is more than all the other types of FPS (which include TLPs, spars and production semi-submersibles) put together.[23]

The storage capacity of vessels that have been impressed into service as FPSO ranges from a low of 47 MB (Crystal Sea) to a high of 2 MMB (Åsgard, Girassol, Petrobrás P31, P32, P33 and P35). FPSO storage capacity has increased markedly in recent years, a reflection of the ever higher production rates achievable in FPSO development projects. Maximum production rates for FPSOs also span a wide range (from 11 MBD for San Jacinto to more than 220 MBD for Norne), with most recently commissioned vessels tending towards the highest side of the spectrum: the large FPSOs mentioned above were all designed for maximum production rates of around 200 MBD, whereas the average maximum production rate for FPSOs put in service before 1994 was only 60 MBD. Over the past decade, the average capacity of new vessels entering service has risen from around 60 MBD to 110 MBD, and it will probably exceed 150 MBD by 2008. For their part, mooring depths for FPSOs go from around 70 feet (Chang Qing Hao) to more than 6000 feet (Seillean, in the Roncador field). Fifty of the active FPSOs are in water depths less than 1000 feet and, until 1995, the average operating water depth was less than 700 feet. Since 1995, though, there has been a dramatic increase in FPSO operating water depth, and this trend will only accelerate in coming years (it is expected to surpass 3000 feet by 2010).

About a fifth of the global FPSO fleet is moored in the North Sea, with the UK accounting for the majority (70 percent) of the vessels. FPSOs also operate in numbers in China, Australia and Brazil, and there is anything from a couple to a handful of vessels in each of Indonesia, Canada and Malaysia, Thailand, Vietnam and the Mediterranean.[24] As of September 2004, a total of 25 FPSO units were on order, and the global fleet is expected to grow to between 230 and 246 vessels by 2010.[25] The key growth market for FPSOs is likely to be West Africa, where the operating number of vessels is set to expand significantly in the short term with the development of a number of deep offshore projects (like Bonga, Dalia, Girassol, Kizomba, Xikomba, Plutonio) in Equatorial Guinea, Angola, Nigeria and other countries in the region.

As can be appreciated, FPSOs have been extensively used worldwide, both in regions where harsh climatic conditions prevail – North Sea, South China Sea, East Canada – as well as environmentally sensitive areas (one FPSO operated 4 miles offshore California from 1981 to 1994).[26] Their environmental record is quite good: after nearly 500 FPSO-years, and the processing of more than 6500 MMB of oil, the most serious incident involving one of these vessels is still the spillage of 3.9 MB in the North Sea from the Texaco Captain FPSO during its start-up.[27] Despite this, environmental groups in the USA have always tended to see FPSOs as a suspect method of producing oil.

GOM is the one major offshore petroleum province in the world where FPSOs have not yet been emplaced (Figure 8.5).[28] The closest thing to an FPSO yet seen there was the converted semisubmersible that Placid installed in 1989 at its unsuccessful Green Canyon GC29 development.[29] Aside from broad environmental concerns, the lack of an FPSO presence in the GOM upstream has been due to a variety of factors. For starters, until quite recently, fields in both the Mexican and the US sectors were to be found relatively close to shore, as well as to major consumption centres. Moreover, the density of production and transportation infrastructure in both sectors (especially the American one) made even very small fields amenable to development with more conventional production methods. Also, the gas prone nature of the GOM Federal OCS militated against FPSOs, as fixed platforms with larger topsides are a more suitable method to develop offshore gas fields (indeed, the rationale for many FPSOs is to exploit oil from fields whose natural gas reserves are, in effect, stranded).

Figure 8.5: Distribution of the World's Active FPSO Fleet

Prospects for FPSOs in the GOM Federal OCS have brightened up considerably, as exploration and production activities in the deepwater have moved to areas that are very remote from existing infrastructure. Indeed, even in areas where the problem of access to infrastructure is not insurmountable, smaller players with lower 'materiality thresholds' have seen in FPSOs a possible means to bypass high trunkline tolls and thereby avoid the prospect of being left with stranded pipeline segments.

In late 2001, after conducting an environmental impact assessment spurred by oil industry interest in FPSOs, MMS approved in principle the use of this production method in much of the Central and Western planning areas.[30] This was heralded as a development that would 'open up bidding, leasing and drilling to many more producers, who would otherwise have stayed out because of the cost stranding situation'.[31] Unfortunately, these high expectations are unlikely to be realised.

Even in those GOM locations that seem ideally suited for FPSO operations (i.e. where oil strikes lie at a considerable water depth, at a long distance from shore and in areas lacking in infrastructure), the minimum profitability threshold for projects developed on the basis of this production method is likely to be rather higher than in other offshore provinces. This is because FPSOs in GOM have to comply with legal requirements that will make their operations more expensive than is the case in other petroleum provinces. FPSOs themselves will not be considered Jones Act vessels and will therefore not be required to be coastwise qualified, unless they are actually used to deliver oil to US ports (which is highly unlikely). This opens up the possibility that companies wishing to develop a GOM field with an FPSO may redeploy a vessel previously in use elsewhere (subject to availability and technical suitability), thereby saving on conversion costs and allowing for tighter production schedules.[32] However, any shuttle tanker carrying oil produced through FPSOs will have to comply with Jones Act requirements and be coastwise qualified. Thus, all shuttle tankers will have to be built in the USA, carry the American flag and employ local crews in US waters, which will significantly increase the costs (fixed and variable) of this shuttle tanker fleet relative to those operating elsewhere. In addition, shipbuilders have stated that a three-year lead time is likely to be the bare minimum for any FPSO development that involves the use of shuttle tankers, chiefly because of the limited capacity available at high-cost US shipyards.[33]

Plenty of firms have expressed an interest, in principle, in operating an FPSO in the GOM region. Moreover, for at least some of these companies, projects requiring an FPSO appear to be the most viable in otherwise unexciting or very daunting development portfolios. So why has it taken so long for such companies to submit concrete FPSO development proposals to MMS? The answer to this question has to do with a problem that FPSO operators generally do not have to contend with: the disposal of natural gas output.

In general, FPSO development projects are predicated on the recovery of oil, not natural gas (the latter tends to be flared and, during the later stages in the development of projects, reinjected into the reservoirs to boost oil recovery). This is because FPSOs are used to exploit reservoirs whose size and location preclude the building of either large diameter high pressure pipelines (the infrastructure best suited to take natural gas to market) or full-scale liquefaction plants. Fields in the ultradeepwater GOM meet the remoteness criteria in full, but the conservation mandate of MMS does not contemplate allowing companies to treat natural gas found in ultradeepwater fields as if it were stranded. MMS will only allow flaring in emergencies or for very short-term tests, and the agency has made it very clear (in response to suggestions that it 'had changed its position with regard to the disposition of associated gas resulting from oil production from floating production, storage, and offloading vessels') that it will not approve any FPSO development plan 'that does not provide for the eventual marketing of produced associated gas under the current lease(s)', even if leaseholders argue that 'it is not technically or commercially feasible to produce and market the gas'.[34]

MMS argues that 'there has been no change in [its] position'[35] regarding the issue of gas disposal from remote leases. This, however, is not strictly true. In fact, the passage of time (and the fact that no ultradeepwater developments have materialised yet) has seen a softening up in the position of the agency. Back in 2001, in the final draft of the environmental impact statement for FPSOs, MMS explicitly incorporated 'pipeline export for associated gas' in its economic assumptions, all the while recognising that the high construction, installation and operation costs that this option entailed would be a significant hurdle for future field developments'.[36] Moreover, in that document, MMS stated that it intended to discourage FPSO operators from re-injecting any produced gas, even if this meant that, in order to dispose of produced gas, leaseholders would have either to build dedicated gas pipelines or else adopt commercially unproven technologies, ranging from the very expensive but seemingly feasible (floating CNG, floating LNG, floating gas to liquids[37]) to the positively esoteric (i.e. trapping of gas in hydrates which would be shipped ashore for subsequent release of the gas). According to the agency, these emerging technologies had been identified in industry studies as being technically and economically within reach and there was therefore no reason why potential FPSO operators should not consider them as alternatives in their decisions about gas disposition.

The literal application of this injunction against gas re-injection would have had a very negative impact on the economic prospects of any FPSO-centred development project. After all, the key attraction of these vessels lies precisely in their ability to operate in the absence of infrastructure, and to dispense with all the complications of taking offshore gas to market, by whatever method of transportation. Moreover, the fanciful technologies that MMS mentioned are emergent to the point that they are nowhere near the working full-scale prototype stage. Thus, the inevitable consequences of the enforced adoption of any of these alternative technologies by MMS would be significant increases in the development lead-times of ultradeepwater FPSO projects, and even greater increases in development and lifting costs for those projects. Fortunately for potential FPSO operators, the MMS *Clarification* regarding the acceptable options for the disposition of associated gas related to oil production now states that

> produced associated gas related to FPSO oil production may be: marketed via pipeline, converted to liquid natural gas (LNG) or methanol (gas to liquid conversion) and marketed; used to generate electricity that is used on or off lease for oil and gas operations and/or marketed; *injected into a producing oil reservoir to increase oil recovery with the expectation that once the oil is*

depleted, the recoverable gas will be produced and marketed; injected into a nonproducing reservoir with the expectation that prior to lease or field abandonment, the injected gas will be produced and marketed.[38]

As can be appreciated, the general thrust of the last two items in this list is totally at variance with the tacit position regarding re-injection that MMS took in the final draft of its FPSO environmental impact statement. This should not be taken to mean, however, that FPSO development projects that depend on gas re-injection will be granted approval as a matter of course. Indeed, there are plenty of reasons to suppose that obtaining such an approval even from a more lenient MMS is still very difficult, which would go a long way to explaining the dearth of FPSO development submissions.

MMS has now expressed its readiness to accept that natural gas 'injected into a reservoir *with no commitment for future recovery* … is not lost and could be recovered by another lessee in the future'. However, MMS has also stressed that it is not 'convinced that this solution will result in the recovery of this valuable resource at a future time, and could easily result in this gas never being recovered'.[39] In other words, at this point in time, MMS appears to be willing to accept in principle only those FPSO development submissions that contain a clear commitment as to how re-injected gas is to be marketed eventually. Few if any OCS lessees will relish preparing such a plan: if the time to depletion or abandonment of a field is long, then the market uncertainty will be too high; alternatively, if the time to depletion is short (probably because of the production profiles characteristic of deepwater fields), then it is unlikely that the gas disposal options will be significantly more attractive than when the decision to develop was taken.

Even assuming that MMS is prepared to accept gas re-injection for a limited period, in exchange for disposal thereafter, it does not follow that the agency will be able to sell this idea to its political masters, much less to the American public at large. After all, how will MMS be able to justify permissions to re-inject gas at a time when the US natural gas market may very well be in the throes of a serious supply crunch? For the same reason, it appears equally unlikely that US politicians will allow American gas from remote ultradeepwater areas to be piped *directly* into Mexico, NAFTA and the constructive suggestions of some oil companies notwithstanding. In any case, the economics of piping gas to Mexico from such locations are likely to be on the marginal side as well (the areas are very remote from suitable landfalls in Mexico, after all).

All of the factors mentioned above point in one direction: despite its prospectivity, the development of GOM ultradeepwater is likely to be more protracted than what conventional wisdom has been suggesting for some time (and, hence, peak GOM deepwater production will come sooner than originally expected). Delays in deepwater FPSO developments are a common occurrence worldwide, of course. For instance, in deepwater Angola during the late 1990s, general industry expectations for project lead times from discovery to first oil were around four years but, as Traynor, Aldridge and Cook underlined a few years later, 'the same set of fields … look likely to have an average seven-year lead time, with an approximate 1% reduction in IRR for every year of delay'.[40] The delays can be attributed to a variety of reasons, most of them technical rather than political in nature. In the GOM ultradeepwater, though, a very different situation obtains.

Although one cannot underestimate the technological challenges facing the oil industry in remote ultradeepwater areas in GOM, their eventual development will depend not so much on the maturity of available technologies but on a satisfactory answer being given to what is a conservation question; namely, what will happen to economically stranded natural gas? This question does not pose too much of a problem worldwide because in most offshore petroleum provinces gas flaring can be practised with gay abandon and nobody bats an eyelid. However, it is a major political problem in the USA, where physical wastage of resources cannot be condoned by an agency like MMS, whose central mandate revolves around resource conservation. Again, the possibility that a technological breakthrough might allow the oil industry to provide a satisfactory answer to this question cannot be discarded out of hand. But, by the same token, one should not underestimate the complex problems and high costs that have thus far prevented technically feasible solutions – like floating liquid natural gas (FLNG) – from leaping off industry drawing boards.

The foreseeable contribution that FPSO operations will make to GOM output in the run-up to the deepwater production peak can be summed up thus: very little, if any. In 2001, after the publication of the environmental impact assessment on the use of FPSOs in GOM cleared the field for this production method, MMS expected to start receiving concrete development proposals at once. As of late 2004, though, such proposals are yet to be received. Given the development and construction lags involved, *and even assuming that natural gas re-injection were to be allowed*, it is difficult to see how fields requiring an FPSO to produce will be in a position to contribute to deepwater output before 2010, if significantly more time were to elapse before the agency starts to receive such development submissions.[41] This would mean that any contribution from FPSOs in these areas will, at best, involve smoothing the post-peak production decline. Of course, if gas re-injection were not allowed, reasonable allowances regarding the maturation of FLNG or CNG technology will delay any contribution of FPSO output to GOM production to an even greater extent. Indeed, it is not inconceivable that by the time such technologies are ready, very high North American gas prices on a sustained basis might have rendered the whole issue of FPSO economics academic, by making it possible even for gas from remote offshore locations to be taken to market via pipeline. Needless to say, this is not an outcome that policymakers are particularly looking forward to, and avoiding it will therefore require a palliative for potential deepwater production declines with a more limited time horizon. Could much improved subsea recovery rates be this palliative?

8.3 Maximising Subsea Recovery

Unlike FPSOs, subsea production facilities form an integral – indeed, a vital – part of the GOM deepwater sector. Even though only small volumes of oil and gas were being produced from subsea wells up until the mid-1990s, the volumetric contributions of subsea wells towards total GOM deepwater oil and gas output grew to 350 MBD and 1.7 BCFD, respectively, by 2005[42] (these figures represent around 34 percent and 50 percent of deepwater oil and gas production, respectively).

As Figure 8.6 shows, subsea deepwater completions steadily increased

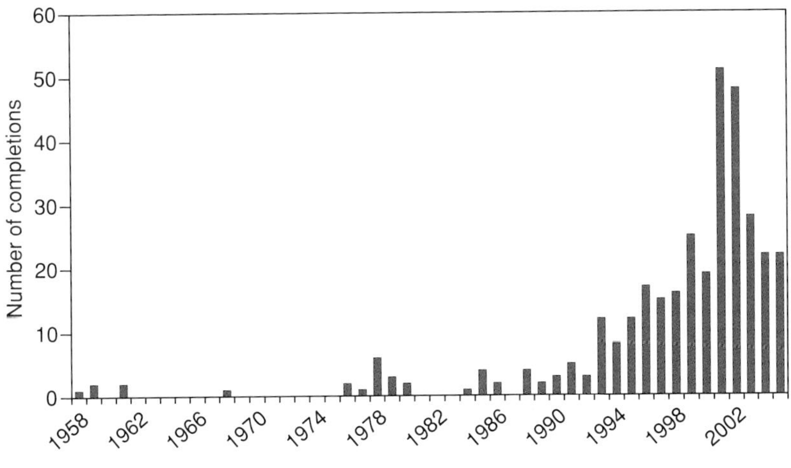

Source: MMS

Figure 8.6: Annual Subsea Well Completions in the GOM Federal OCS 1958–2003

from the mid-1990s onwards, and by 2003 they numbered 164 (compared to 131 subsea completions in the shallow water). In terms of depth, progress has been even more remarkable: the deepest subsea completion lay beneath 350 feet of water until 1988, when a subsea completion for Placid's unsuccessful Green Canyon GC29 project took the record to 2243 feet (Figure 8.7). In 1996, a Mars well was completed at 2956 feet. The current depth record stands at 7216 feet, for the Camden Hills development (however, 70 percent of the subsea completions are still located in water depths of less than 2500 feet, as Figure 8.8 shows). GOM subsea operations have also been trendsetters on a global scale in terms of the length of tiebacks between subsea wells and host surface piercing platforms. Although most subsea wells are within a 10 mile radius of their respective host platforms (Figure 8.9), GOM also boasts the two longest subsea tiebacks in the world, in Mensa (at 62 miles) and Canyon Express (at 55 miles). These tieback

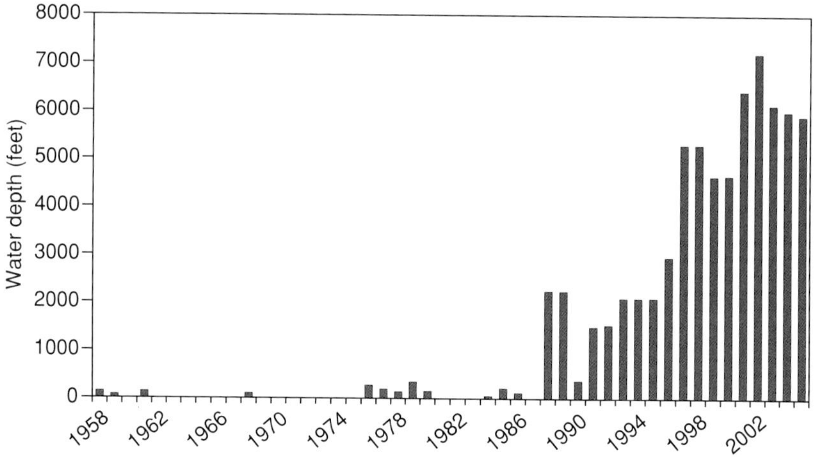

Source: MMS

Figure 8.7: Maximum Depth of Subsea Well Completions in the GOM Federal OCS 1958–2003

lengths represent increases of 206 and 183 percent, respectively, over the erstwhile record of 30 miles, held by the Troll Oseberg Gas Injection project in the Norwegian sector of the North Sea.

Subsea technology is proven and cost-effective, and even though GOM subsea oil production has increased at a far slower rate than deepwater production as a whole, it is reasonable to assume that a significant share of future output will come from subsea wells. After all, already there are a significant number of discoveries whose development

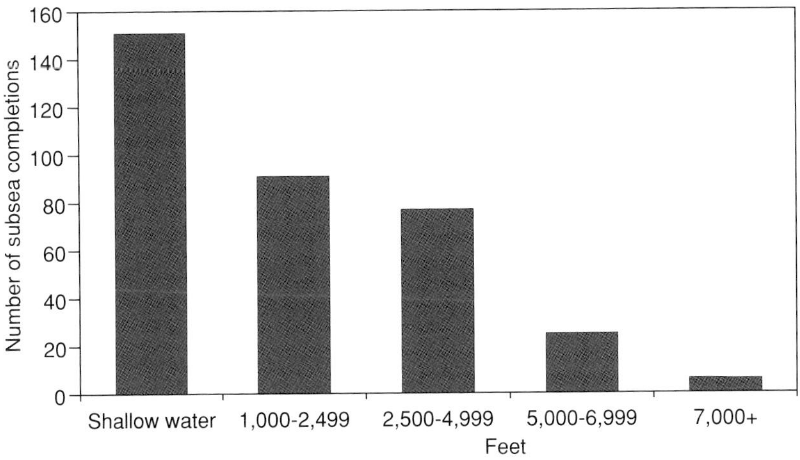

Source: MMS

Figure 8.8: Subsea Well Completions in the GOM Federal OCS, by Depth Bands

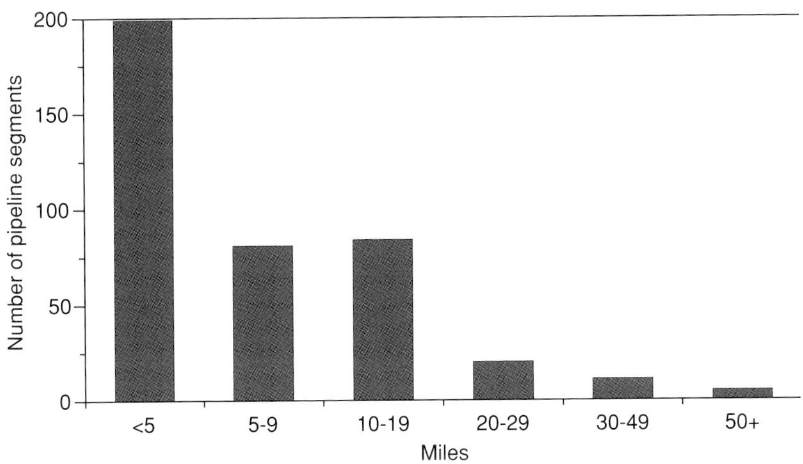

Source: MMS

Figure 8.9: Subsea Well Completions in the GOM Federal OCS, by Tieback Length

with current technology as subsea satellites would be viable if only they were located closer to potential host platforms. Given the industry's track record in these matters, it is safe to assume that technological advances will neutralise the flow assurance problems that currently preclude tiebacks approaching or exceeding 100 miles in length. This, together with the development of multiphase flow and subsea separation technology, will allow the commercial exploitation of hitherto submarginal prospects, without requiring a proportionate – and very expensive – increase in the density of host platforms and trunklines.

The installation rates for subsea facilities will continue to be primarily a function of discovery rates. Thus, it follows that if subsea technology is to make a contribution to production over and above that attributable to exploratory success, this can only come in the form of increased recovery factors. There is, moreover, plenty of scope for improvement in this direction. The most thorough study on subsea recovery factors, carried out jointly by Statoil, Norsk Hydro and the Norwegian Petroleum Directorate (NPD), found recovery from subsea wells to be 15–20 percent lower than that for wells with direct platform access. This is a reflection of the fact that accessing completed subsea wells is far more difficult and costly than for wells drilled from a fixed installation, not least because vessels or mobile rigs have to be used even for minor jobs. Although in recent years a significant effort has been made to expand the fleet of mobile units that undertake this sort of work, the number of light intervention vessels or rigs available remain limited, and below requirements. This often translates into suboptimal maintenance for subsea wells, as well as insufficient collection of reservoir management data, both of which impinge negatively on recovery factors. These problems are compounded in subsea wells with long tiebacks (i.e. the type of subsea well that is bound to become more and more common in GOM) by the considerable difficulties inherent in maintaining a large enough pressure over time to maintain tail production.

Improving subsea recovery rates (and thereby prolonging the lives of mature fields) will require a collaborative effort between oil companies and equipment and service suppliers, who have carried the brunt of the development work (and risk) up to now, but are no longer in a position to continue doing so. There is a need not only for improved quality control during development and fabrication but also for the deployment of more robust materials, leading to greater reliability in all subsea equipment, flow lines, risers and umbilicals. In addition, there should be a greater degree of standardisation in both materials and design. These goals seem to be well within the reach of the considerable capabilities of oil companies, service companies and technology and equipment providers. Hence, it seems entirely reasonable to assume that in the not too distant future, recovery rates from subsea wells will indeed be comparable to those from conventional wells.

Unfortunately, the impact of this improvement in recovery rates will be felt chiefly at the level of ultimate recovery, as opposed to sustainable production at peak. In other words, improved subsea recovery will greatly prolong the longevity of some fields but, by the same token, it will not allow them to increase their pre-decline production rates to any great extent. Improved recovery, then, is not to be the panacea for GOM deepwater output. But is it possible that the goal of significantly higher output may require nothing more complicated than a significantly more lenient and flexible fiscal regime?

8.4 Can Fiscal Incentives Sustain the Deepwater Boom?

From 1993 onwards, the deepwater GOM has been by far the hottest exploration play worldwide: proved developed and undeveloped reserves additions plus discoveries over this period have amounted to nearly 9 MMBOE. During this time, a large proportion of the worldwide fleet of deepwater rigs has been committed to GOM (Figure 8.10 shows the number of rigs operating in the sub-province), and these rigs have drilled a most impressive number of wells, both exploratory and development (Figures 8.11 and 8.12). In addition, more than 4000 miles of deepwater oil and gas pipelines have been installed (Figure 8.13). To sustain these intense levels of activity during the most frenzied period of the GOM deepwater boom (1998–2003), the oil industry spent

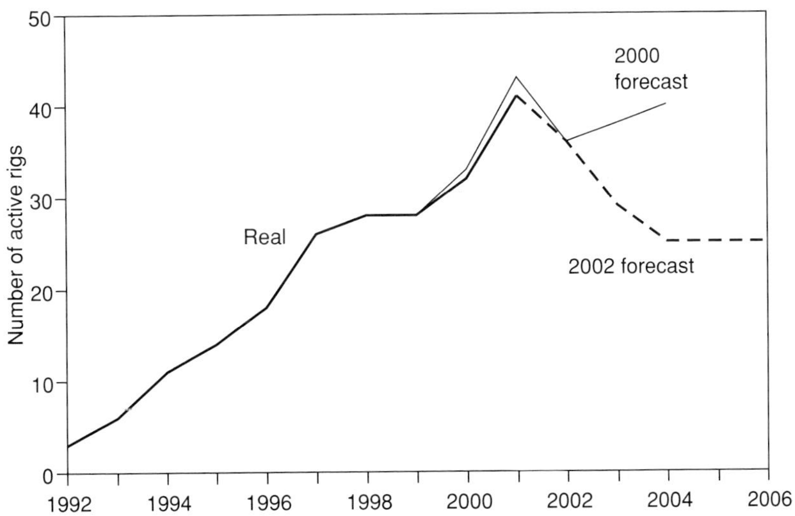

Source: MMS

Figure 8.10: Number of Rigs Operating in the GOM Federal OCS Deep Waters, 1992–2006

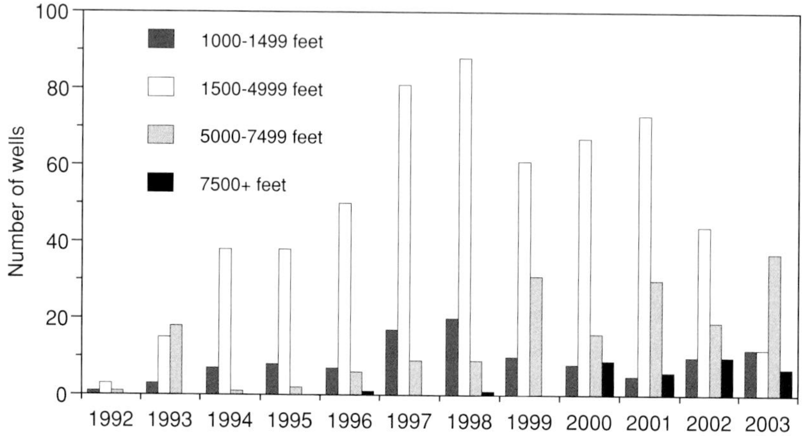

Source: MMS

Figure 8.11: Exploratory Wells Drilled in the GOM Federal OCS, by Year and Depth Range, 1992–2003

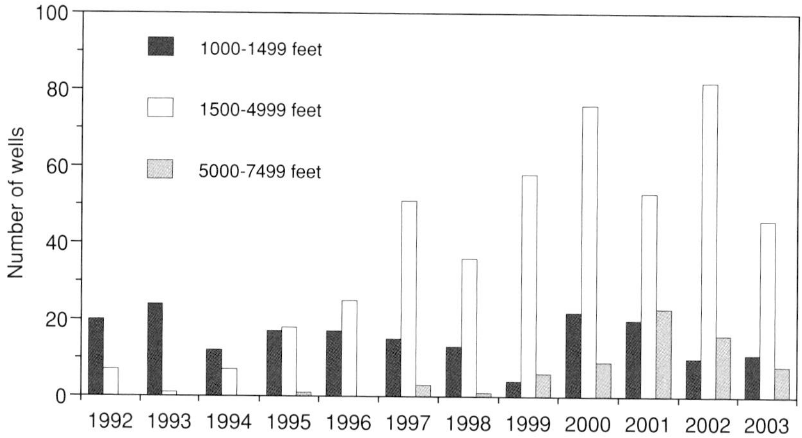

Source: MMS

Figure 8.12: Development Wells Drilled in the GOM Federal OCS, by Year and Depth Range, 1992–2003

around USD 12 billion, a figure equivalent to nearly 50 percent of the cumulative global spending in deepwater upstream activities.

Conventional wisdom regarding the future direction of deepwater activities considers that, up to 2008 inclusive and very possibly beyond, about 50 percent of the deepwater rig fleet will remain active in

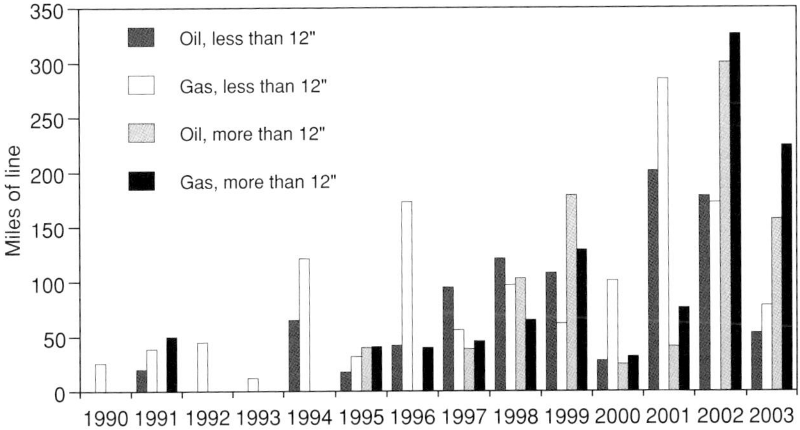

Source: MMS

Figure 8.13: Deepwater Pipelines Installed in the GOM Federal OCS, By Diameter and Type, 1990–2003

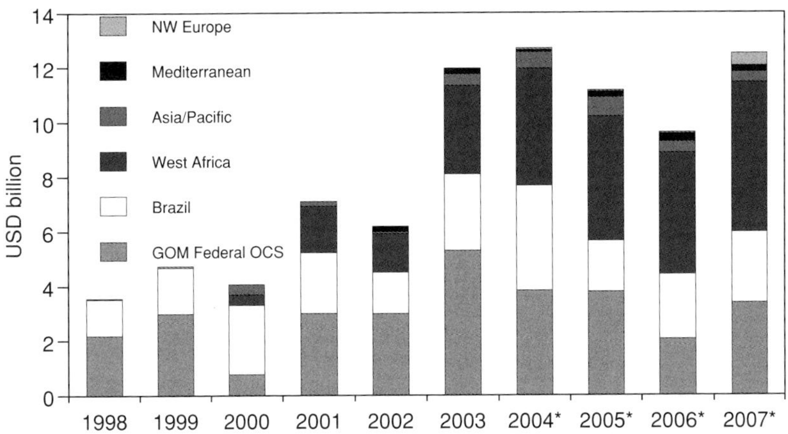

* Estimates

Source: Robertson and McFarlane 2004

Figure 8.14: Worldwide Investment in Deepwater Development Projects, by Province, 1998–2007

GOM (Figure 8.14). Despite the rise to prominence of other provinces (especially West Africa) and the marked slowdown in drilling activity seen over the 2002–2004 period, the GOM Federal OCS is still seen as accounting for around a quarter of worldwide expenditure in deepwater activities by the end of the present decade.

This ability of the GOM region to hold its own in the investment league, against more prospective provinces, has been put down to a variety of factors (political and institutional stability, proximity to market and ancillary services, and so forth). However, many observers have singled out for praise the various branches of the US government, crediting them for the way in which they have striven to make overall terms in the deepwater more attractive for oil companies, in order to attract investment capital to the region. Indeed, looking towards the future, the willingness on the part of the US government to go the distance on the fiscal front is widely seen as being perhaps the best indicator of the future health of the GOM upstream sector. In other words, no matter what the outcome of exploration efforts in frontier areas and FPSO regulatory controversies might be, it is taken as read by many observers that the GOM fiscal regime will be adjusted in a way that will make it worthwhile for oil companies to continue to invest in deepwater oil and gas, thereby ensuring that GOM output continues to expand. But to what extent is this trust in the power of tax breaks to 'buy' increased output justified?

8.4.1 Deepwater Royalty Relief

Deepwater royalty relief is often cited as the supreme example of the unwavering disposition of the US government to do what has to be done in order to give the maximum incentive possible to domestic oil production, chiefly because it represents the explicit abandonment of the principle that public mineral property (seen as a capital accumulated by Nature) should never be surrendered or conveyed to private parties without fair and proper compensation. In oil circles everywhere, the deepwater royalty relief initiative has been touted as being responsible in no small part for the sharp increase in the number of deepwater blocks receiving bids in lease sales held after 1995. The following lines, penned by MMS officers, succinctly express the view held by this agency as to how royalty relief has contributed to keeping the exploration momentum going in the deepwater sub-province:

> in the 2 calender [*sic.*] years preceding passage of the [royalty relief act], bidding for newly issued leases in deepwater was modest at best: 78 tracts in 1993, and 71 tracts in 1994. Subsequently, bidding on deepwater tracts exploded: 334 in 1995, 877 in 1996 and 1,280 in 1997. Clearly, the mandated royalty suspensions available to new fields, *regardless of economic need*, played an important role in this outcome.[43]

Although deepwater royalty relief has been hailed as a major policy innovation, in fact it has a long pedigree. The 1970s vintage OCS legislation granted the Secretary of the Interior the statutory right to reduce the royalty rate for a field if its abandonment was imminent. However, as Mead explains, no Secretary of the Interior ever dared to exercise this authority because of their fear that 'the public might not understand the economic reasoning indicating that society would benefit by such action and might instead see it as a "give away" to oil companies'.[44] Indeed, before the passage of the 1995 Deepwater Royalty Relief Act (DWRRA), the best that the US Federal government had been able to come up with in order to improve the profitability of supposedly marginal fields as well as fields in frontier areas was to ensure that deepwater tracts only attracted the minimum royalty rate set by law (12 ½ percent). In this sense, it is worth remembering that barely two years before the passage of DWRRA, a very similar initiative put forward by Senator J. Bennett Johnston of Louisiana was easily defeated in the Senate, amidst strident denunciations that it amounted to nothing more than 'corporate welfare'.[45] After this failure, steering the deepwater royalty relief initiative through the halls of Congress required supporters of the bill to put into play all of their legislative wiles to secure its passage (and, even then, the bill survived mainly because it was appended to another legislative vehicle that enjoyed quite broad support: the repeal of the ban on Alaskan oil exports).[46]

The truly novel aspect of deepwater royalty relief is that it has been successfully enacted for the benefit of what everybody acknowledges is not a marginal play whose prospects would be stunted if not carefully nurtured with tax breaks, but rather a highly profitable world class hydrocarbon province where large oil companies enjoy an overwhelming presence, and cash-strapped small companies do not form a part of the picture.[47] Moreover, the path blazed by deepwater royalty relief has been widened to cover also the deep gas sub-province, which again would be an attractive play even in the absence of tax incentives. To that extent, DWRRA (which President Clinton signed into law as Title III of the Alaska Power Administration Sale Act, S. 395) deserves to be seen as a major milestone in the evolution of the institutional framework governing GOM oil activities.

DWRRA was meant to encourage exploration and production by exempting all fields found in deepwater leases issued after 28 November 1995 from royalty payments, dispensing with any administrative process of economic evaluation of need.[48] The automatic waiver would apply until such time as output reached a predetermined target, which varied according to depth (17.5 MMBOE for leases lying at water depths between 600 and 1200 feet, 52.5 MMBOE for leases lying between 1200 and 2400 feet, and 87.5 million barrels for leases located beyond

2400 feet). These 'stipulated minimum royalty suspensions volumes', which MMS readily confessed to be 'large',[49] were to apply 'to the fields upon which … leases reside, not to each individual lease on the field',[50] regardless of how many OCS blocks a given field might straddle.

Threshold prices based on NYMEX contract settlements were also established for both crude oil and natural gas: full royalties would be payable whenever the daily price for both commodities pierced either or both thresholds (price thresholds applicable to subsequent calendar years would be determined by multiplying the base year price times the GDP implicit price deflator). The base (1994) oil and gas prices set in DWRRA were exceptionally generous, and up until 2004 inclusive, observed NYMEX prices had only rarely penetrated either threshold (Figures 8.15 and 8.16). Thus, the royalty waivers extended under the original DWRRA provisions have proven very valuable for those companies whose development projects came on stream in time to enjoy this benefit.[51]

The original provisions of the royalty relief bill expired at the end of 2000, and a number of changes were introduced to the scheme in 2001. Firstly, probably on legal grounds, it was determined that, henceforth, royalty suspensions would only apply to specific leases, regardless of the field to which they might be assigned.[52] Secondly, the automatic royalty suspension volumes were phased out and, in their stead, MMS was given the faculty to set waivers (again applicable to leases, not fields)

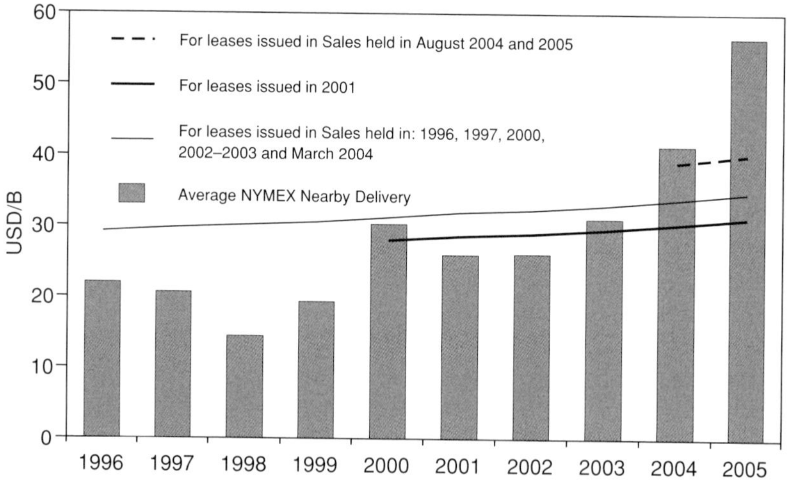

Sources: MMS, NYMEX

Figure 8.15: Crude Oil Price Thresholds for Deepwater Royalty Relief, and Observed NYMEX Oil Prices, 1996–2005

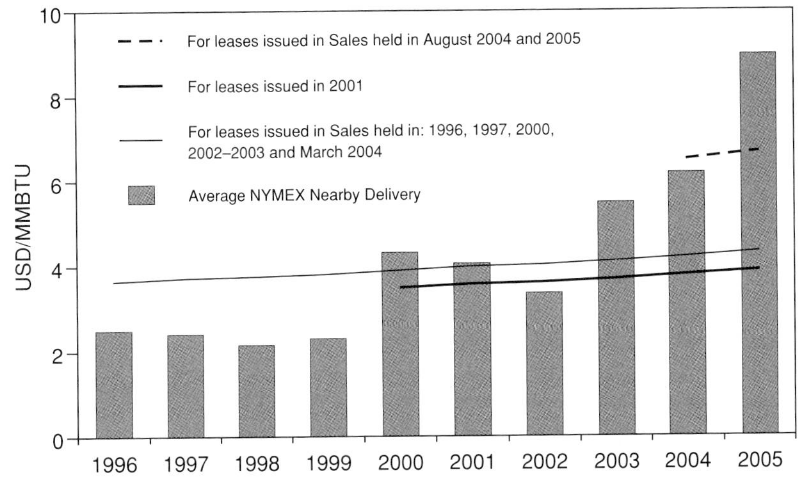

Sources: MMS, NYMEX

Figure 8.16: Natural Gas Price Thresholds for Deepwater Royalty Relief, and Observed NYMEX Natural Gas Prices, 1996–2005

for every lease sale thereafter. The post-1999 strength of both oil and gas prices persuaded MMS to go for slightly less generous suspension volumes (after Lease Sale 178, waivers for sales held in the Central and Western planning regions have been limited to 5 MMBOE in water depths between 1200 and 2400 feet, 9 MMBOE in water depths between 2400 feet and 4800 feet, and 12 MMBOE in water depths greater than 5300 feet). In addition, lower price thresholds for both crude oil and natural gas were introduced but, in the event, these have only ever applied to leases assigned during 2001.

The fact that the threshold prices set in 2001 have been pierced more often than those established for leases held during previous and subsequent years is symptomatic of how delicately poised US oil and gas markets have been since that year. It should certainly not be seen as a reflection of a move by the US government to tighten up the fiscal screws in the OCS. In fact, both Congress and the Federal Executive have gone out of their way to make OCS fiscal incentives permanent, even after the industry weathered the 1998 price crisis. As a representative from the state of California put it in 1998:

> Congress was wrong in 1995 when we provided a royalty free-ride on the first 88 million barrels of oil and gas extracted from ... deepwater tracts ... *[T]he holiday was not needed since the boom was already on and the technology available to make the investment more cost-effective*. But, small consolation though it was, we were assured that if the holiday turned out to be too generous to the

> oil industry, the royalty rates could be adjusted so we could recoup some of the loss in the future. However, once it became known that ...[DOI] was considering raising those post-holiday rates, the Congress quickly stepped up to protect its special interest supporters again by adding report language to the Emergency bill to prohibit the Interior Department from raising the rates on royalties from deepwater leases.[53]

Likewise, MMS stopped using the 2001 threshold prices in subsequent sales as soon as it realised that, unless it did so, companies would henceforth receive very little royalty relief at all (again, this is why the comprehensive energy bill approved by the US Senate in May 2005 did not specify threshold prices).

DWRRA also allowed the Secretary of the Interior to waive royalty payments temporarily, so long as it could be demonstrated that production from leases issued prior to the passage of the Act would be uneconomic – or would cease – without the relief requested (all 1996–2000 and post-2000 leases in water depths greater than 670 feet are also eligible to apply for discretionary and end of life relief, if necessary). However, neither type of relief can be granted either to leases that produced prior to the passage of the Act[54] or to leases not found in waters at least 670 feet deep and lying wholly west of 87 degrees, 30 minutes West longitude. In addition, DWRRA reaffirmed the faculties of the Secretary of the Interior to reduce/eliminate royalties in order to promote development on non-producing leases or to encourage primary, secondary or tertiary recovery from marginal GOM leases. Importantly, the legislation provides for automatic approval of any petition of royalty relief in the event that the Secretary of the Interior fails to act upon it within a certain time-frame (180 days).[55] The administrative procedure whereby discretionary royalty relief can be obtained is quite convoluted[56] but there is no need to dwell at length on it, because it will never have more than a negligible economic impact (the monetary value of automatic deepwater royalty relief is far greater).

The avowed objective of deepwater royalty relief was to convince oil companies to explore more aggressively than they might otherwise have done, by giving them the added security of a royalty waiver in case they made a discovery whose development on a forward-looking basis would be uneconomic after the inclusion of sunk costs. Things have not worked out quite like that, though. The names of many of the projects that have qualified for royalty relief thus far (Horn Mountain, Nansen, Einset, Typhoon, Aconcagua, Camden Hills, Gunnison, Black Widow, Morpeth, Klamath) read like a veritable who's who of cutting edge deepwater technology. It is highly unlikely that their development would not have been undertaken in the absence of royalty relief.

Indeed, the timetables of the earlier starters among these projects show that they were already well underway by the time they qualified for royalty relief (to cite an example, in January 2000, MMS extended its approval for the development of Typhoon, but the field only received its royalty waiver 11 months later). Small, truly marginal, fields are certainly conspicuous in the group of projects receiving royalty relief, but by their absence.

The above strongly suggests that deepwater royalty relief has not fulfilled the stated objective of its designers. In a nutshell, royalty relief has not prompted oil companies to develop truly marginal prospects, and it has had no discernible effect on their exploration budgets (what extra E&P dollars there have been have come about in response to prices and, as we have argued before, these larger budgets have in no way been proportional to price increases). Thus, the incremental output that can be ascribed to royalty relief is negligible to the point of non-existence. The main effect of royalty relief has been to enhance the profitability of projects that would have been attractive anyway on an ex-waiver basis.

The desultory output response that has followed the enactment of deepwater royalty relief would not surprise John Mitchell, who rightly observes that 'fiscal terms affect supply only marginally ... They do not greatly affect the general level of activity, which is more influenced by overall price levels and exploration attractiveness'.[57] Nevertheless, there is no scarcity of observers who insist that the increase in the number of deepwater blocks receiving bids since 1995 would not have been quite so marked in the absence of the relief measures.[58] However, the upwards and downwards variations in bids submitted and tracts bid upon since 1995 as a rule can be much better explained in terms of the behaviour of oil prices during the period preceding acreage auctions (see Figure 8.17). Indeed, in moments of candour, companies themselves downplay the importance of royalty relief. As an official in a large independent company pointed out, 'the discoveries that you are likely to make [in deepwater] are much larger than in shallower waters. That's the real attraction. The royalty holiday is an enhancement, but it's not the reason for deepwater drilling.'[59]

8.4.2 Can Royalty Relief Succeed Where Areawide Leasing Failed?

Although the deepwater royalty relief initiative is invested with an image as a major innovation, in actual fact it is merely the most recent manifestation of a longstanding fiscal incentive policy pursued by every US presidential administration from Ronald Reagan onwards. Indeed,

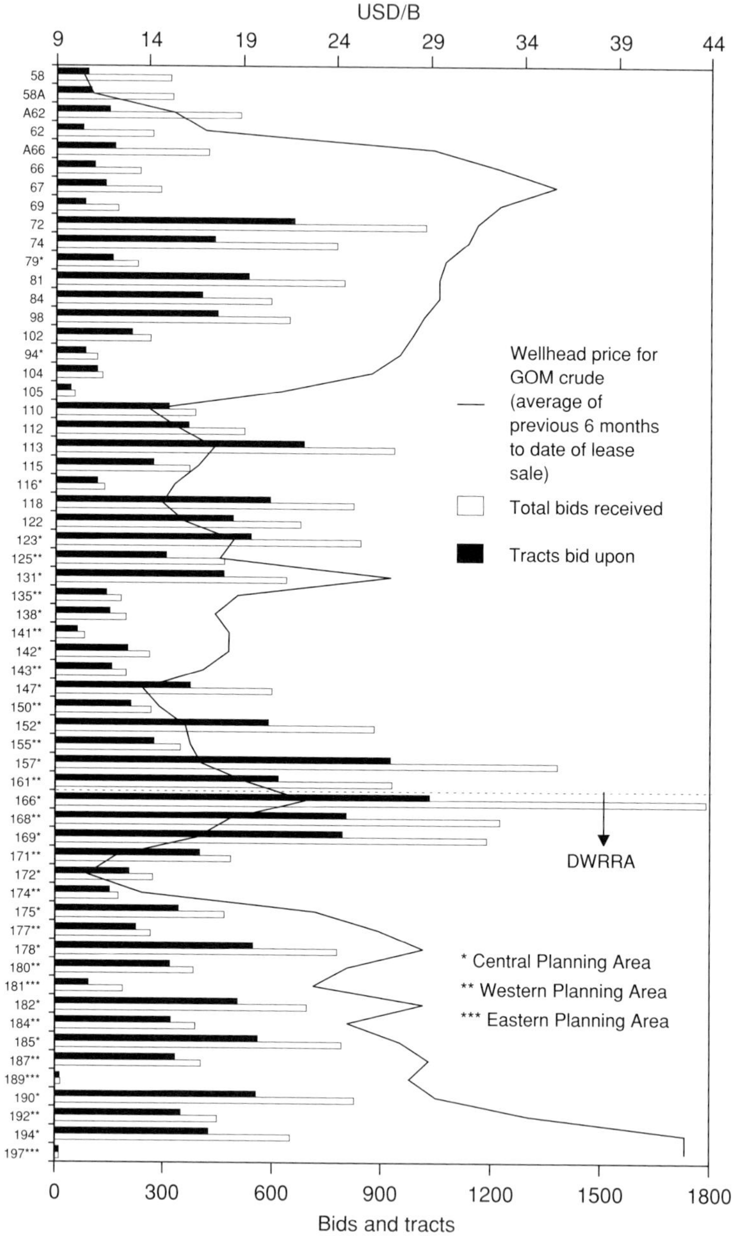

Source: MMS

Figure 8.17: Behaviour of Key Indicators in GOM Federal OCS Lease Auctions, by Sale Number, 1979–2004

the fiscal regime on the US Federal OCS was radically relaxed in 1983, precisely in pursuit of fast output growth. However, this is not something that has registered in the public consciousness at all, either within or outside the USA, due to a lack of awareness or understanding in terms of the central role that signature bonuses play in the idiosyncratic OCS fiscal regime.

In almost all countries, licensing rounds are relatively rare occurrences, so proceeds from acreage auctions actually constitute a form of extraordinary income. In contrast, in the Federal OCS, licensing rounds are very frequent and, given the absence of any petroleum-specific taxes, bonuses are the sole means whereby the US government obtains an income from its ownership rights on the petroleum resources found in its land (while royalties ensure the government's ongoing participation in the benefits of oil discoveries on public property). *It cannot be stressed strongly enough that, contrary to popular belief, GOM signature bonuses are* ***not*** *payments in exchange for a 'right to drill'.*[60] Neither does a signature bonus represent the fair market value of the resources present in a given tract (as this value can only be determined *ex post*). Rather, the bonus represents 'the fair market value of the rights and opportunities conveyed by the lease to explore, develop, produce, and sell whatever resources might be present on a particular tract'[61] (net of taxes and royalties, and at a *given* point in time, it should be added). In practical terms, in the Federal OCS, signature bonuses fulfil a role that governments in other parts of the world usually entrust to so-called resource rent taxes (RRTs), like Britain's Petroleum Revenue Tax (PRT) or Australia's Commonwealth Resource Rent Tax (CRRT). The key difference between signature bonuses and RRTs as vehicles for excess profit taxation lies in the fact that the latter collect Ricardian rents only after they have actually materialised, whereas the former are meant to capture *expected* Ricardian rents.

The fiscal take on oil activities is generally defined as the sum of a variety of contractually defined payments (royalties, rentals) plus the taxes (both petroleum-specific and general) levied by local and national governments. Due to the unique form of subsoil tenure that prevails in the USA (with ownership of subsoil resources being vested in the possessor of the surface),[62] the revenues that make up the government fiscal take in the Federal OCS actually come from two sources that are worlds apart from one another in conceptual and statutory – if not necessarily practical – terms.

On the one hand, there are those revenues that the Federal government receives in its capacity as a sovereign fiscal authority from the federal income tax obligations – and, between 1980 and 1988, from the

Crude Oil Windfall Profit Tax (COWPT) liabilities – that oil companies incur as a result of their profits from operations in the Federal OCS (state and local governments cannot tax OCS activities, and the only revenues that coastal states receive from them come from congressional appropriations under the Coastal Zone Management Act and/or the environmental clauses of OCSLA amendments).[63]

On the other hand, there are those lease payments (royalties, surface rental fees and signature bonuses) that are due to the Federal government purely in its capacity as the landholding party to mineral lease contracts governed by private law, and for which lessees are liable in order 'to compensate the general public for the market value of the resources that ... [they] remove from public lands'.[64] The rules governing the collection of GOM lease payments are specified in law – in the SLA and OCSLA and their various amendments, as well as the Emergency Petroleum Allocation Act (EPAA) of 1973, and the Federal Oil and Gas Royalty Management Act (FOGRMA) of 1982 – just as genuine taxes are. However, as the Office of Management and Budget of the US Congress points out, OCS leasing is one of those activities where 'the Government, not acting in its capacity as sovereign, is leasing or selling goods or resources, or is providing a service ... under ... business-type conditions'.[65] In a word, lease payments are undertaken as private or commercial acts (*acta jure gestionis*), whereas taxes are compulsory contributions levied by the authority of a sovereign power (and hence qualify as *acta jure imperii*).[66]

Figure 8.18 shows the evolution of the overall fiscal burden on GOM since 1977.[67] It is evident that there has been a significant fall in this indicator from 1983 onwards: the fiscal takes for the periods 1977–82 and 1983–2001 average 50.9 percent and 27.5 percent, respectively, including COWPT liabilities.[68] The 50 percent fall in the incidence of Federal income tax (net of Investment Tax Credits until 1986) on gross OCS income from the mid-1980s onwards is attributable in its entirety to the effects of President Reagan's 1986 Tax Reform Act (which did not single out the oil industry for special treatment). This means that the abrupt decline in the OCS fiscal burden after 1983 is almost entirely due to the behaviour of lease payments, in general, and signature bonuses, in particular. Up until that year, lease payments amounted to 55 percent of GOM cumulative gross income. From then on, the ratio of total lease payments against GOM gross income has never again come close to such percentage. It declined to its lowest point yet (16.66 percent) in 1992, and even during the banner years of 2000 and 2001, it only managed to reach 17.23 and 17.02 percent, respectively. Indeed, for the period 1983–2001 as a whole, total lease

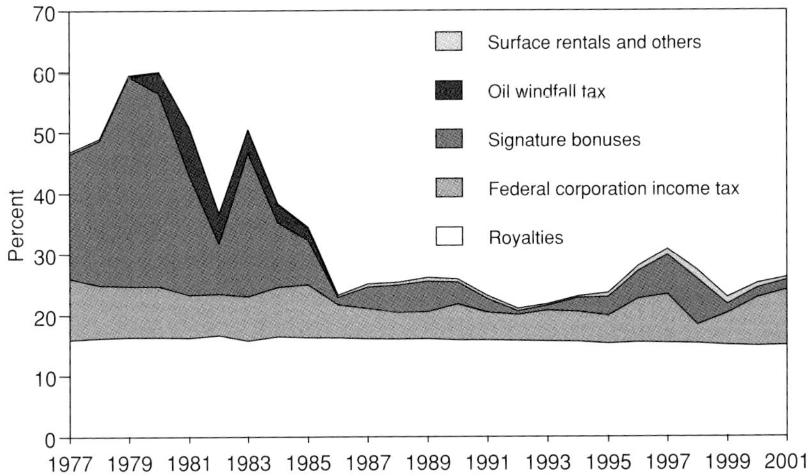

Sources: MMS, DOE

Figure 8.18: GOM Federal OCS. Lease Payments and Federal Income Taxes as a Percentage of Gross Income, 1977–2001

payments represent only 21.5 percent of a total GOM cumulative gross income of USD 352.5 billion.

As Figure 8.19 shows, GOM lease payments peaked in the early 1980s (reaching their maximum of USD 8.2 billion in 1983). During that period, the wellhead price for GOM crude also reached its post-OPEC revolution apex (averaging USD 35.12/B in 1983, a record that stood unchallenged until 2004). In the eyes of the Reagan administration, this undesirable outcome reflected the fact that, up until that point, the OCS leasing programme was used 'to meet short-term budget needs rather than focusing on the consumer'.[69] In other words, the enormous bonuses paid for OCS acreage during the late 1970s were conceptualised as being symptomatic of an underlying pathology in the market for offshore acreage, caused by the restrictive conditions of access to acreage that DOI had imposed on the oil industry throughout the history of the OCS leasing programme. These restrictions were seen as an important obstacle in the path of the exploration effort necessary for US dependence on oil imports to be reduced and the impact of the 1973–81 oil price increases to be mitigated (not least because the frenzied atmosphere whipped up by the OPEC revolution raised the prospect of American oil firms bidding themselves into financial oblivion).

The urgency of the US government to shake off OPEC's perceived stranglehold on the country's economy led it to embark on an all-out

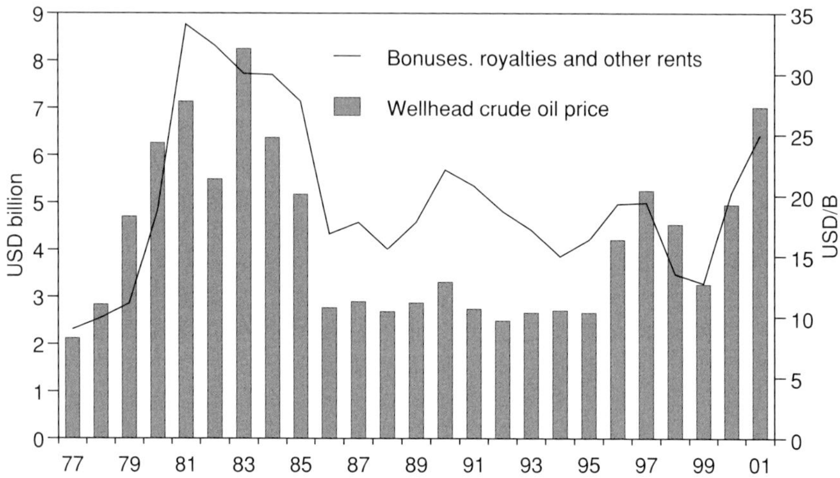

Sources: DOE, MMS

Figure 8.19: GOM Federal OCS. Total Lease Payments Versus Wellhead Crude Oil Price 1977–2001

effort to increase the frequency of lease sales, to offer more tracts for lease in each sale, and to streamline the bid acceptance and pre-sale planning processes. This effort crystallised into an extraordinarily ambitious five-year leasing programme that hinged upon offering the industry nothing less than the entire extension of the Federal OCS, by means of 41 lease sales. This programme, and the policies that gave form to it, came to be known under the uninspiring name of areawide leasing (AWL). This moniker was due to the fact that the cornerstone of the programme was to offer entire OCS planning areas at a time (each one up to 50 million acres in extension), in preference to the method used until that point, which consisted of only offering tracts that had been specifically nominated by firms.

The political ferment that AWL provoked in coastal states like California and Florida led to vast swathes of the Federal OCS being effectively put out of bounds by drilling and leasing moratoria. Despite this, AWL did succeed in greatly increasing the OCS acreage offered and leased. During the first 18 months of the programme's existence, DOI (through the newly created MMS) offered 265 million OCS acres, and leased 13 million acres (of which 10.46 million were in the GOM region).[70] From 1983 to the end of 2004 inclusive (i.e. up to Lease Sale 192), a total of 1.3 billion OCS acres have been offered (many tracts have been offered repeatedly) and 85 million acres have been leased. In contrast, during the previous 29 years of OCS leasing, the figures for total acreage offered and leased came to only 58 and 23 million acres, respectively. Indeed, under the leasing procedures in use until 1982 inclusive, assigning the 13 million acres leased during AWL's first 18 months of existence would have taken 98 months – until June 1991, that is – assuming a leasing rate similar to that achieved during the final days of the previous system[71] (OCS acreage offered and leased had already increased by 238 percent and 197 percent, respectively, between 1973 and 1979).

At first glance, one might think that the introduction of AWL did not have any discernible effect on bonuses. After all, bonus payments peaked in 1983 at USD 6.65 billion (of which USD 4.9 billion were paid for GOM blocks). GAO conducted an investigation into the first ten areawide leasing sales and reached the conclusion that 'the increased pace of offshore leasing through the area-wide programme decreased competition (in terms of the number of bids received for each tract) and reduced government revenues (in terms of the amount of high bids for individual tracts)'. According to GAO, the 'significant negative relationship between areawide leasing and the number of bids received for each tract'[72] meant that the USD 8.9 billion that the Federal government received in bonuses in the first ten areawide sales was – in nominal terms – USD 7 billion less than what it would have received had 'the slower pace of the prior tract-selection programme … been followed'.[73] Although the magnitude of this shortfall caused surprise and consternation in some circles,[74] the direction of the movement in signature bonuses certainly did not: at the time of the adoption of AWL, DOI openly announced its expectation that, by 1985–1986, bonuses would have declined to a level of USD 2–3 billion per year (down from the USD 5 billion that they averaged in the lease sales held in the run-up to and the immediate aftermath of the Iranian Revolution).

The decline in revenues from bonus payments from 1984 onwards has exceeded these expectations by an order of magnitude. As conditions in the international oil market deteriorated and crude oil prices entered into a period of acute decline, signature bonuses followed suit, falling by no less than USD 5.6 billion between 1983 and the *annus horribilis* of 1986, although proceeds from lease sales had already slipped well below the USD 2 billion mark (to USD 1.56 billion) by 1985. The stabilisation of oil prices after the netback crisis did not lead to any great improvement in the situation, and bonus payments reached a nadir of USD 84 million during 1992. Thus, the shrinkage in bonus payments since the introduction of AWL has greatly exceeded the decline in international oil prices over the same period. Indeed, since 1986 and up to the end of 2004 inclusive, yearly proceeds from

OCS lease sales have only exceeded the USD 1 billion mark (i.e. *half* the minimum level originally predicted by DOI) on three occasions – 1988, 1997 and 1998 – after having averaged USD 5.3 billion over the 1979–81 period. Furthermore, the superficially impressive signature bonuses receipts from auctions held during the mid-1990s only reflect the vast surface leased.

Signature bonus receipts have remained relatively static even as oil prices have reached record levels and activities in deepwaters have become an everyday – indeed, an essential – part of the remit of more and more oil companies. Thus, even though a lot has been made of the record-breaking nature of the GOM acreage auctions that took place during the middle to late 1990s, this success is palpable only if it is measured in non-monetary indicators (like the total number of bids submitted or the total number of tracts bid upon). Even after being deflated, the USD 1.4 billion in bonuses that the Federal government obtained in the much vaunted lease sales of 1997, for instance, compares unfavourably with the USD 4.9 billion figure recorded during 1981 (Figure 8.20). On a per acre basis, the contrast is even more striking (Figure 8.21).

Figure 8.22 shows, in a condensed form, the behaviour of OCS lease payments before and after the introduction of AWL. This graph makes it abundantly clear why, in practical terms, AWL constitutes one of the most aggressive tax cuts ever enacted for the benefit of the oil industry, comparable in this respect even to the policy measures adopted in the

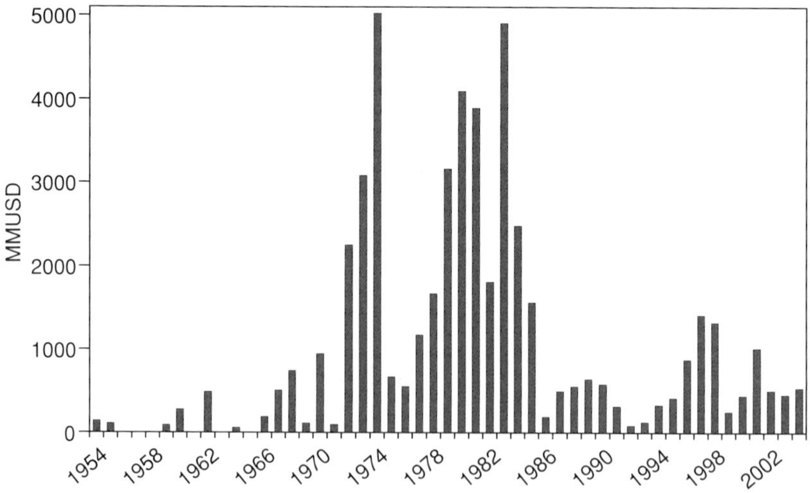

Source: MMS

Figure 8.20: GOM Federal OCS Signature Bonus Payments, 1954–2004

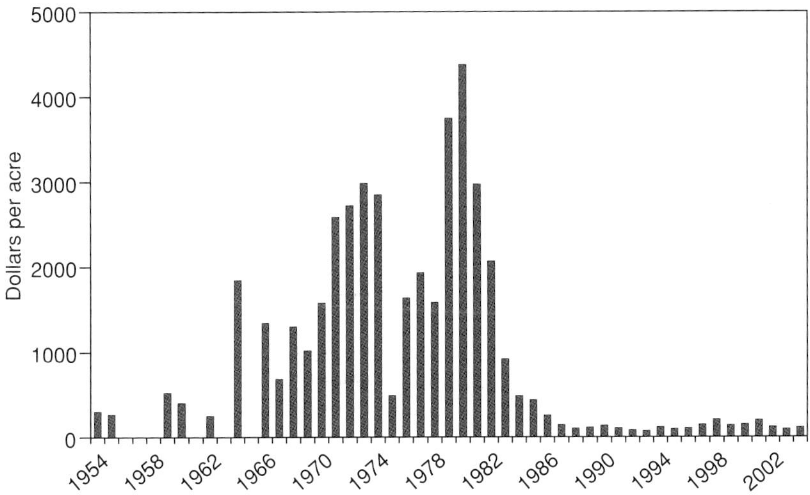

Source: MMS

Figure 8.21: GOM Federal OCS Signature Bonus Payments by Unit of Surface, 1954–2004

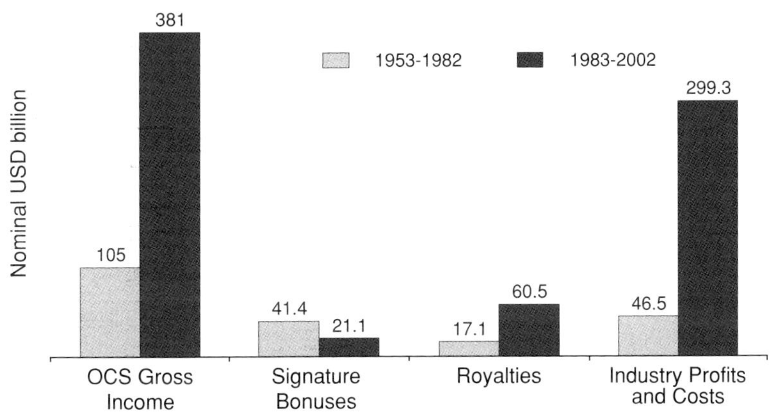

Period	*Acres Offered (Millions)*	*Acres Leased (Millions)*	*US Dollars per acre*
1953–82	34.4	16.8	1,275
1973–82	23	10.1	2,253
1983–02	1,235	84.4	207
1993–02	513	46.9	136

Source: MMS

Figure 8.22: Behaviour of Total OCS Lease Payments, by Period

British North Sea by Margaret Thatcher's Tory government and its successors. Therefore, it stands to reason that if the alleged connection between oil output and more flexible and generous fiscal terms were valid, GOM output should have expanded significantly throughout the mid- to late-1980s.

With the introduction of AWL, the US Federal government expected royalty receipts to increase dramatically (USD 3–4 billion per year by 1986, from an average of less than USD 1.3 billion for the period 1970–1982), in line with rising production. Due to the behaviour of oil prices during the late 1980s, these estimates proved pure fantasy: royalty payments surpassed the USD 3 billion mark for the first time only in 1996 and, up to 2000, they have only managed to repeat this feat twice (during 1997 and 2000). However, the important aspect of these post-AWL royalty receipts estimates is the gross output figures that they imply. If one assumes that increased production was to have led oil prices to fall by 25 percent for the period 1983–1986 (and such a fall was the key objective sought by AWL), DOI's royalty estimates suggest that the agency expected to see a 50−60 percent increase in output to between 1.8 and 2 BBOE per year for the entire OCS (with GOM accounting for around 75 percent of the total).

GOM output for the 1983−2000 period averaged only 1.2 billion BOE per year (elsewhere in the OCS, drilling moratoria put a cap on output). Indeed, GOM output only reached the annual target of 1.4 billion BOE during the deepwater boom period (1993−2000). This means that AWL failed to elicit any output response from the industry during the first 12 years after its adoption. It is true that the current GOM output (for both oil and gas) is much higher than it was during the 1980s (and even during its 1970s peak). It is also true that the declining trend in GOM production has not only been arrested but actually reversed (notwithstanding the harsher economic conditions that the industry has encountered as it moved into ever deeper waters). Furthermore, the bulk of the incremental GOM output has come from fields lying in blocks offered to the industry after the adoption of AWL (and particularly during the early years of the programme). Is this not suggestive of the existence of a direct – albeit lagged – link between AWL and the surge in deepwater output, a link which would implicitly justify the fiscal sacrifices that AWL entailed?

The answer to this question is negative: the nature of this apparent link is *sequential* and not *consequential*, because the real driving force behind the renaissance in GOM production has been technological progress, rather than ease of access to prospective acreage. In other words, the length of time that had to elapse before output reacted to the incentives allegedly provided by AWL suggest that the causal link joining them is tenuous to the point of irrelevance. After all, even large oil companies (which amassed impressive deepwater lease portfolios during the 1980s) were unable to prevent a long time from elapsing between the dates of lease assignment and first oil, simply because developing their deepwater prospects profitably was a task that lay beyond the possibilities of 1980s offshore technology. Thus, development of these blocks only got underway as the oil industry gradually acquired the capability of working in 2000 feet of water and beyond in places like the North Sea and Brazil, a technology-led process in which AWL played no part whatsoever.

To sum up: AWL had virtually no impact on output, either on a medium- or even a long-term basis (in the latter case, because deepwater production would have come about in its own time, with or without AWL). The magnitude of the fiscal incentive afforded to the industry by AWL vastly exceeds that which will be made available to it as a result of royalty relief, and all the more so if the value of both programmes is expressed on an NPV basis (and this is really the best way to do it, given the front-loaded character of signature bonus payments). Thus, it seems fair to conclude that royalty relief will be of no consequence to the short- or medium-term deepwater production path, or even to the ultimate recovery achievable in the province. Indeed, since royalty relief was introduced, the increase in the pace of leasing has been twice as large as the increase in the pace of exploration, leading MMS to propose – out of sheer frustration, one presumes – the adoption of a sliding scale structure of rental increases 'to encourage exploration drilling … earlier in the lease term'.[75] If adopted, this system will not make a blind bit of difference, of course, not least because even for companies with very extensive leaseholdings, rental costs are but a fraction of drilling and exploration costs, on the one hand, and royalty payments, on the other.

Plenty of deepwater oil and gas remains to be found and produced, of course, but not nearly enough to sate the insatiable US demand. Furthermore, although cumulative production in the shallow and deepwater provinces may possibly turn out to be comparable, the aggregate production profile in both provinces will be very different, with the deepwater having a markedly shorter lifespan. The deepwater era in GOM will resemble a straw fire: a conflagration that burned with singular intensity but did so, alas, only too briefly. Indeed the prediction can be hazarded that deepwater oil and gas output has already peaked, given what it will take for the sub-province to recover from the colossal damage wrought by hurricanes Rita and Katrina.

It is ironic that fellow party members of the fiscal arch-conservatives who devised and implemented AWL should also have been responsible for the subsequent enactment of the Budget Enforcement Act, because the pay-as-you-go provisions of that legislation would have made it virtually impossible for a programme with the budgetary impact of AWL to be approved today. Thanks to AWL, the effective government take rate in the GOM Federal OCS has already gone down to one of the lowest levels worldwide, broadly comparable to that found in the British North Sea (see Figure 8.23).[76] Nevertheless, in spite of the exorbitant cost of AWL and DWRRA and the modesty of their achievements, there is every likelihood that the GOM fiscal take will be further rolled back in years to come (indeed, the principle of royalty relief has been extended to deep gas, and there are moves afoot to reform royalty measures come on stream). Indeed, the only reason why the effective tax rate in the UK North Sea is lower than in the Federal OCS is that royalties have been scrapped in the UK (but, as shown in a subsequent chapter, the US Federal government is hard at work on getting rid of royalties, and if this ever comes to pass, the GOM Federal OCS will probably become the petroleum province with the lowest taxation levels, bar none). In the meantime, the world watches with bemusement as the Federal government harvests some of the consequences of its fiscal largesse, such as not receiving any royalties from producing leases issued under the DWRRA aegis in 1998 and

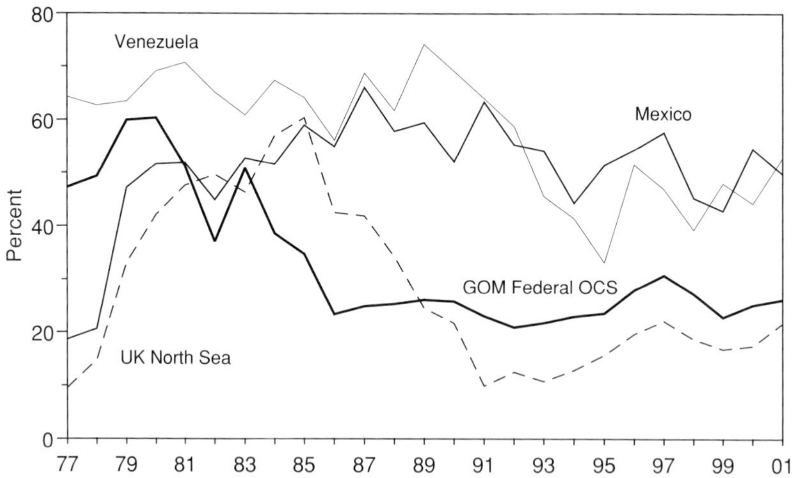

Sources: DOE, DTI, MEM, MMS, PDVSA, PEMEX

Figure 8.23: Total Government Take of Upstream Oil Industry Income for Selected Fiscal Regimes, 1977–2001

1999 amidst record prices for both oil and natural gas, due to the 'unwitting' removal by a clerical employee of a provision setting a price threshold for such leases. The potential cost of this omission has been estimated at USD 10 billion.[77]

Of course, one cannot conclude solely on this basis that the GOM fiscal take is too low or, for that matter that oil companies in the region have been at the receiving end of a bonanza financed by the American taxpayer. After all, very little has been said thus far about deepwater costs. These are, of course, the factors that ultimately determine the share of gross revenues that is available to be divided between the government take and profits. Depending on how this division is done, some fiscal regimes may be classified as unduly generous, others as too harsh. So, before assigning a place to the GOM fiscal regime in this continuum, the deepwater cost structure has to be examined and understood, and it is to this issue that we now turn our attention.

NOTES

1 In 2005, within a six-month period, eight hurricanes catastrophically disrupted GOM offshore activities.
2 French *et al.* 2006: 7.
3 The year 2004 saw the installation of the world's deepest (5610 feet) dry-tree spar in the Devil's Tower project, the beginning of production from the world's largest spar (at Holstein) and truss spar (Mad Dog), the beginning of production from the world's first cell spar at Red Hawk, the emplacement of the world's largest semisubmersible production unit at Thunder Horse, and the successful installation of the world's deepest (4320 feet) TLP at Marco Polo.
4 Sale 190 (March 2004), for instance, for the first time saw bids (six in number) being entered for blocks in Amery Terrace.
5 Anderson and Boulanger 2002: 4.
6 *Hart's E&P*, April 2000: 60–1.
7 *FT*, 25 April 2002: 29.
8 After ConocoPhillips' costly disappointment in the Magnolia project, the head of the company's upstream operations conceded that 'there are quite a few fields there [in the deepwater GOM] where we would probably not want to be involved in' (*PON*, 18 November 2004: 4).
9 *PON*, 16 March 2005: 1.
10 *PON*, 10 May 2002: 1–2.
11 *PON*, 26 August 2004: 1. The hub project will involve the following finds: Merganser, Vortex (both Kerr-McGee), Spiderman, Atlas, Atlas Northwest, Jubilee (Anadarko) and San Jacinto (Dominion). It will be operated by Anadarko, and the gas transportation pipeline will be built and operated

by Enterprise (formerly GulfTerra El Paso Energy Partners).

12 Young *et al.* 2000.

13 Unocal's Sardinia well – drilled in block KC681 – penetrated an extensive interval of reservoir-quality sandstones harbouring non-commercial amounts of hydrocarbons, in a horizon analogous to that encountered by the St. Malo discovery well in Walker Ridge (*PON*, 1 September 2004: 1).

14 *Hart's E&P*, April 2000: 60.

15 Marathon drilled two wells in blocks WR30 and WR165, the latter to a depth beyond 18,000 feet. Texaco drilled the first ever Walker Ridge wildcat in 1999 on block WR70 (the well was temporarily abandoned at a total depth of 8000+ feet, but nothing has come of Texaco's plans to drill it up to the 26,000 foot mark).

16 Hewitt 1995. See also DiMarco *et al.* 2004.

17 In May 2003, for instance, BP suffered a riser separation problem when drilling the ninth development well at Thunder Horse (*PON*, 27 May 2003: 3). In January 2004, Kerr-McGee and Devon Energy finished drilling one of the most expensive wells in GOM ever (at USD 86 million) in their Yorktown prospect in Mississippi Canyon block MC886. One of the reasons why the well was so expensive was that work had to be suspended on a number of occasions due to the strong eddy-generated currents. On that occasion, Anadarko also had to interrupt drilling at its Atlas prospect, while installation of production infrastructure in the Matterhorn, Devil's Tower and NaKika development projects was also disrupted (*PE*, October 2004: 15). Drilling at the Thunder Hawk prospect (located in 5716 feet of water in Mississippi Canyon block MC734) likewise had to be prematurely suspended in April 2005 on account of currents.

18 Toledo found an estimated 750 MMCF of gas, a non-commercial volume given its location. Tobago has reputedly found 75 MMB of oil.

19 Anderson and Boulanger 2002: 10; italics ours. The Baha structure is a faulted, simple, four-way closure that straddles ten OCS blocks.

20 In the event, although Shell was targeting Mesozoic sediments at more than 22,000 feet, the Baha–1 well reached a depth of only 11,254 feet.

21 The Toledo well cost USD 72 million to drill.

22 Argyll and Duncan produced 57 MBD at peak, and had a lifespan of 17 years.

23 Robertson and McFarlane 2004.

24 IMO 2002.

25 *Hart's E&P*, September 2004: 43.

26 This was the Offshore Storage and Treatment (OS&T) vessel, a 55,000 DWT tanker serving the Hondo Field and Platform Hondo (located in 490 feet of water in the Pacific Federal OCS). The vessel processed oil from a fixed platform and offloaded the oil to dedicated shuttle tankers. It was decommissioned after installation of a pipeline to shore.

27 MMS 2001c: I–7. The Terranova FPSO off Newfoundland spilled about 1000 barrels in August 2004, but the incident caused great consternation because of particularly high seabird mortality.

28 PEMEX operates one Floating Storage and Offloading (FSO) vessel in the southern Gulf of Mexico (Bay of Campeche). The ta'Kuntah FSO has a storage capacity of 2.342 MMB of oil, and is capable of receiving up to 800 MBD of crude. The FSO is chiefly used to provide standby storage capacity in the event of inclement weather.

29 In 1995, Enserch Exploration Inc. used much of Placid's concept and even some of Placid's equipment to develop its Garden Banks GB388 discovery, located in 2190 feet of water.

30 MMS will not allow the use of FPSOs in a 471-block zone (just off the continental shelf from Galveston to New Orleans), which forms part of the U.S. Coast Guard lightering-prohibited areas.

31 *Offshore*, June 2000.

32 Consultants Douglas-Westwood estimate that as many as 39 FPSOs may become available for redeployment during the 2002–2006 timeframe (Robertson and McFarlane 2004: 30).

33 *Hart's E&P*, July 2001: 39. The three-year estimate is rather optimistic, furthermore. As Traynor and Slorer (1998: 11) indicate, 'nearly every [FPSO] project has ... faced cost or time delays compared with original estimates', with most problems arising from a combination of 'overly ambitious timetables ... aggressive bidding by inexperienced engineering contractors ... unable to correctly assess the complexity and costs of construction and outfitting of an FPSO [and]... problems related with sub-sea equipment and installation contractors'.

34 *Clarification on the Disposition of Associated Gas Related to Oil Production from a Floating Production, Storage, and Offloading (FPSO) in the Gulf of Mexico Region.* MMS news release, 21 October 2002.

35 *Ibid.*

36 MMS 2001c.

37 The cost of a gas-only FPSO geared to CNG exports has recently been estimated at USD 235 million, with a further USD 93.6 million in annual operating costs. A production and processing vessel configured for pipeline export would cost USD 531.4 million, but its annual operating costs (USD 38 million) would be much lower. A floating LNG vessel would have both high capital costs (USD 868 million) and high annual operating costs (USD 116 million). See *Hart's E&P*, July 2003 deepwater supplement: 38. These figures do not include the cost of the dedicated CNG or LNG carriers, or of appropriate terminal facilities onshore. According to NPC (2003, vol. V: AD–4), the cost of a standard LNG carrier is around USD 160 million, and the cost of a CNG carrier 'is expected to exceed that of a standard sized LNG carrier'. Costs for LNG or CNG carriers in GOM would be increased by a substantial margin by the requirement to comply with the Jones Act.

38 MMS, '*Clarification...*'; italics ours.

39 *Ibid.*, italics ours.

40 Traynor, Aldridge and Cook 2002: 25.

41 Devon Energy is thinking about the possibility of emplacing a FPSO if

one or more of its Walker Ridge prospects turn out to be a commercial find. First oil would supposedly occur in 2008 at the very earliest (*PON*, 6 August 2004: 1). A more likely date would be 2010.

42 Richardson *et al.* 2004: 91.

43 Rose, Farndon and Fraser 1998: 2, italics ours. Consultants from the defunct Arthur Andersen considered that, 'after 1995, when Congress reduced royalties on certain deepwater leases, the pace of leasing reached fever pitch ... [again setting] records for the number of tracts bid on and the number of bids submitted as energy companies battled for the right to exploit the deepwaters of the Gulf' (Riddle, Snyder and George 2001: 4).

44 Mead 1993: 239.

45 Seydlitz, Sutherlin and Smith 1995: 36.

46 The account of the passage of the relief initiative through Congress is a salutary lesson into the tortuousness of the legislative process in the USA (see Davis and Neff 1996).

47 Royalty relief for GOM leases had been studied as an option before, but only to promote Enhanced Oil Recovery (EOR), on a strictly project-by-project basis (see GAO 1985b). To date, such projects have been conspicuous by their absence in GOM.

48 Rose, Farndon and Fraser 1998: 2.

49 The royalty exempt volume for leases lying at 2400 feet of water or greater exceeded by 10 MMBOE the average size of all GOM deepwater finds.

50 *Ibid.*

51 In December 2000, Chevron's Typhoon (lying under nearly 2500 feet of water in Green Canyon blocks GC236 and GC237) became the third field to be granted royalty relief under DWRRA. The first field to benefit from automatic royalty relief under the act was Walter Oil & Gas Corporation's East Breaks block EB168, originally assigned in August 1996.

52 A January 8, 2003 United States District Court ruling (*Santa Fe Snyder Corporation et al. v. Norton et al.*) held invalid the regulation assigning royalty suspensions to fields rather than leases. The verdict is under appeal, but if it were to stand, leases issued under the DWRRA in sales held from 1996–2000 will be lease-specific rather than field-based royalty suspension.

53 George Miller, in US Congress 1998: 40, italics ours.

54 Royalty relief for existing leases would have triggered the pay-as-you-go provisions of the Budget Enforcement Act.

55 No royalties will be due for any new production achieved after the date of enactment, pursuant to a Development Operations Coordination document approved by the Secretary of Interior (Davis and Neff 1996: 46).

56 Applicants for deepwater royalty relief for leases sold prior to late 1995 use an MMS cash-flow model called the Royalty Suspension Viability Programme (RSVP). Applicants describe the risk of the proposed venture by specifying the uncertainty in the geologic, engineering, and cost inputs as ranges of possible values and/or measures of the likelihood of occurrence for each possibility, and these data are used to simulate the prospective net

present value before royalties and sunk costs for any given field. If this simulation results in a positive value, the field is proven to be economically viable. MMS then incorporates royalty payments and eligible sunk costs in another simulation. If the resulting figure is positive or zero, the field does not qualify for relief. If the present value is negative, MMS has to calculate the volume of royalty free production required to take the value to zero. If the required waiver is equal to or less than the minimum mandated for the block's water depth category, the applicant receives the minimum suspension volume. If the required waiver exceeds the minimum suspension volume, the applicant receives the former.

57 Mitchell *et al.* 2001: 49–50. This conclusion echoes that of GAO (1990: 63); 'petroleum taxes and regulation do not appear to be the most important factors in determining the location of petroleum investments. While foreign tax policies and other inducements can be contributing factors, favourable geologic characteristics ... appear to be the main factor behind the preference ... to explore and develop ... petroleum resources ... Taxes were neither generally the most important influence on the location of petroleum investments, nor were taxes responsible for the decline in US domestic drilling activity'.

58 Riddle, Snyder and George 2001: 10. As these authors see it, 'after 1995, when Congress reduced royalties on certain deepwater leases, the pace of leasing reached fever pitch ... [again setting] records for the number of tracts bid on and the number of bids submitted as energy companies battled for the right to exploit the deepwaters of the Gulf' (*ibid.*: 4).

59 *Inside FERC.'s Gas Market Report*, 4 October 1996: 4.

60 See *WSJ*, 4 April 1996: A–1.

61 MMS 1983: 98.

62 Inexcusably (or perhaps understandably, in view of her devotion to the idea that natural resource ownership is an anachronistic and irritating irrelevance in a globalised world), Susan Strange makes constant references to 'concessions', 'governments' and 'concessionaires' in her overview of the evolution of American oil in the greater context of the international oil industry (Strange 1998: 198). No mineral concession as such has ever been granted in the USA, of course, and governments (whether at a local, state or federal level) only figure in lease contracts in their capacity as lessors. That did not put off a renowned American legal consultant from (oxymoronically) saying that the USA constitutes an extreme example of a concessionary system where individuals have the original dominion over mineral rights (Moore 2000). Mark Kaiser and Allan Pulsipher take this American confusion surrounding mineral property regimes even further. According to them, a concessionary regime is one in which the government or land owner will transfer title of the minerals to a lessee/licensee, which then becomes subject to the payment of non-negotiable and transparent royalties and taxes, specified in the country or state's legislation (Kaiser and Pulsipher 2004: 5). In fact, the key characteristic of concessionary regimes is the collective property of subsoil resources, whose exploitation

is granted in concession to investors (hence the name) but whose property is never transferred to the latter unless and until the resources in question are severed from the subsoil. However, these authors labour under the delusion that 'contractual systems derive from the Napoleonic era and are based on the French legal concept that mineral resources should be owned by the state for the benefit of all citizens' (*ibid.*: 32). This idea, of course, is actually at the heart of *concessionary* regimes. The American system of petroleum leasing (which they view as concessionary in nature, in common with Moore) is, in contrast, an authentic *contractual* system, governed by private mercantile law.

63 As explained more fully in a subsequent chapter, OCS oil leasing is the sole federal programme authorising the leasing, sale or disposal of public resources in which no provision is made for the sharing of revenues with states in whose territory the land from which these resources were extracted is located or with states affected by the development of these mineral resources. This is noteworthy because the rules governing the collection of bonuses and production royalties on public lands specifically enjoin the Federal government to distribute 'a share of those receipts to the states to help state and local governments meet their costs of supporting development activities' on such lands (CBO 2000: 3).

64 *Ibid.*

65 *Federal Register*, 65 (187), 20 September 2000: 57771, italics ours. The case *Winstar Corp. v. United States*, (64 F3.d 1531, Fed. Circuit 1995; affirmed 518 US Supreme Court 839, 1996) elaborates on the circumstances in which the United States government is liable for breach of contract despite the sovereign acts doctrine.

66 According to Christopher (1953), the legislative power intended to exercise both the proprietary powers of a landowner and the police powers of a sovereign in OCSLA. Thus, OCS lessees only acquire limited property rights, which are in any case subject to the government's regulatory authority (in much the same way as a tenant that acquires a lease in Texas is still beholden to the Railroad Commission of Texas, for starters, as well as other regulatory instances). US courts have always recognised the dual nature to OCS leasing activities, and on occasion have interpreted even a quintessentially sovereign act like the passage of legislation as a breach of contract. In 1992, for instance, Conoco and other companies responded to the *Outer Banks Protection Act* (*OBPA*) of 1990 (about which see Fitzgerald 2002: 49–52) with a suit alleging that *OBPA* and other Congressional and Executive directives 'breached the contracts at issue, frustrated performance thereof, rendered such performance impracticable, or constituted a taking in violation of the Fifth Amendment of the US Constitution (*Conoco Inc. v. United States* 1996). The US Court of Federal Claims agreed with the plaintiffs, but the Federal Court of Appeals reversed the decision, only to see the US Supreme Court in turn reverse this verdict (incorrectly and unsoundly, in Fitzgerald's view; Fitzgerald 2002: 52).

67 See Appendix 1.

68 COWPT was a strictly temporary excise tax aimed solely at capturing the windfall gains realised by producers of crude oil resulting from the decontrol of domestic crude prices (effective 1 June 1979). See Fiske *et al.* 1982: 15–2 and US Congress 1981.

69 Bradley 1996, v.1: 306.

70 GAO 1985a: 9. The total acreage offered includes blocks previously sold but returned to the MMS, and it counts separately 35 million acres in the GOM that were offered on two occasions. Thus, the net new acreage submitted to the consideration of the oil industry to the end of 1984 came to 346 million acres (OTA 1985: 135).

71 GAO 1985a: 66.

72 During the period covered by the GAO study, the overall number of bids per tract declined to 1.65, from a figure of 2.44 before the adoption of AWL; in the GOM region, the average number of bids received per tract fell even more steeply, from 2.67 to 1.56.

73 GAO 1985a: 18.

74 Interior challenged GAO's calculations in a response to Congress, which turned it over to GAO for consideration. GAO showed that Interior's critique was methodologically flawed, and insisted that its own calculations continued to provide the best estimate of the initial fiscal effects of areawide leasing (GAO 1986: 13–7).

75 *PON*, 3 March 2005: 4. Under the proposed system, if a lease is drilled within the first five years of its initial period, escalation fees would be avoided entirely, and the rental rate would stay the same until the start of royalty bearing production or relinquishment (as applicable). The sliding scale would be as follows: USD 9.50/acre for the first five years, USD 10.50/acre in year six, USD 12/acre in year seven, USD 13.75 /acre in year eight, USD 15.50/acre in year nine and USD 17.50/acre in year 10.

76 As Gerking explains, given 'the myriad of exemptions, incentives, different tax bases, special features and frequent changes in tax laws' entail 'considerable complexity in understanding and tracking of tax law over time'. Comparison across different oil provinces does not require an itemisation of all the minute tax code details, though. Instead, one can compute what are called effective tax rates, expressed as the ratio of taxes, royalties and so forth collected from a particular tax to the value of production. As Gerking explains, 'the calculation of effective tax rates fully account for all tax incentives granted against all types of taxes faced by oil and gas industry. Also, use of a common denominator (value of production) makes it easier to compare tax burdens between states' (Gerking 2005: 3–4).

77 *PON*, 16 June 2006: 4. Notwithstanding the ludicrousness of royalty relief in a 70 USD + price environment, oil companies have insisted that the government has no right to change conditions on these leases, and Kerr-McGee filed suit to this effect in the US Federal Court at Lake Charles, Louisiana, on 20 March, 2006.

CHAPTER 9

DEEPWATER ECONOMICS

Two distinct perspectives permeate the abundant specialised literature on GOM deepwater economics. One stresses the various risks and enormous financial outlays that are characteristic of deepwater projects, and which supposedly place operators perpetually on a razor's edge between commercial survival and ruin. The other, more exuberant view dwells on the upside that technology offers to make the deepwater a very attractive, if not superior, investment and, therefore, the source of prodigious volumes of oil and gas. The first view might be labelled commercial conservatism; the second, technological optimism.

When confronted by views that are polarised to this extent, neutral observers tend to assume that the truth probably lies somewhere between the two extremes. However, it is usually the case that diametrically opposed views like these are held by groups with very divergent interests and compositions. In the case at hand, though, both strands of thought are embraced by the same fairly broad coalition of industry and governmental interests, with its members simply changing their tune according to whom they might be addressing. Thus, when the target audience is composed of domestic policymakers, risks and upfront costs are stressed. This message is meant to raise the spectre of unnecessarily low production sometime in the future, hence discouraging its recipients from entertaining dreams about increasing rent extraction (i.e. oil taxes). In contrast, cost abatement and the inexorable march of technology are the foci of choice when the target audience is foreign policymakers, who need to be constantly reminded about a number of axioms; to wit, that multinational oil companies have a great variety of attractive investment choices open to them; that market share will incrementally accrue to oil extracted from higher cost provinces where output responds favourably to high technology; that investment will flee from areas whose fiscal regimes are not 'internationally competitive', even if they have a low cost base; and, finally, that all of the above factors translate into an imperative for countries that cling to outmoded conceptions of territorial sovereignty to liberalise access to their petroleum resources.

The objective of this chapter is to probe beyond the rhetoric and dissembling that surround deepwater production economics, in order to pin them down analytically. To do so, we will first examine the cost structure in the province, and its evolution through time. In terms of finding and development (F&D) costs, the main thrust of the analysis will be to come to grips with three key questions; namely, just how far these costs have fallen, why they have fallen and, finally, what their behaviour is likely to be in the future in light of recent discovery trends. As far as lifting costs are concerned, the characteristics that set GOM apart from other deepwater provinces will be highlighted, and an assessment offered as to whether they are advantageous or otherwise. Thirdly, the revenue generation capabilities of GOM deepwater fields will be assessed in the light of this cost structure. Finally, these various strands will be woven together into an account that explains why the GOM deepwater has probably been the most profitable petroleum province in the world for the past decade or so.

9.1 Finding and Development Costs

The undeniable drawbacks and obstacles that deepwater activities pose to companies are a good starting point for any exercise purporting to quantify and explain the behaviour of deepwater costs through time. As Merrill Lynch analysts have rightly pointed out,

> the challenges of deepwater are unlike anything else the industry has encountered in the last 100 years when developing platforms and land based assets. The complexities of deepwater environments – deepwater currents, weather conditions, production components operating in suspension, remote operation of equipment, limited or no well accessibility post completion, leave no room for compromise'.[1]

In the GOM deepwater province, the problems that characterise deepwater projects in general are compounded by the imaging and drilling difficulties to which we have often made reference throughout this study, albeit in more extreme manifestations. Take, for instance, the issue of high pore pressure variability in areas with extensive shallow faulting (and, hence, weak fracture gradients).[2] Under such conditions, routinely encountered all over the GOM deepwater sub-province, the margin between pore pressure and fracture gradient translates into an extremely narrow drilling envelope that not only requires closely spaced extra casing strings to maintain control in the shallower parts of wells but may also lead to small hole size above prospective target intervals. Indeed, this is one of the reasons why quantifying with any certainty the volumes of the discoveries made and understanding reservoir

Table 9.1: Development Delays for Selected Deepwater Projects Worldwide, 1986–2005

Project	*Operator*	*Discovery year*	*Original Onstream Date*	*Delay*	*Reasons*
Atwater Valley 1	Shell	1986	2004	2010	Subeconomic
Mensa	Shell	1987	1997	Months	Technical issues with sub sea wells
Atwater Valley 8	ConocoPhillips	1991	2004	2009	Subeconomic
Garden Banks 302	Walter Oil	1991	2003	2008	Subeconomic
Schiehallion	BP	1993	1998	Months	Problems with riser
Macaroni	Shell	1994	1999	Months	Tie-in infrastructure issues
Foinaven	BP	1994	1997	2 years	Technical issues with sub sea equipment
Caratinga	Petrobrás	1994	1997	Ongoing	Electrical problem resulted in tilting of FPSO (P34)
Girassol	TFE	1996	2001	1.5 years	Redesign of subsea equipment and FPSO construction delays
Bonga	Shell	1996	2003	2 years	FPSO delays; drilling of appraisal wells: 50% cost inflation
Roncador	Petrobrás	1996	2001	Ongoing	Platform sank (P36)
Green Canyon GC82	ExxonMobil	1996	2002	2011	Subeconomic
Garden Banks 386	EEX	1997	2002	2007	Not known
Mississippi Canyon 162–3	BP	1998	Not known	2011	Not known
Mississippi Canyon 443	EEX	1998	2000	2007	Subeconomic
Nnwa (Nigeria)	Statoil	1999	2005	2007	Not known
Neptune	BP	1999	2004	Abandoned	Written off*
Ikja (Nigeria)	ChevronTexaco	2000	2005	2009	Not known
Ekoli (Nigeria)	Statoil	2000	Not known	2009	Not known
Thunder Horse	BP	2000	Jul-05	2 Q. 2008	Major delays. First almost sunk by hurricane (repairs 250 MMUSD); then issues around export infrastructure integrity

* Field development taken up by BHP Billiton

Sources: Merrill Lynch 2002; Sankey 2005

dynamics are tasks that have proven to be fraught with pitfalls. Many projects throughout the world have run afoul of the constraints that deepwater operations entail (Table 9.1), with consequences ranging from the merely irritating albeit quite expensive – as in BP's weather-related delays in Foinaven – to the catastrophic and even more expensive – as in the sinking of the P36 platform in the Roncador field, or the near capsizing of the Thunder Horse platform after the passage of hurricane Dennis.

Well integrity and flow assurance (i.e. maintaining a steady flow through the pipelines and risers that take oil and gas from the ocean floor to production facilities at the surface) are particularly complicated affairs, because low temperatures and high pressures promote the build-up of solids (waxes and gas hydrates) within tubes, and might end up by blocking them entirely. Likewise, strong deepwater currents may curtail drilling at times through conventional single-gradient mud systems and marine risers, because of high riser loads.[3] The connection of equipment at great depths for deepwater sub-sea completions has also been revealed as a minefield (especially in ultra deepwaters).

As a general rule, this sort of difficulty becomes exponentially more severe at greater depths (and, as Figure 9.1 shows, oil companies in GOM have been drilling to ever more extreme depths in recent years). It is for this reason that deepwater development project plans often have a rather nebulous quality about them. Many projects have frequently

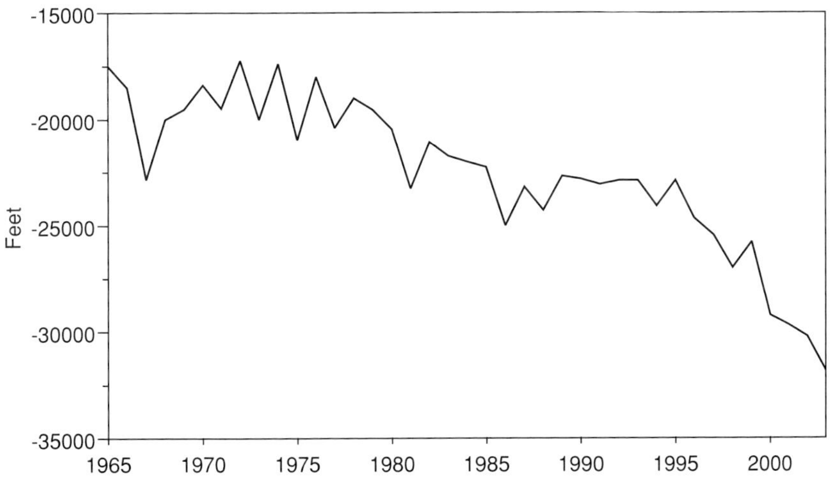

Source: MMS

Figure 9.1: Maximal True Vertical Depth Achieved by Any Well Drilled in the GOM Federal OCS, 1965–2003

suffered from significant (sometimes even dramatic) changes in both scope and cost.

Take, for instance, the BP-led Mad Dog development. In the words of the general manager of the project, Mad Dog was 'all designed and ready to go', but then it turned out that 'the reservoir was different than … initially thought, and additional appraisal wells were required', which forced the development team 'to stop the process, throw most things out and start all over again'.[4] ConocoPhillips' Magnolia, located in Garden Banks blocks GB783 and 784, provides an even better example in this regard. When this project was sanctioned (2002), it was meant to develop 150 MMBOE in reserves, with a peak production rate of 75 MBOED, and at a total cost of USD 600 million. By mid-2004, costs had climbed to around USD 1 billion, while the reserve estimate had been cut in half and the peak production rate by one quarter. Indeed, at a time when international oil prices were posting record highs, ConocoPhillips submitted a royalty relief request for Magnolia, which led some analysts to conclude that the company was not entirely sure about the prospects of ever recouping its investment even in the midst of such a favourable price environment.[5] Moreover, the type of problem encountered at Magnolia has not been restricted to new developments. Even supposedly well-understood deepwater fields have managed to spring some unpleasant surprises on their operators. An example of this phenomenon involves the post-2001 production decline at the Auger field (Figure 9.2).

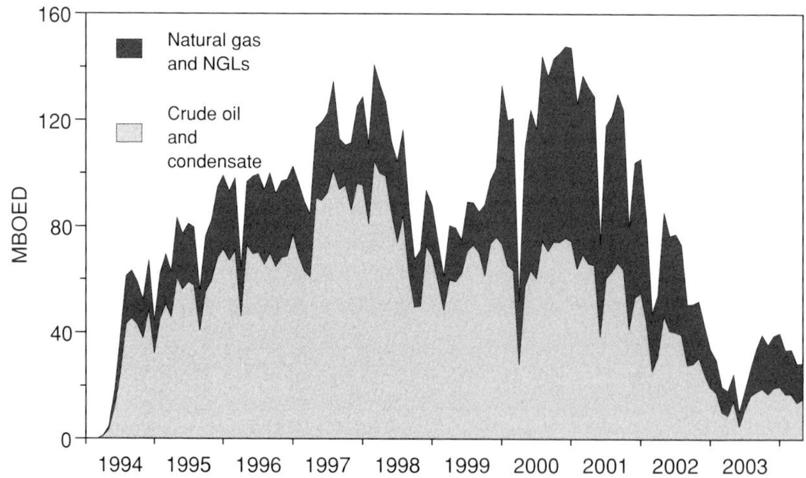

Source: MMS

Figure 9.2: Production Profile of the Auger Development, 1994–2004

Another annoying geological peculiarity of the GOM deepwater province is the frequency with which shallow water flow sands have been encountered at water depths beyond 1700 feet. Since 1984, 'water flow[s] from an overpressured shallow aquifer occurring above the first pressure-containing casing string' have been encountered in seventy deepwater leases, and such flows have had a significant impact on drilling and cementing practice and, on occasion, have led operators to change drilling locations and even to lose wells through 'loss of integrity, plus buckling or collapse of shallow casing strings'.[6] Not without reason, shallow water flow events have been called 'the single most costly and dangerous hazard in the deepwater exploration and production business'.[7]

The hazards do not end there, though. Gas hydrates are particularly abundant on and in the GOM sea floor, and this is yet another factor that greatly complicates drilling, because the rotation of the drill bits generates sufficient heat to decompose the hydrates. These tend to dissolve into and saturate the drilling mud, thereby compromising well control. The process of hydrate decomposition can also lead to sea floor subsidence (which in turn can undermine the foundation support for offshore platforms or pipelines) and, potentially, it could conceivably lead to the release of colossal large volumes of free methane gas (with dire safety and even environmental implications). Likewise, strong sea currents in certain areas (notably the Mississippi Fan Belt, the focus of much of the industry's most recent exploration effort) cause furrows in the muddy sediment, which complicate both drilling and the installation of pipelines and other fixed structures.

In the light of the factors detailed above, it is hardly surprising to find that drilling costs in the GOM deepwater can be eye-watering. A run-of-the-mill well in any of the areas where most exploration efforts are being directed will probably have a true vertical depth of 20,000 feet or more, and will cost between USD 40–60 million, assuming everything goes according to plan. For instance, Unocal's St. Malo well (located in 6900 feet of water and spudded on July 6, 2004) was drilled to a true vertical depth of 29,066 feet in a very short time (100 days), without any major complications or setbacks. Nevertheless, the well ended up costing USD 62 million. It beggars belief to think how frightening the final bill might have looked like had any part of its drilling programme gone seriously awry. Also, one begins to understand why, for development to take place any target will have to contain fairly large minimum recoverable reserves of oil (and even larger reserves of gas).

A further key difference in drilling fundamentals between the GOM deepwater, on the one hand, and more traditional exploration plays, on the other, is aptly summarised in the words of a high E&P official at Kerr-McGee: 'when we drill a well onshore or on the shelf, if the first well is a discovery, that is cause for tremendous celebration. In deepwater, because of the necessary economic size needed for an accumulation to be economic, the first well is significant ... but the second, or third or even fourth well is the one that is a cause for celebration'.[8] Indeed, there have been cases when four successful wells have merely been the prelude for further heavy expenditure. Between 2002 and 2004, for instance, BHP Billiton and its partners found themselves over USD 420 million out of pocket after drilling a string of successful wells at their Neptune (AT)[9] prospect in Atwater block AT575, and they were still in no position to decide if the find was commercial or not (as the reservoir turned out to be far more compartmentalised than originally expected). Thus, the main thing that this company had to look forward to was further heavy outlays associated with a formal appraisal programme (and, to add insult to injury, all these exploratory wells have had to be plugged and abandoned, as they cannot be converted into production wells). Finally, in July 2005 (and with oil prices at all time highs), BHP and its partners duly announced that they would proceed with the development of the field by means of a TLP, at a total cost of USD 850 million (this figure does not include USD 100 million worth of oil and gas export pipelines, to be built by Enbridge).[10]

Drilling costs constitute the largest single item of expenditure in GOM deepwater F&D costs, usually accounting for 45–60 percent (depending on water depth) of the total F&D outlays. Even though the GOM deepwater is more complex in geological terms than either West Africa or Brazil, F&D costs across the three provinces are roughly comparable, not least because GOM projects tend to be run on tighter timescales. The post-1996 GOM project lead time of 4.3 years (see Figure 9.3) compares favourably with a global average of 7.6 years. In contrast, the times from discovery to first oil (cycle time) in parts of West Africa, in particular, have proven acutely disappointing. On the whole, in no frontier oil province has the downward displacement along the learning curve been quite as quick and as steep as in the deepwater GOM (Figure 9.4).

Whenever the issue of F&D costs arises in any discussion, technological optimists will take the opportunity to display a cost curve resembling that shown in Figure 9.4. The message that this picture is meant to convey is unequivocal: the relentless advance of technology – the latest embodiments of which are riserless drilling, slender wells and expandable casings, to name but a few – will, slowly but surely, bring

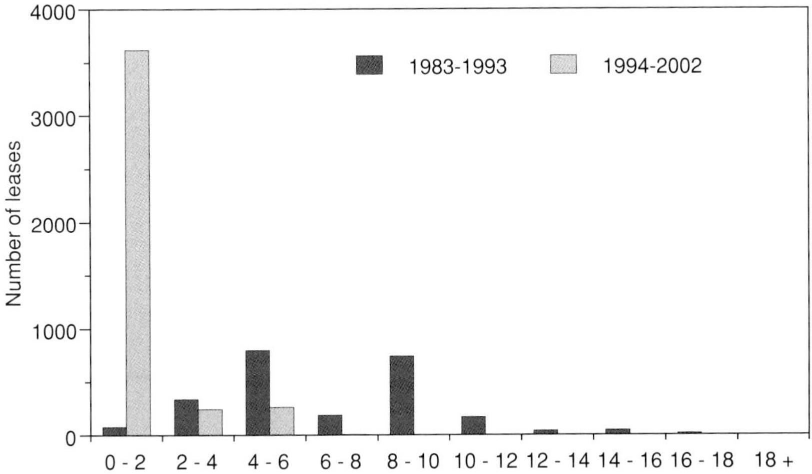

Source: MMS

Figure 9.3: Years Elapsed between Lease Assignation and First Oil For Successful Deepwater Projects in the GOM Federal OCS, 1983–2002

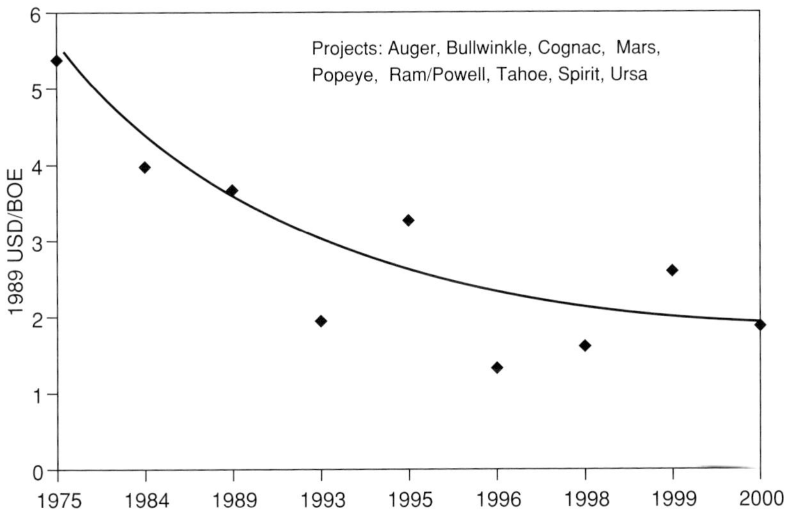

Source: Shell Offshore

Figure 9.4: Finding and Development Costs for Selected Projects in the GOM Federal OCS Deep Water, in Real Terms, 1975–2000

more and more deepwater resources into play, by means of increasingly robust projects with ever lower breakeven economics.

The data that underlie a cost curve like the one shown in Figure 9.4 are not representative, though: the projects incorporated in the graph are either much larger than the average or otherwise lend themselves to be fitted closely to a downward sloping curve. However, if one constructs this same graph on the basis of a broader (although still very selective) group of projects, a significantly different picture emerges (Figure 9.5). The decrease in F&D costs, for starters, is nowhere near as dramatic as in the previous graph, chiefly because, as far as smaller finds are concerned, these costs resolutely refuse to be pushed below the USD 4.50/BOE mark (the global average F&D cost in deepwaters is currently around USD 3.5/BOE). This has to do chiefly with the fact that costs of finding and appraising a deepwater field are not strictly related to prospect size.[11] Thus, it is obvious that a key contribution to lowering costs in the province has come from economies of scale achievable in large projects (i.e. similar costs in absolute terms distributed among a larger number of barrels).

This is not to suggest that this cost abatement has been illusory; after all, these projects have accounted for most of the cumulative output in

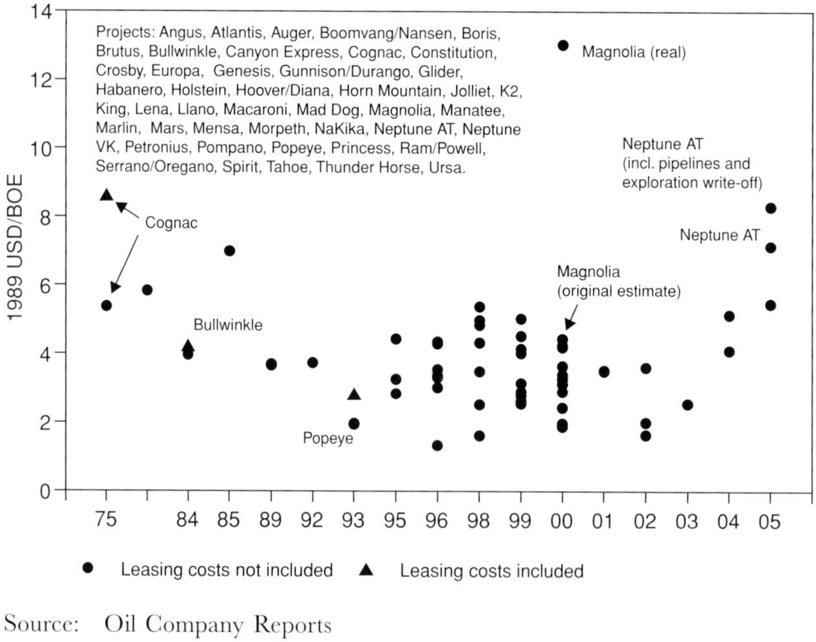

Source: Oil Company Reports

Figure 9.5: Finding and Development Costs for Selected Projects in the GOM Federal OCS Deep Water, in Real Terms, 1975–2005

the deepwater province. But, by the same token, it means that continuous cost abatement at the rate seen since 1996 will only materialise if largish finds continue to be made, an assumption that runs foul of the lognormal distribution of field sizes. True to form, finds have been getting smaller over time, and large finds have become conspicuous chiefly by their absence, both unmistakable signs of maturity (in 2003, for instance, only 330 MMBOE was discovered in the deepwater GOM, a desultory performance that led some analysts to postulate that 'the late 1990s' feast is being closely followed by a period of famine'[12]). Again, the finds that could prove such pessimistic appraisals groundless might be just around the corner, but it is more likely that they are not.

There is another good reason to doubt that the reduction in F&D costs will continue at a steady pace. This has to do with the development contracts that many international oil companies were able to sign with service contractors during the latter part of the 1990s, with fixed costs and delivery dates, meaning that 'contractors carry the capital risks in case there are cost over runs or/and changes in project specs'.[13] Unfortunately for service companies, losses associated with such contracts have really piled up. To cite but one example (by no means isolated), Halliburton is facing a potential loss of USD 434 million associated to delays and cost overruns for development work done on Petrobrás' behalf in the Barracuda/Caratinga field complex.[14] Naturally, this type of charge has done nothing for the balance sheet health of the service companies, leading Merrill Lynch analysts to conclude in 2002 that 'further F&D cost improvements at the expense of contractors [are] unsustainable going forward'.[15] After that year, moreover, there has been a phenomenal rise in the world price of steel, so the passage of time can only have reinforced the economic arguments underlying these conclusions.

Cost curves like that from Figure 9.4, in other words, tend to overstate the extent to which deepwater F&D costs can be reduced further. At the same time, by excluding acreage acquisition costs from the overall picture, these curves also manage to downplay the contribution that the fiscal element has made towards the profitability of deepwater projects. Indeed, thanks to the negative impact that AWL had on signature bonus payments, *this element has probably made a greater contribution towards the cost competitiveness of the GOM deepwater province than all other factors put together.*

The reason usually put forward to justify the exclusion of acreage acquisition costs from F&D cost calculations sounds plausible enough: the magnitude of signature bonuses is a function primarily of the oil price level at the time the blocks were acquired and, given the volatility that characterises the international oil market, this may lead to wide cost discrepancies across projects which have nothing to do with their underlying economics. Hence, goes the argument, excluding signature bonuses from these calculations is a methodologically sound step that allows both analysts and companies to compare like with like, and draw the relevant conclusions.

The weakness of this argument, though, lies in the fact that, regardless of the date when specific blocks might have been leased, acreage acquisition costs represent an immaterial proportion of F&D costs for all deepwater projects. For instance, the signature bonus payments for the six blocks that harbour the Mars field account for less than one-half of 1 percent of this project's F&D costs. Acreage acquisition costs for deepwater projects tapping reserves located in more 'expensive' blocks will vary between three- and four-fifths of 1 percent and, very exceptionally, they may reach the 1 percent threshold. For the purposes of analysing the economics of specific projects, this degree of variance is clearly irrelevant. To put it into its proper context, one need only recall that acreage acquisition costs at Cognac (the sole deepwater development project undertaken before the advent of AWL) amounted to 63 percent of total F&D costs!

9.2 Lifting Costs

While it is true that F&D costs in GOM have declined very substantially in recent years, lifting costs on the whole have not. If anything, they have increased slightly. For instance, Merrill Lynch estimates that, while deepwater lifting costs are around USD 4.7/BOE in Brazil, and USD 5/BOE in Angola and Nigeria, in the GOM region they can top USD 7/BOE (Figure 9.6). The culprit behind this disparity is not too difficult to pinpoint: relative to the other provinces, the GOM deepwater looks overcapitalised in terms of expensive transport infrastructure (i.e. deepwater pipelines).

On the face of it, this is somewhat counterintuitive. After all, recondite as the GOM deepwater fields might appear in the context of domestic US oil operations, they are literally thousands of miles closer to their target markets than their counterparts in Brazil, and even more so West Africa. To be sure, the type of processing, compression and transportation equipment that can cope with the extreme conditions encountered in GOM is not cheap, not only because it is required to work for more than a decade on end without any maintenance but also because of its colossal energy needs: the Holstein spar, for instance, uses

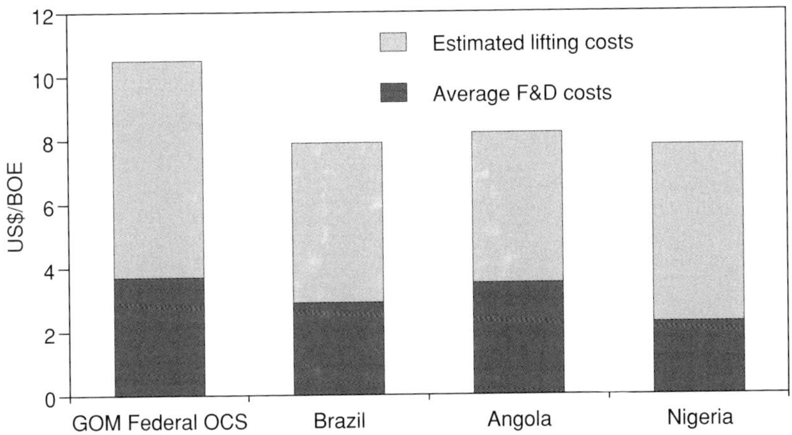

Source: Merrill Lynch 2002

Figure 9.6: F&D and Lifting Costs for Deepwater Fields in Selected Oil Provinces

as much electricity as the entire city of Tulsa (2004 population: 3.45 million souls). Having said that, the same type of equipment is also used in Brazil and West Africa and moreover, there can be no doubt that mobilising manpower, technology and financial resources is considerably less problematic offshore Louisiana than, say, offshore Angola. It is true that the relatively small field size in GOM does translate into more risers, flowlines, drilling and, in general, costly pipe. But the decisive factor in this regard is probably the domestic orientation of GOM projects compared to those in other deepwater provinces.

Outside the USA, deepwater projects tend to be oil-centred and export-oriented. The transportation infrastructure requirements of such projects, therefore, are defined by whatever is needed to take oil to a suitable export point. Usually, this is not very much: if crude is recovered by means of an FPSO, the production, processing and export facilities are one and the same, after all. As far as gas is concerned, deepwater projects only have to devise the best way of stripping it of liquids, and then decide on how to dispose of it (i.e. whether to flare or to re-inject it). However, in GOM, projects are oriented towards large domestic markets for both oil and gas, with pipelines being the preferred method to supply these markets from offshore locations (not least because, as explained above, conservation strictures mean that the exploitation of gas may not be carried out wastefully or even, in the case of associated gas, indefinitely postponed). Given the long distance to shore of deepwater development projects, this translates into very high tariff and amortisation expenses, and it is these two elements that inflate GOM lifting costs well above those of the other deepwater provinces.

As regards amortisation expenses, it has been pointed out that one of the reasons behind the very high cost of some deepwater pipelines lies in the fact that their capacity significantly exceeds the throughputs achievable on the basis of production from the fields they were intended to serve. It certainly appears to be the case that some sizing decisions have been taken on the basis of optimistic assumptions as to the development of satellite reserves, most of which are yet to be found. However, tariffs on many of the key transportation systems in the deepwater GOM have been calculated on the basis of relatively low throughputs, thereby allowing companies to achieve a reasonable return on their investment without necessarily having to accommodate oil or gas produced by others.

Tangible proof for this assertion is available in the form of the proceedings of a recent Federal Energy Regulatory Commission (FERC) hearing in which BP Exploration & Oil Inc. protested against a tariff filed by the ExxonMobil Pipeline Company (EMPCO) for tariffs for the Hoover Offshore Pipeline System (HOOPS).[16] In this hearing, BP (a full partner in the project served by the line) argued that the proposed tariff structure would result in HOOPS receiving revenues well in excess of its cost of providing service, to the detriment of shippers whose costs would consequently be higher than they need have been. BP convinced the regulator that there was no good reason for the HOOPS tariffs to be calculated solely on the basis of Diana/Hoover production, as it was likely that substantial volumes of crude oil would be produced in as yet undiscovered fields located within the catchment area of HOOPS throughout this pipeline's useful life.

As far as tariff expenses go, these are again very much on the high side but it is not advisable to read too much into this fact. The deepwater GOM is one of those areas where lack of infrastructure has acted not so much as a hurdle but as an opportunity to add to field development value by turning pipelines and processing facilities into levers to gain strategic control of basin development. Offshore infrastructure corridors in remote areas have indeed proved to be very effective to extract rents from less advantaged developers forced by natural monopoly economics to use this infrastructure. Recall that McKinsey estimates that Shell has managed to add more than 60 percent to its original field development values in some GOM fields through third-party pipeline and processing fees.[17]

In a very real sense, the high GOM tariff expense reflects the quasi-monopolistic pricing of access to transportation and processing facilities.

To that extent, it cannot be seen as a handicap for the competitiveness of the province as a whole, given that the majority of GOM deepwater infrastructure is in the hands of the same companies that also account for a large part of the output and to whom the bulk of the infrastructure monopoly rents accrue. Thus, notwithstanding high nominal tariff rates, companies like Shell, BP or ExxonMobil (all of whom have extensive deepwater lease holdings) are in a position to develop even quite small finds, as their development decisions can be evaluated on an integrated basis. The development prospects of smaller fields discovered by new entrants, though, are more hampered by the market power of the incumbents in the infrastructure game, especially since there does not appear to be a straightforward regulatory remedy for it: offshore gas gathering facilities do not fall under FERC's regulatory umbrella,[18] and the agency cannot enforce the Interstate Commerce Act (ICA) with respect to deepwater lines, because the latter are located *wholly* within Federal jurisdiction and cross no state lines (see Chapter 10 below).

In sum, hefty lifting costs are far less of a problem in GOM than they are in other provinces. To be sure, they are symptomatic of a high level of capital intensity that translates into commensurately high capital charges. However, the pipeline infrastructure associated with this capital allows companies not only to develop much smaller pools than would be feasible in other provinces but also to monetise natural gas finds which would be left stranded elsewhere. Moreover, GOM lifting costs include most of the costs that a company incurs to take oil and gas to market, whereas in other provinces lifting costs constitute but a fraction of this total (one has to add shipping and handling costs, import duties, storage and pipeline costs at the destination, and so on). Thus, the differences between GOM lifting costs and lifting costs in other provinces become trivial when one expresses project economics in terms of the all-in delivered cost of supply to key markets. In GOM (as in Brazil but unlike West Africa), high lifting costs may still leave better netted back prices for operating companies, not least because they can sell their natural gas output at attractive prices (instead of having to flare or re-inject it), without having to go to the added expense of turning this gas into LNG (Figure 9.7).[19]

9.3 The Income Dimension

Having dissected costs at length, we can now turn to the positive dimension of GOM deepwater operations, namely, their remarkable revenue generation capabilities. Weimer, Rowan, McBride and Kligfield

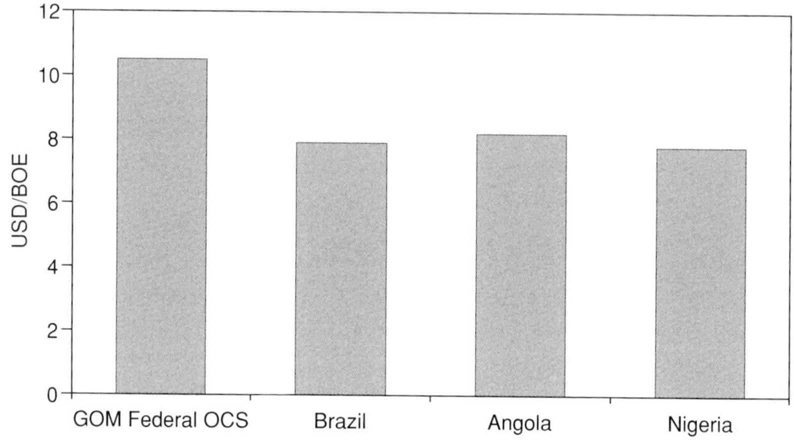

Source: Merrill Lynch 2002

Figure 9.7: All-in Delivered Costs of Supply of Crude to USGC Refineries for Deepwater Fields in Selected Oil Provinces

expressed the secret behind these capabilities thus: 'the petroleum industry has been pleasantly surprised by the extremely high sustainable rates of production from ... [GOM deepwater] reservoirs'.[20] This is quite the understatement. In fact, the production tests that showed that Bullwinkle wells could produce at double the rate originally expected qualify as some of the most momentous events in the history of the US oil industry over the past thirty years, given the way in which they transformed the economics of the deepwater discoveries that Shell had already identified (Auger in 1987, Ram/Powell in 1985). Recall that Shell's earliest plans for Auger contemplated the drilling of a total of thirty wells, a drilling programme that would have entailed an expenditure of around USD 825 million. Actual drilling costs, however, came to less than half this figure in nominal terms (around USD 385 million), because the number of wells required was much smaller (initially 14, going up to 17). Moreover, each one of these wells was capable of sustaining very high output rates, with the result that peak production capacity for the project was more than double that originally expected.

Figures 9.8 and 9.9 show the highest daily averages from a full month of production achieved at any one GOM deepwater oil or gas well, respectively (together with comparable figures for shallow water fields – which include deep gas fields, whose impact on productivity is readily apparent from 2000 onwards). As can be appreciated, during the early 1990s, the highest deepwater oil production rate was around 5

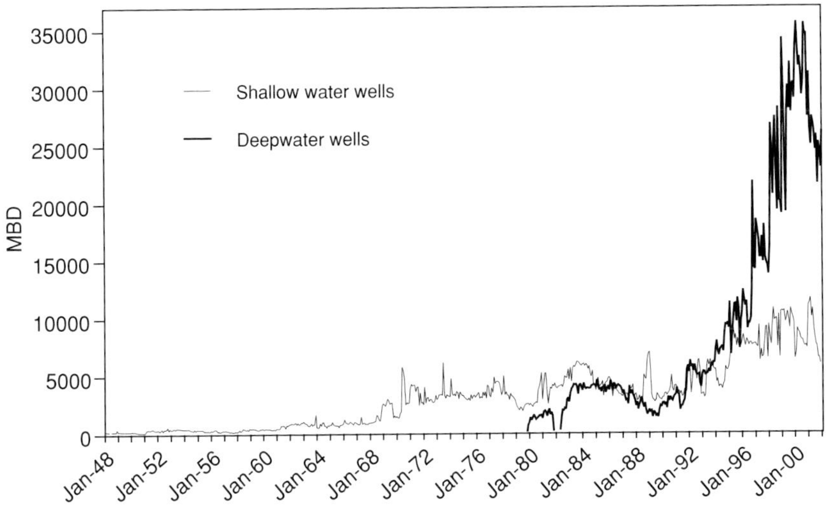

Source: MMS

Figure 9.8: Maximum Monthly Production Rate Achieved at Any Oil Well in the GOM Federal OCS, 1948–2001

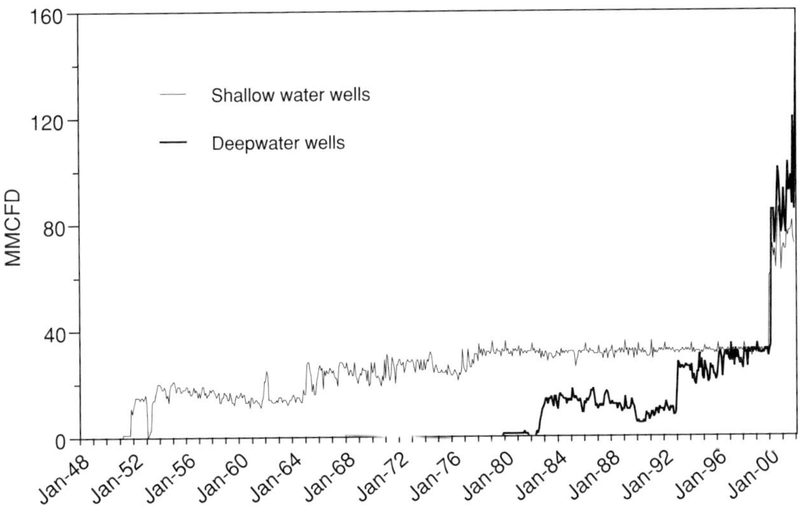

Source: MMS

Figure 9.9: Maximum Monthly Production Rate Achieved at Any Gas Well in the GOM Federal OCS, 1948–2001

MBD for a Bullwinkle well. However, when Auger came on stream, this record was shattered. From 1994 through 1999, maximum deepwater oil production rates continued to climb steadily, as new wells located in water depths between 1500 and 5000 feet came on stream. From 2000 onwards, though, maximum oil production rates, while still very respectable, have declined. As far as gas is concerned, throughout the early 1990s, maximum deepwater gas production rates hovered around 25 MMCFD (a figure not unlike those easily achievable at shallow water gas wells). Then, a well in Shell's Popeye field quadrupled the monthly deepwater gas production record and ever since, deepwater gas wells have been achieving even higher maximum production rates. Currently, the record oil and gas production rates (for a single well, on a single day) stand at 41,532 MBD at the Troika field and 145 MMCFD at the Mica field,[21] respectively. Overall, as of 2005, the average deepwater oil completion produces at about 25 times the rate of the average shallow water oil completion, while the average deepwater gas completion produces at about eight times the rate of the average shallow water gas completion.

While deepwater fields are blessed with high productivity, they are cursed with very short times to peak output. This characteristic makes deepwater projects as highly front-loaded on the income side as they are on the investment side. In other words, a significant proportion of the cash flow that a deepwater project is going to create over its useful life is generated during its first few years in operation, in a manner that allows for the quick recouping of the admittedly high capital costs associated with deepwater operations. Cash flows generated during the tail-end years of a given project are discounted heavily in net present value (NPV) calculations used to see whether a given investment option might be worth pursuing, so early life cash flows have a disproportionate effect in terms of the estimated profitability of projects, and also in determining their degree of robustness vis-a-vis unfavourable changes in oil and gas prices.

For the reasons given above, from an oil company perspective, deepwater field economics are considerably more attractive than the economics of many conventional projects in more tractable areas (both on- or offshore), notwithstanding the greater capital intensity of the former. Having said that, the fiscal environment in the GOM deepwater is also very favourable, and this begs the question of which particular element has made the greatest contribution to post-tax investment returns in the deepwater province. In order to answer this question, we will compare the economics of two landmark deepwater development projects, Cognac and Mars (Table 9.2), both carried out

Table 9.2: Comparison of the Economics of Major Deepwater Projects in the GOM Federal OCS, in 1975 (1993) Dollars

Cognac		*Mars*
1975	Discovery date	1989
1975	Project go–ahead	1993
1979	On–stream date	1996
1,000*	Water depth (feet)	2,940*
200	Initial reserves (MMBOE)	500
464 (1,100)	Development costs (USD million)	490 (1,161)
783 (1,855)	Total costs (USD million)	492 (1,166)
2.43 (5.76)	Development costs (USD/BOE)	1.07 (2.53)
3.91 (9.26)	Total costs (USD/BOE)	1.07 (2.53)
110	Peak production (MBOED)	220
62	Number of wells	24
2.5 (5.92)	Drilling cost per well (USD million)	11.6 (27.48)
>2	Average production per well (MBOED)	9
295 (699)	Leasing costs (USD million)	2.2 (5.21)
1.48 (3.51)	Leasing costs (USD/BOE)	0.004 (0.009)
13,020 (30,844)	Leasing costs per acre (USD)	64.6 (153)
MC108, MC151, MC194, MC195	Block(s)	MC108, MC151, MC194, MC195
Shell	Operator	Shell
Amoco, Barber Oil, Conoco, Drillamex, Koch, Murphy, Ocean Oil, Ocean Production, Sonat, The Offshore Co., Unocal**	Other stakeholders	BP

* World depth record at the time

** Current partners are Shell, BP, Agip, El Paso, ChevronTexaco, Unocal, Murphy, Phillips and Koch

Sources: Shell, MMS

under Shell's leadership. Whereas the former project was undertaken before the onset of both AWL and the deepwater boom, the latter was begun after the reforms to the GOM institutional framework had been in place for some time.

Cognac, it will be recalled, was the first genuine deepwater development in the world, and the project set a number of records at the time of its completion, not least that for water depth (at 1000 feet). Like Cognac, Mars was also a record-breaker, in physical as well as in financial terms: the Mars TLP was installed in May 1996 at a water depth of 2940 feet, breaking the GOM record for a permanent drilling and production platform by around 100 feet. The main difference between both the projects lies in the fact that Mars is a much larger field than Cognac: initial recoverable reserves for the former were estimated at 500 MMBOE, against 200 MMBOE for the latter.

Although much is made in the trade press about the phenomenal ante that Shell had to put up in Mars, the fact is that Cognac was just as heavily front-loaded in terms of investment. Development costs at Mars look much larger in nominal terms, of course, but not so much in real terms. Moreover, again expressed in real money, Cognac's full costs (i.e. development plus exploration plus lease acquisition costs) are higher than those of Mars, which means that the latter project was actually less heavily front-loaded than the former.

The effect that high well productivity has on project profitability can be readily appreciated by comparing drilling costs across both projects. The 24 wells at Mars cost USD 625 million in all (i.e. 56 percent of a development expenditure of USD 1.1 billion). Drilling costs at Cognac (USD 155 million) only accounted for 33 percent of the development expenditure at this field. Drilling costs per well at Mars were therefore 4.6 times as large as those of Cognac, but since the latter project required the drilling of almost three times as many wells, total drilling costs at Mars were 'only' 80 percent greater than those of Cognac (with all magnitudes expressed in real terms). Furthermore, since the Mars wells were also five times as prolific, drilling costs at Cognac were significantly *higher* than those of Mars, *on a unit production basis.*

As can be appreciated in Figure 9.10, Cognac was by all accounts a successful, profitable project. It turned cash positive in 1984 (1983 if leasing costs are ignored) and had achieved an internal rate of return (IRR) of 7.78 percent – a figure more or less equivalent to Shell's cost of capital in the mid-1970s – by 1984. The Cognac IRR grows to 16.30 percent if leasing costs are not taken into account. The IRR over the life of the project is 14.26 percent (22.48 percent without leasing costs). And while it is true that Cognac benefited greatly from the very high prices prevailing in the world oil market during the early years of its operational life, it is also true that the project had to weather a longish period of very low oil prices. Over the life of the project, the ratio of government take (including Federal income taxes[22]) to gross income, on the one hand, and profits, on the other hand, has been 27.66 and 42.80 percent, respectively. The bonus payment for the blocks accounts for 18 percent of the total government take in nominal terms, with royalties and Federal Income Tax making up the remaining 39 and 43 percent, respectively. However, bonus payments account for 45 percent of the value of total government take on present value basis

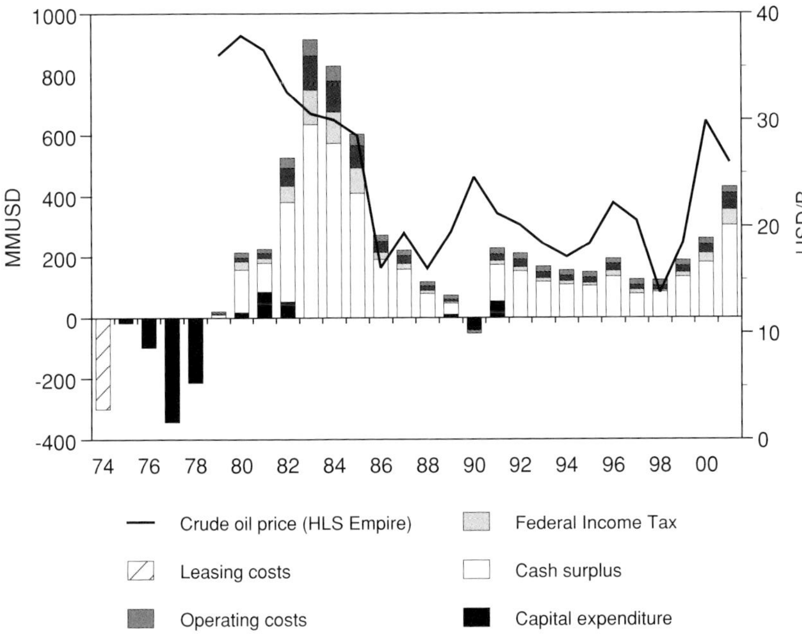

Sources: MMS, Platt's, Shell Offshore

Figure 9.10: Net Cash Flow Breakdown for the Cognac Development Project, 1974–2001

(using a 10 percent discount rate), with royalties and Federal income tax accounting for 30 percent each.

If it is compared to a profitable project like Cognac, Mars can only be called a runaway success (Figure 9.11). The project was expected to pay for itself only one and a half years after first oil, but that milestone was reached in March 1998, even though at that point international oil prices had already entered into an accelerated decline that would take them to the lowest levels seen (in real terms) since the First Oil Shock. At 26.31 percent (27.22 percent without leasing costs), the IRR as of 2001 for Mars is 3.4 times higher than the one achieved by Cognac at the same stage of its life. Indeed, since first oil in 1996 and up until 2002 inclusive, the free cash flow generated by the Mars project is equivalent to 87 percent of the cumulative free cash flow generated by Cognac over a much longer period of 23 years.

The ratio of government take versus gross income, on the one hand, and profits, on the other hand, is also significantly more favourable to the Mars partners: 22 and 36 percent, respectively. This is hardly surprising when one considers that Shell paid USD 5.3 million in

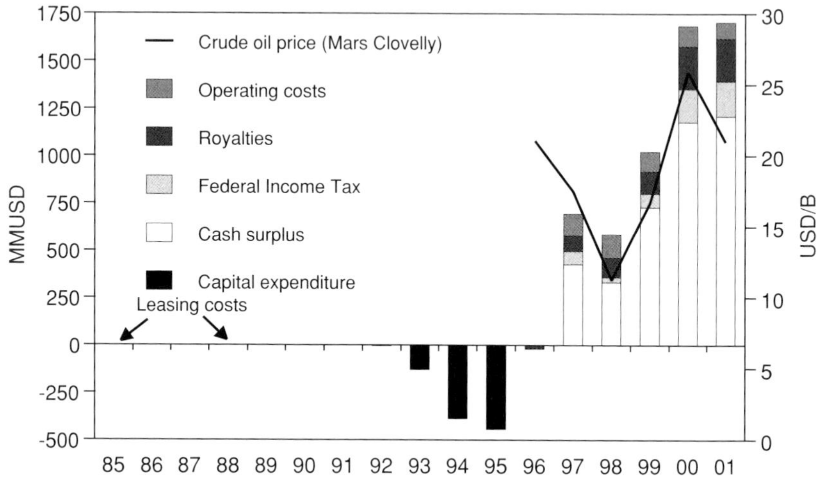

Sources: MMS, Platt's, Shell Offshore

Figure 9.11: Net Cash Flow Breakdown for the Mars Development Project, 1985–2001

bonuses for the six blocks that straddle the Mars field (equivalent to USD 0.004/BOE of reserves), whereas the various working interest owners at Cognac had to pay the equivalent of USD 1.48/BOE of reserves (and USD 13,000 per acre). In other words, the Mars leases cost Shell only USD 2.2 million (USD 64.6 per acre) in 1974 money, an astonishing 134 times less (or 201 times less on a per acre basis) than the amount that the company and its partners paid for the considerably less prolific (and, for their time, equally challenging) Cognac blocks. The Mars bonus payments account for 1.1 percent of the value of total government take over the period 1985–2001 on a NP basis (using a 10 percent discount rate).

Intuitively, one would think that the larger size of the Mars field should account for most of this disparity, but this explanation founders upon consideration of other deepwater projects: Auger (380 MMBOE), Brutus (324 MMBOE) Hoover/Diana (370 MMBOE), Petronius (110 MMBOE), Ursa (300 MMBOE). These projects are more similar to Cognac in terms of reserves than Mars, but the IRR figures estimated for them by Arthur Andersen consultants in 2001 (19.15 percent, 53.02 percent, 29.35 percent, 28.07 percent and 29.17 percent, respectively) are also significantly higher than the Cognac IRR.[23] Clearly, factors other than field size seem to be at work here.

In order to highlight what these factors are, the 'spider diagramme' in Figure 9.12 presents the summary of a sensitivity analysis on the

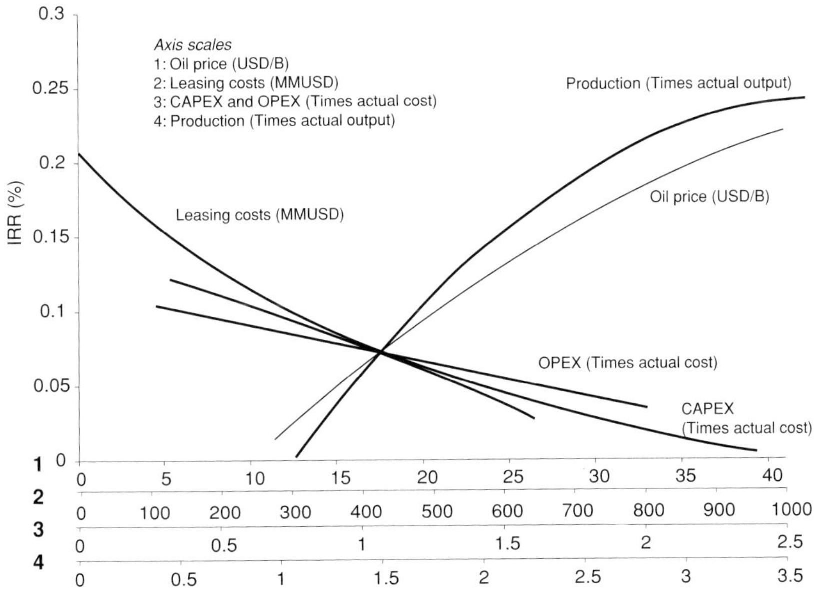

Sources: DOE, MMS, Riddle, Snyder and George 2001

Figure 9.12: IRR Sensitivity Analysis for Mars Development Project

economics of the Mars project, on the basis of observed production and price data for the 1996–2002 period, observed capital expenditure and operating costs,[24] a royalty rate of 12.5 percent and a *marginal* income tax rate of 34 percent (a rate much higher than the one Shell has had to face, as explained below). In the graph, IRR lines are plotted for each of the following (with all the other ones held constant): capital expenditure (including drilling costs), leasing costs, operating costs, oil production rates and, last but not least oil prices. As can be appreciated, the midpoint elasticities of rate of return with respect to leasing costs are greater than those for either capital expenditures or operating costs. Indeed, among the key variables selected, only prices and production rates have a greater impact on project IRR than variations in leasing costs.

This result is consistent with the role of signature bonuses as the vehicle for excess profit collection in GOM. By the same token, the very attractive returns achieved at deepwater projects strongly suggest that, in the post-AWL world, signature bonuses have been fulfilling this role far from adequately. In other words, although it is true that GOM deepwater geology has been very favourable, the fact that 'the levels of economic rent generated from the deepwater are high, more than twice the industry's cost of capital'[25] is *fundamentally* attributable to the benevolent fiscal environment created by AWL.

A similar fiscal generosity prevails in other deepwater provinces around the world. According to Merrill Lynch analysts, this is a reflection of the fact that 'in the early 1990s when fiscal terms were negotiated ... governments (and the industry) underestimated the pace at which technology would move during the mid-1990s. Essentially, technological advancements allowed deepwater projects to be developed faster and sooner than had been envisaged in the various fiscal regimes. What was originally intended as a mid-teens rate of return ended up being a high-teens rate of return for the industry.'[26] As explained in greater detail in the concluding chapter to this study, there are reasons to doubt that oil companies would have been content with a 'mid-teens rate of return' in other oil provinces (they clearly wanted much more than this). But, in any case, Merrill Lynch's conclusions do not really apply to the case of GOM. *In this province, rates of return have certainly been very attractive, thanks to radical changes to the prevailing fiscal framework (i.e. the adoption of AWL). Crucially, though, these changes were put in place much before the early 1990s, at a moment when no avant garde deepwater development prospects appeared to be in the offing.*

The conventional interpretation for the factors that determine bonus payments is that they depend on the level of interest that it manages to arouse in competing bidders, which in its turn 'is more directly related to an area's resource potential than to the method of leasing'.[27] According to this view, the Mars blocks were leased for peanuts because, aside from Shell, no one – including MMS – could see any sense in tying down money in acreage whose odds for profitable development seemed infinitesimal, given the costs involved and the prevailing oil price expectations at the time the blocks were being offered (Shell had an idea, a concept, that cost them a great deal in terms of the geo-scientific expertise that they developed, maintained and brought to bear on this province, and these costs translated into success against the competition). Conversely, and notwithstanding the steepness of its costs, Cognac attracted much more bidding interest than Mars because the expectations for the oil price in 1974 were extremely bullish (there can be no question that the bonuses paid for the Cognac blocks would have been much smaller had these come up for offer during the middle- to late-1980s).

It is a truism to say that the NPV of a proposed development project (calculated by different companies' factoring in the different price expectations going around, as well as cost estimations at various reserve probabilities) will limit the *maximum* amount that a seller will be able to obtain for acreage. For this reason, too, achievable prices will be a function of *when* the acreage is offered. But, in addition to this, the *manner* in which the acreage is put on the market will also be instrumental to determine whether this maximum is achieved or not, as different selling methods will influence both the nature and the intensity of competition for offshore oil and gas leases by allowing (or not, as the case may be) players with differing capabilities – technological, managerial and financial – to form an idea about an area's resource potential, and to tailor their bids accordingly.

On the strength of the above, one can claim without unfairness that, throughout the 1980s, MMS leased most of the choice deepwater acreage prematurely, foregoing the chance to collect massive windfall profits when these began to materialise from the mid-1990s onwards. Given Mars' cash generation capabilities (even on the basis of conservative oil and gas price assumptions: USD 23/B for crude, and USD 3/MMBTU for natural gas), the six blocks harbouring the field could very well have been leased (in 1990, say) for a *combined* total payment of USD 600 million (around USD 20,000 per acre), and the project would still have achieved an IRR significantly higher than Shell's cost of capital (and comparable to the return achieved at Cognac).

By the mid-1980s there existed a fair amount of 2D seismic and other geo-scientific information (DSDP) on the deepwater province and some companies (above all, Shell) were using it to develop ideas and geological concepts, well in advance of the technology necessary to carry them to fruition. Obviously, when deepwater acreage started to become available, the companies with the more advanced ideas and concepts would have been in a position to take the best prospects first. This does not amount to a 20/20 hindsight suggestion that MMS *missed* seeing these future rents or, for that matter, that Shell somehow knew these rents would be there for the taking. Rather, the policy failure that lies at the heart of the enormous fall in value of OCS acreage since 1983 is that AWL gave companies in the geo-scientific vanguard the means to put a corner on deepwater acreage way before offshore technology was sufficiently mature to develop it. A very different outcome would have obtained had MMS doled out deepwater leases in a more controlled fashion, akin to that in which the Norwegian government has made offshore acreage available through time. The pace of Norwegian licensing has not been dictated by the fact that clever officials have been able to 'see' and quantify future rents. Rather, their actions and policies have been motivated and energised by the intuitions underlying the trenchant criticisms that Stephen McDonald levelled in 1979 against Nixon's accelerated leasing initiative in the Federal OCS:

> leasing could with benefit be accelerated if the build-up were gradual, if the new higher rate ultimately to be achieved were known in advance, and if, in the light of the total area ultimately to be leased, the new rate could be sustained for a long enough period of time to satisfy additional capacity in the supply industries. The recent, very sharp increase in the rate of leasing of OCS lands seems not well calculated to conform to these conditions and therefore to increase unambiguously the present value of the expected economic rent … *We do not mean to suggest that [DOI] attempt to measure … the pure economic rent available and then insist upon using the figure as a minimum payment by the lessees as there is no practical way it could do that. Rather we meant to say that [DOI] should try to create and maintain leasing conditions which are conducive to the desired result. Thus, for example, conditions that reduce lessee uncertainty, that increase competition for leases, that increase efficiency of resource extraction.*[28]

The key problem with AWL (a problem that explains the long-term collapse in GOM acreage values) is that, as a policy, it failed (and continues to fail) miserably on all these counts.

9.4 Just How Generous Is the GOM Fiscal Regime?

The analysis presented above strongly suggests that, although it is true that GOM deepwater geology has been very favourable, the fact that 'the levels of economic rent generated from the deepwater are high, more than twice the industry's cost of capital'[29] is *fundamentally* attributable to the benevolent fiscal environment created by AWL. In other words, the phenomenal success of the GOM deepwater would seem to owe far more to taxes (or more precisely, the absence thereof) than to technology.

Derman and Johnston cogently dispute this characterisation of the GOM Federal OCS fiscal regime as unduly generous. The way they see it, GOM terms 'are good but certainly no give-away. The [US] Government is not leaving any money on the table and for the petroleum industry there is no windfall in sight.'[30] According to these authors, calculations such as those presented in this chapter are flawed because they 'are based upon the division of profits from an undiscounted (nominal) and un-risked point of view',[31] a simplification that – while understandable – is all the more unsustainable when it is used to analyse the production economics of a province such as the deepwater GOM, 'where costs and lead times are greater as is the difference between technical success and commercial success'.[32] These authors also object to the fact that 'bonuses paid on blocks where no discovery is made do not get collapsed into typical take statistics. In a country/region or particular play type where the success probability (chance factor) is on the order of say 20 percent, 4 out of 5 bonuses do not get factored-in to the take calculations'.[33] As if this were not enough, most analyses make no allowances for the time value of money, even though 'typically, there is a lag of several years between the point at which a bonus is paid for rights on a GOM shelf block and the resulting discovery if there is one'.[34]

Derman and Johnston estimate that the value of the timing differences between bonus outlays and royalty receipts are such that the latter outweigh the former to the tune of 2 to 1, on an NPV basis, 'even though the typical bonus paid in the US OCS is insignificant when compared to the total revenues generated by a producing field'.[35] Furthermore, on a province-wide (i.e. macro) basis, they estimate that acreage acquisition costs have ended up by absorbing a very significant 20 percent of *discounted* gross revenues[36] (rather than a *nominal* 3–5 percent). Indeed, they reach the conclusion that 'from this perspective the Government Take [in GOM] is over 70 percent [of profits] – tougher than world average from an oil company point of view'.[37]

The points that Derman and Johnston raise are valid, and forcefully argued. However, from a project-specific (i.e. microeconomic) and – even more importantly – *retrospective* viewpoint, their estimates of fiscal take (Table 9.3) are significantly overstated, even if one accepts their – reasonable enough – estimate that the capital and operating costs associated with generating OCS revenues amount to approximately 35 percent of gross revenues over the life of a given field.

Table 9.3: Comparison of Per Barrel Government Take Methodologies for GOM Federal OCS, Full Cycle Basis

Item (in percentages)	*Bonuses not considered*	*Derman and Johnston*	*Retrospective analysis*
Gross revenues	100	100	100
Royalty payments	-16	-16	-12.5
Bonus impact	0	-20	-7
Revenues, net of royalties and bonuses	84	64	80.5
Total costs (estimated)	-35	-35	-35
Taxable income	49	29	45.5
Corporate Income Tax	-17.15	-10.15	-5.8*
Company net income	31.85	18.85	39.7
Company Take (Company net income/ (Gross Revenues – Costs))	49	29	61
US Government Take (1 - Company Take)	51	71	39

* Effective Federal Income Tax rate for 1990–2001, calculated as per Appendix 1

Source: Derman and Johnston 1998.

On the royalty front, the problem lies in Derman and Johnston's use of a 16 percent royalty rate, when in fact deepwater production attracts the minimum royalty rate set by law (12½ percent), unless DWRRA provisions apply.[38] As far as Federal income tax liabilities are concerned, because the aim of their analysis is to illustrate the attractiveness of *future* investments and investment incentives, their calculations are based on a *marginal* rate of 35 percent (this is because the marginal tax rate is a forward-looking measure). However, the objective of this section is to evaluate the profitability of *past* rather than future investments, which means that a different, backward-looking, measure is called for; namely, the average tax rate.[39] The substitution of the average tax rate for the marginal tax rate in the calculations makes a significant difference because, due to the peculiarities of the US fiscal regime (notably the fact that GOM upstream activities are not ring-fenced for income tax purposes), the former is significantly lower than the latter. For illustrative purposes only, Figure 9.13 plots the behaviour through time of the average effective and the marginal statutory Federal income tax rates. The former is calculated on the basis of *gross* income,[40] so the yawning gap between both rates is not proportionate to the extent by which past OCS income tax liabilities calculated on the basis of the latter (and only meant to apply to *net* taxable income) will have been overstated.

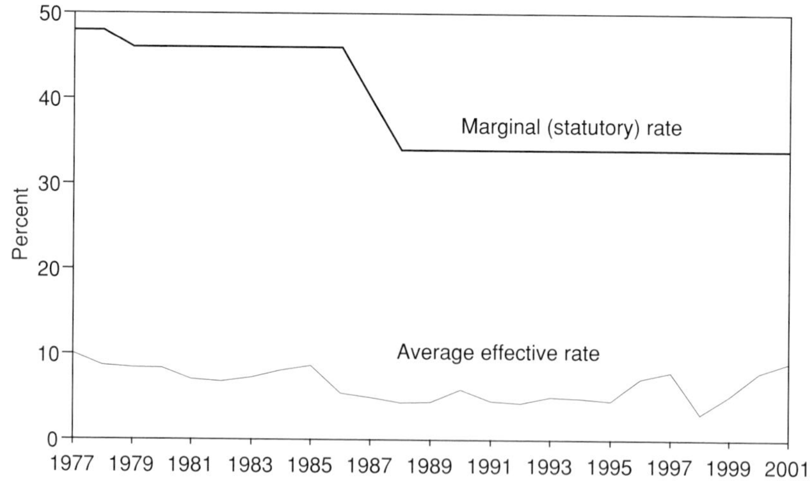

Sources: DOE, MMS

Figure 9.13: Comparison of Federal Income Tax Rates for Upstream Income Generated in the GOM Federal OCS, 1977–2001

Derman and Johnston's figures regarding the proportional magnitude of signature bonuses also appear too high, but in this case this is due to their underestimation in discounted revenues. In the case of Mars, for instance, given this project's production profile over its lifetime and conservative price assumptions,[41] it turns out that Shell would have had to disburse around USD 925 million in signature bonuses for these payments to be equivalent to 20 percent of the project's discounted gross revenues. This figure exceeds the observed bonus payment by 125 times.

Obviously, Shell did not have a firm idea of what the eventual revenues from Mars would be (and it is safe to say that absolutely no one expected oil prices to behave as they have from 2000 onwards). Furthermore, the comparison as it stands also makes no allowance for bonuses paid for dry or unexplored blocks (and, again, Shell did not know in advance which particular blocks in its lease inventory would turn out to contain a commercial find). Having said that, aggregate dry hole expenses and the carrying costs of a lease inventory are corporate-wide costs for any company engaged in E&P activities and, there is no non-arbitrary way to allocate such costs among successful projects. Suffice it to say, though, that the estimated bonus figure for Mars exceeds by USD 100 million the total discounted bonuses paid by Shell over the 1983–2001 period. And since Mars accounts for only 20 percent of Shell's gross cumulative deepwater output over the same timeframe, it is obvious that there is no way that Shell's total discounted bonus payments have been anywhere near 20 percent of its province-wide discounted revenues.

These calculations suggest that, even after all the relevant risking and discounting has been done, the GOM fiscal regime has indeed proved to be the stuff from which corporate windfalls are made, at least as far as Shell is concerned. Shell is by no means a typical GOM deepwater operator, though. And it is very interesting to see that when one carries out these cost calculations for other – more representative – firms, the results are quite different to those outlined above, *in a way that makes Derman and Johnston's estimates of fiscal take look conservative*. For instance, Marathon's discounted deepwater bonus expenditure over the period 1983–2003 actually comes to 30 – rather than 20 – percent of its discounted gross revenues. Moreover, because of this company's disappointing deepwater exploration programme so far, its unit finding and development costs are considerably higher than Shell's.[42] Thus, the fiscal take on this company's profits looks to have been consequently higher, very likely surpassing the average estimated by Derman and Johnston.

9.5 The Most Profitable Petroleum Province in the World? Only for Colossi...

In light of the evidence presented in the section above, Derman and Johnston's contention in the sense that the GOM fiscal regime is fair (rather than unjustifiably generous) seems untenable. However, the contraposition between the profitability of the deepwater operations of Shell, on the one hand, and Marathon, on the other, suggests that there is little additional value to be derived from asking to what extent the GOM deepwater has been the most profitable oil province in the world since 1996 (because it is obvious that it has consistently been in first place or thereabouts). Far more interesting is to ask *who* has been at the receiving end of this profitability and, even more importantly, *how* they got themselves in that position.

Answering the first of these questions is not difficult: it is a well-attested fact that participation in large low-cost deepwater fields is entirely restricted to a small number of major oil companies. Most other participants, in contrast, have seen comparatively larger outlays in bonuses and dry holes buying them a disappointingly small share in overall deepwater output. Thus, the fiscal regime has worked out to be highly advantageous to a few players and neutral to poor for the rest of the players in the province (not to mention the US Federal government and the US taxpayer).

Figure 9.14, based on the very useful ranking system for world fiscal regimes developed by Daniel Johnston,[43] seeks to underscore this point by placing the GOM Federal OCS fiscal regime within this continuum, *but as seen from the different perspectives of Shell, on the one hand, and Marathon, on the other.* It is worthwhile to point out that the allocation of places for the different fiscal systems in the graph depends on the figures obtained for marginal government take on net income, whereas the places for Shell and Marathon reflect effective government take on gross income. It can be readily appreciated that, as far as Shell is concerned, the GOM Federal OCS fiscal terms are as attractive as those of the fiscal regime that tops Johnston's list, Ireland (but whereas there is plenty of petroleum in GOM, Ireland is pretty much as dry as a bone). And even though Shell has to contend with an effective royalty rate of 12.5 percent, its terms also look much better than the representative GOM terms, as calculated by Johnston on the basis of a zero royalty rate (because of DWRRA) and a 34 percent tax marginal income tax rate.

In contrast, the GOM terms that Marathon effectively has faced are much tougher than Johnston's estimations, and are quite near the world

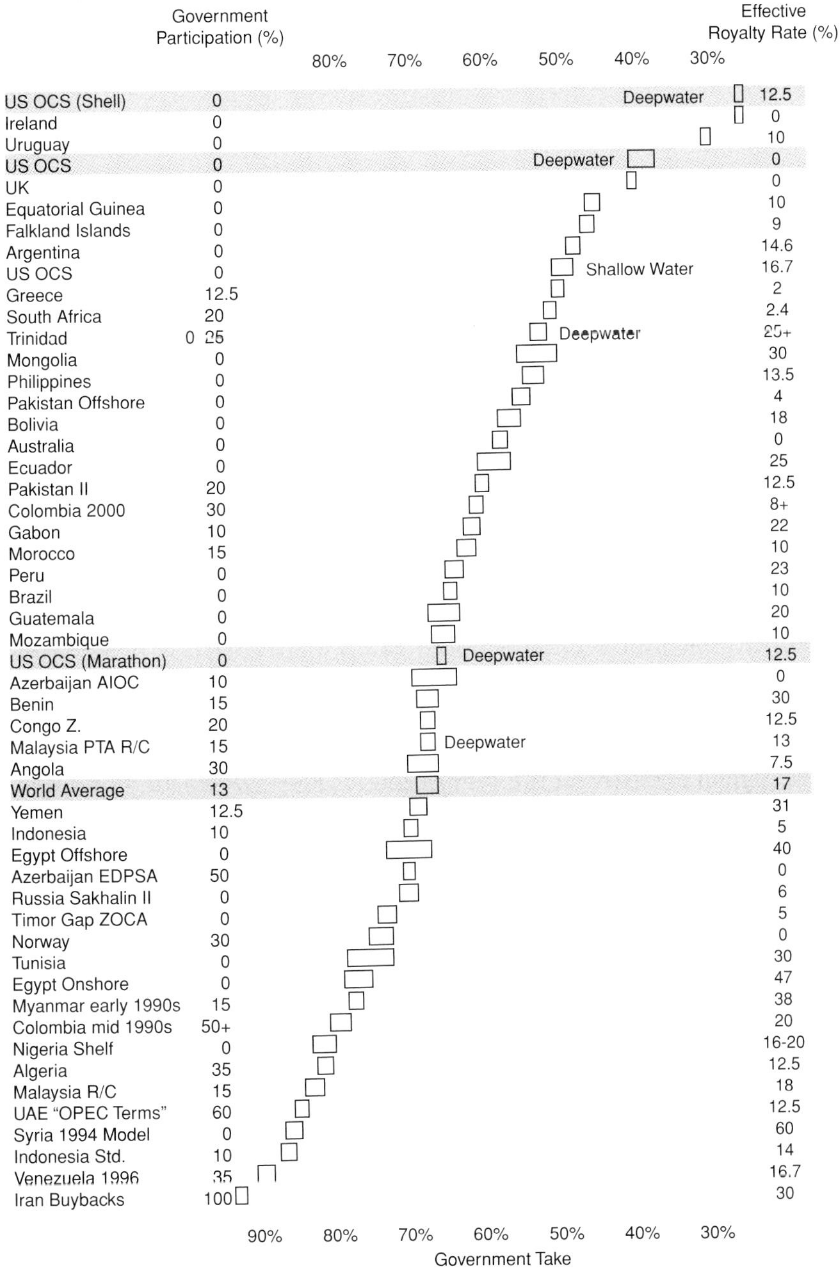

Source: Daniel Johnston & Co

Figure 9.14: Worldwide Fiscal Take Comparison

average for total government take. Since GOM deepwater costs exceed the global average, this strongly suggests that activities in this province may have been as much of a value-destroying proposition for Marathon as looking for oil in Guatemala would have been. But whereas failure in Guatemala would have quickly been met by a decision to cut its losses, Marathon has persisted in its expensive efforts to strike deepwater oil in GOM, tantalised by the successes achieved by others.

Analysts usually explain away the deepwater dominance by companies at the very top of the size rankings by invoking the unassailable technological leadership that such companies supposedly enjoy (an argument that conveniently overlooks the crucial role that the service companies played in developing a lot of this technology). Large majors are presented as the only players able to marshal the enormous financial and human resources necessary to tackle the development of frontier projects that are fraught with all kinds of uncertainties and technological exigencies (as all the early GOM deepwater projects undoubtedly were). Because they spent vast sums on developing leading edge production technology, the argument goes, firms like Shell, BP and Exxon were able to stake claims for the best deepwater acreage first, thereby carving out a dominant position for themselves and pre-empting players who had to wait for the technology to mature, in order to buy it off the shelf.[44]

This explanation overstates the degree of technological and project-management superiority that companies like those named above supposedly enjoy. After all, to disprove it one need only look at the many offshore upstream projects posing unique and enormous challenges (often at the cutting edge of technology) which smaller majors or first-tier independents have managed to tackle nonetheless (Cook Inlet, the North Slope, Ekofisk, the Campos basin, the Athabasca Tar Sands). Even if one acknowledges the very different nature of the challenges posed by the deepwater GOM and early operations in the Norwegian North Sea,[45] there is no *a priori* reason to doubt that the likes of Pennzoil, Amerada or Kerr-McGee would have been able to rise to the occasion in the same way as Phillips did in Norway or Richfield and Arco did in Alaska, and for much the same reason: the colossal magnitude of the potential rewards. Furthermore, even after the 1986 price crisis, the availability of finance through the junk bond market meant that, up until the early 1990s, the availability of capital was not even a binding constraint for many of the larger independent firms (recall, for instance, that Pennzoil came close to mounting a takeover bid for Chevron with the settlement money it got from Texaco in connection with Pennzoil's stillborn merger with Getty Oil.

In the specific case of GOM, independents had long proved themselves perfectly capable of punching above their weight in the project management and technology league. A statistical study on drilling effort and outcomes conducted by Pulsipher, Iledare and Baumann, for instance, found 'little evidence to support the speculation that ... [GOM] resources would be less aggressively or efficiently developed should independents continue to play a progressively larger role in the search for and development of hydrocarbon resources'.[46] As a matter of fact, their data showed that independents had become progressively more willing to assume GOM exploration risks than the majors, no doubt because the dearth of prospects elsewhere affected the former more acutely than the latter. Over the 1986–96 period independents accounted for 53 percent of cumulative wildcat permits and nearly 70 percent of total exploratory wells drilled (with small independents accounting for nearly 30 percent of total wells drilled). Furthermore, more than 50 percent of the drilling effort by independents was concentrated on exploratory wells (compared to 22 percent for the majors).

Independents were not only more willing than majors to assume exploration risks in GOM, they were also marginally more successful at this game. Between 1983 and 1992, independents added an average of 265 BOE per successful foot drilled (compared to 227 per successful foot for majors), and 111 BOE per total feet drilled (compared to 106 BOE for majors). More tellingly, the wildcat success rate for independents (33 percent) over this period significantly exceeded that of majors (25 percent), while the increasing maturity of the GOM exploratory rate had a more negative incidence on the reserve additions per foot of well completed for the majors. All of which led Pulsipher, Iledare and Baumann to conclude that there was 'no statistical evidence of significant differences in the responsiveness of the gross find rate of hydrocarbons reserves to technical progress among firms of different sizes operating on the OCS'.[47]

In light of the above, to assert that medium-sized companies would have been both willing and able to undertake largish deepwater projects (even at a relatively early stage in the development of the province) clearly does not amount to misplaced optimism in their capabilities (not least because, as of 2006, the number of deepwater discoveries by non-major companies has surpassed that by major companies). Furthermore, their capabilities would have expanded enormously as the more esoteric elements of offshore technology gradually became more widely available. The negative effects of a greater involvement on the part of relatively smaller companies during these early stages would probably have been nothing more serious than slightly longer lead times for those projects in which such companies assumed the leading role. Arguing counterfactuals is fraught with pitfalls, of course, but it is probably fair to conclude not only that the resurgence in GOM output would still have taken place had independents not disappeared off the radar screens after Cognac, but also that it would have proceeded along roughly similar lines.

The Cognac development can be cited in support of this counterfactual, to challenge the assertion that large companies are the only entities in a position to handle exposure to the complexities and costs of deepwater operations. As has already been mentioned, medium-sized firms like Murphy and Unocal had a small equity participation in Cognac. Together with small independents, they held around a quarter of the working interest in the project.[48] At first sight, and in the light of the supposed technological superiority of the majors, the presence of firms like these in the Cognac development appears baffling. After all, they were certainly in no position to assure the success of a project that was very much at the 'bleeding edge' of technology in those days. So why then did the majors bring these smaller fry on board?

The answer revolves around the enormous costs of not only developing but – primarily – *leasing* the Cognac blocks. Simply put, very large upfront outlays of this magnitude (*circa* USD 300 million) made even major oil companies keen to spread both costs and risks, a purpose for which the money of small companies was as good as that of anybody else. Indeed, the money of relatively smaller firms was arguably better than that of other companies, because as Hendricks and Porter pointed out in their study on joint bidding in OCS acreage auctions, parties wishing to reduce their outlays on attractive tracts 'may not want to share [their] knowledge or expertise with firms that could use that information in bidding for other leases', and hence they will have a clear incentive 'to seek partners with no desire to explore the region themselves'.[49]

The case of Cognac suggests that, even in the extreme case that mid-sized or smaller companies had indeed proved incapable of taking on their own the operating leadership of early deepwater projects, they could nevertheless have carved out a respectable output position for themselves in the GOM deepwater. For that to happen, however, it would have been necessary for large firms to have an incentive to offload part of their leasing and development costs on smaller partners. This incentive disappeared with AWL, which among other things led to order-of-magnitude reductions in acreage acquisition costs that allowed majors not only to carry large deepwater lease inventories on their own but also to evaluate them in relative leisure.[50]

While the reduction in acreage costs might have eliminated the need for majors to engage in cooperation, it does not explain why smaller players did not play any meaningful part in deepwater lease sales, with the result that production operations were entirely in the hands of a few majors until 1997–1998 (when Kerr-McGee's Neptune/Thor, Marathon's Arnold and Amerada Hess' Baldpate projects came on stream). After all, part of the benefit from the decline in bonus payments could have accrued to smaller players, especially since AWL was a policy supposedly enacted on their behalf (as an MMS bureaucrat put it, keeping bonus bids high 'limited competition and kept the [OCS] drilling activities in the hands of the very few major oil companies that could offer such bids ... [whereas the Reagan's] administration's philosophy does not cater to the big corporations but, rather seeks to stimulate competition').[51] In theory, therefore, smaller players could conceivably have built a large deepwater lease position themselves, just as long as they found ways to pool their capital and resources, in order to mobilise them more effectively. But therein, precisely, lies the rub.

In the following chapter, we demonstrate that the decision by most industry players to leave the field in the hands of majors for more than a decade is a consequence of the unsoundness of the Federal government's leasing policy after 1983. AWL prevented the government from capturing the Ricardian rents generated in the deepwater, to the great benefit of precisely the most affluent companies. But in addition to this, *in an entirely predictable fashion*, it rendered smaller companies unable to compete with their larger peers and, therefore, needlessly sacrificed their enterprising dynamism (responsible for opening up GOM to oil activities, no less) on the altar of facile supply-side economic dogma.

NOTES

1 Merrill Lynch 2002: 228.

2 The fracture gradient in a formation is determined by measuring the pressure at which fluid losses begin to occur in the hole section of a well, and then converting this downhole pressure into its equivalent in drilling mud weight.

3 This problem can be countered, at great expense, through the use of dual-gradient systems, which divert mud from the drilling riser to separate riser return lines, replacing it with seawater.

4 *Hart's E&P*, April 2005 special supplement: 13–14. Mad Dog threw up another very nasty and unexpected surprise in the form of highly mobile tar intervals in the reservoir. The tar was encountered in one of the pre-drill programme wells, which not only ended up with three well bores but

also had to be sidetracked and temporarily abandoned.

5 *PON* 27 August 2004: 5. Anadarko's Marco Polo provides a less extreme example: at its peak (in October 2004), the field produced around 8 MBD less than expected. Further, production pressure held up for a disappointingly brief time (production started in July 2004, but output had declined to only 24 MBD by year end), probably because the field is more compartmentalised and geologically faulted than initially thought. The Marco Polo platform was designed to handle up to 120 MBD of oil and 300 MMCFD of gas, figures which will probably never be reached even if the development of all the discoveries around Marco Polo (Genghis Khan, K2, K2 North) come up to expectations. Similarly, Kerr-McGee's Gunnison development did not manage to come anywhere near its design capacity (40 MBD of liquids and 200 MMCFD of gas).

6 Smith 2002: 11. Safe drilling through shallow water-flow areas requires the use of additional casing strings and quick-setting foam cements, as well as sophisticated – and expensive – monitoring and measurement techniques.

7 *Hart's E&P*, July 2003 deepwater supplement: 15.

8 *Hart's E&P*, August 2000 deepwater supplement: 15.

9 The AT suffix differentiates this project from a Kerr-McGee development with the same name, located in Viosca Knoll block VK826, and henceforth called Neptune (VK). BP had already sunk USD 85 million in the Neptune (AT) prospect, drilling a well that it had declared to be non-commercial. As of the end of 2004, BHP had drilled seven wells.

10 *PON*, 1 July 2005: 3.

11 Oil companies consider that the complexity of GOM geology means that, on average, they will spend three times as much in finding and appraising a deepwater discovery in GOM in comparison to Angola, say (*PON*, 31 May 2005: 4).

12 McMahon 2004: 1.

13 Merrill Lynch 2002: 235.

14 *PON*, 23 June 2003: 1–4.

15 Merrill Lynch 2002: 235.

16 Hearing order: 91 FERC 61,182 (2000) for ExxonMobil Pipeline Co.

17 Conn and White 1994: 66.

18 Such facilities, under 1(b) of the Natural Gas Act (15 U.S.C. 717), are free from FERC's regulatory jurisdiction. The agency considers that any facility collecting gas at depths of 200 metres or greater is a gathering facility, 'up to the point or points of potential interconnection with the interstate pipeline grid' (Shell Gas Pipeline 74 FERC 61,896). From that point, the Commission applies a so-called 'primary function' test to determine (on the basis of a sliding scale that broadens the definition of gathering pipelines in terms of length and diameter as a function of the distance from shore and the water depth of the offshore production area) whether a given facility falls within its regulatory jurisdiction or not. The tests are explained in *Williams Gas Processing-Gulf Coast Company, L.P. v. FERC*, 1998; the sliding

scale in Amerada Hess, 52 FERC 61,268.

19 Turning Angolan deepwater gas into LNG will cost anything between USD 3 and 5 billion.

20 Weimer, Rowan, McBride and Kligfield 1998: 865.

21 At a price of USD 4/MMBTU, this would pay out a USD 60 million well in little over a year. BP expects to achieve flows of 50 MBOED with some of the wells that will tap the Thunder Horse field.

22 See Appendix 1.

23 See Appendix to Riddle, Snyder and George 2001.

24 As reported in *ibid.*

25 Merrill Lynch 2002: 224.

26 *Ibid.*

27 OTA 1985: 154.

28 McDonald 1979: 1; italics ours.

29 Merrill Lynch 2002: 224.

30 Derman and Johnston 1998: 9.

31 *Ibid.*: 8.

32 *Ibid.*: 9.

33 *Ibid.*: 1.

34 *Ibid.*: 5.

35 *Ibid.*: 8.

36 *Ibid.*: 7.

37 *Ibid.*: 8.

38 In subsequent analyses, Johnston always used the lower royalty figure. His latest estimates, whose objective is to assess *expected* project returns, set the royalty rate at zero to reflect the impact of DWRRA provisions on new investment.

39 See DOE 1991: 4.

40 See Appendix 1.

41 Riddle, Snyder and George 2001: appendix.

42 Marathon's so-called 'high impact' deepwater drilling programme in GOM, which started in mid-2001, experienced a 63 per cent failure rate. The company's development costs over the period 1996–2001 were about USD 12/BOE (*PON* 18 January 2002: 1–5). The drilling programme was completely rejigged during late 2002, but major successes continue to elude it.

43 See Johnston 2002: 11 (and also Johnston 1994).

44 As a Research Coordinator of the Shell Group put it, 'if you decide to buy this know-how, you will find yourself in the same queue as your competitors. Much of our research is done to keep ahead of our competitors' (Conn and White 1994: 63).

45 Ekofisk, for instance, was located in only 200 feet of water.

46 Pulsipher, Iledare and Baumann 1996: 1.

47 *Ibid.*: 2.

48 Currently Shell holds a 34.87% working interest in Cognac. Its partners in the venture are as follows: BP with 21.8%; Agip with 16.5%; Sonat with

10.7%, Texaco with 6.9%, Unocal with 4.7%, Murphy with 2.4%, Conoco Phillips with 1.2% and Koch with 1.1%.

49 Hendricks and Porter 1992: 510.

50 In a study carried out by the Louisiana Coastal Marine Institute on behalf of MMS, a number of officials from major oil companies were asked their opinions about AWL in a series of structured interviews. Unsurprisingly, they were very keen on AWL, and credited the policy with all sorts of benefits, ranging from 'justif[ying] extensive 3D use in the GOM and ... other technology purchases', all the way through to 'allowing ... [companies] to put together multiblock prospects', as well as making operations more convenient by exposing companies 'to a wider area ... [that] allows ... [them] to look at GOM in a regional context rather than a block by block basis' (Seydlitz, Sutherlin and Smith 1995: 73–4).

51 MMS 1983: 26.

CHAPTER 10

COMPETITION IN THE MARKET FOR GOM LEASES: THE GREAT CASUALTY OF AREAWIDE LEASING

The previous chapter established that the single most important contribution to the profitability of GOM deepwater operations has come from the dramatic reduction in signature bonus payments following the radical innovations to the fiscal framework introduced by the Reagan administration. These innovations, it is worth repeating, were aimed at putting downward pressure on entry costs, with the main designated beneficiaries of such a reduction being small and medium-sized firms. By means of these reductions, in turn, AWL was supposed to induce a significant expansion in GOM output, on the one hand, and to intensify the degree of competition prevailing in both the upstream sector and the market for offshore leases (through the elimination of regulatory burdens which were seen to be stunting the dynamism and entrepreneurial drive of American oil firms), on the other.

The sanguine expectations attached to AWL on the output front went unrealised (despite superficial appearances to the contrary), because up to the early 1990s little production was forthcoming from deepwater blocks. Thus far, though, not a great deal has been said about the effect of AWL on competition in both the market for offshore leases and the upstream oil industry, although the concluding remarks to the previous chapter do not appear too favourable. The discussion in this chapter seeks to ascertain, first of all, whether this negative preliminary impression is borne out by the facts. It will also try to explain the reasons behind any changes in the intensity of competitive forces at work in this market in the wake of the adoption of AWL.

10.1 The Official Verdict on AWL and Competition: a Critique

The general consensus in government and oil industry circles alike is that AWL has been a resounding success on the competition front.[1] In order to demonstrate its reinvigorating effect on competition, MMS, oil analysts and the oil industry alike always point to the post-1983 behaviour of two key indicators: the number of bids submitted and the number of tracts receiving bids.

AWL certainly had a beneficial effect on both these indicators in comparison to the equivalent figures recorded in the last sales held immediately before the adoption of AWL (Figure 10.1). However, these gains look rather less impressive when compared to the level of participation achieved in sales held during the late 1960s and early 1970s. But what truly undermines this conventional wisdom argument about the beneficial effects of AWL is that neither of these two indicators constitutes a sensible yardstick for measuring either the competitive soundness or the allocative efficiency of the offshore leasing programme. After all, the submission of a large number of bids for a similarly large number of tracts says absolutely nothing about whether an auction might have been driven by genuinely competitive (that is to say, adversarial) interaction between potential lessees. Consider the following. To say that AWL was good because of the bids received is analogous to a real estate agent telling a home owner/seller that he put the latter's house into a citywide sale and received a stupendous number of bids in the sale but, alas, only one for the seller's home. Is the owner supposed to believe therefore that there was an effective auction that tested the market for the value of his house?

An altogether sounder method for gauging the intensity of competitive rivalry consists in focusing on the number of bids drawn by individual items in auctions. In two studies on GOM leasing published in 1993 and 1994,[2] Walter Mead (a self-confessed admirer of AWL) argued that the fact that each tract leased over the 1954–1981 period drew an average of 3.3 bids provided strong *prima facie* evidence of the vigour of competition in the market for offshore petroleum leases. Unfortunately, Mead somewhat undermined the credibility of his positive assessment of AWL by not examining the figures for post-1982 sales, despite the availability of at least ten years' worth of additional statistics. Had he done so, he would have seen (as in Figure 10.2) that the average number of bids per tract leased in all AWL sales up to the end of 1992 was an unimpressive 1.38 (an average that remains essentially unchanged if the period under consideration is extended to the end of 2004). Furthermore, as Figure 10.3 shows, he would also have encountered plenty of AWL lease sales where the ratio of winning bids against total monies exposed has approached 90 percent. In such lease sales, winning bids were pretty much the *only* bids that MMS has received.

If one looks in greater detail at the lease sales covering the deepwater boom (Table 10.1), it is obvious that the blocks assigned in such sales were not subject to any bidding genuinely driven by inter-firm rivalry. It is most revealing to compare these statistics with data compiled for

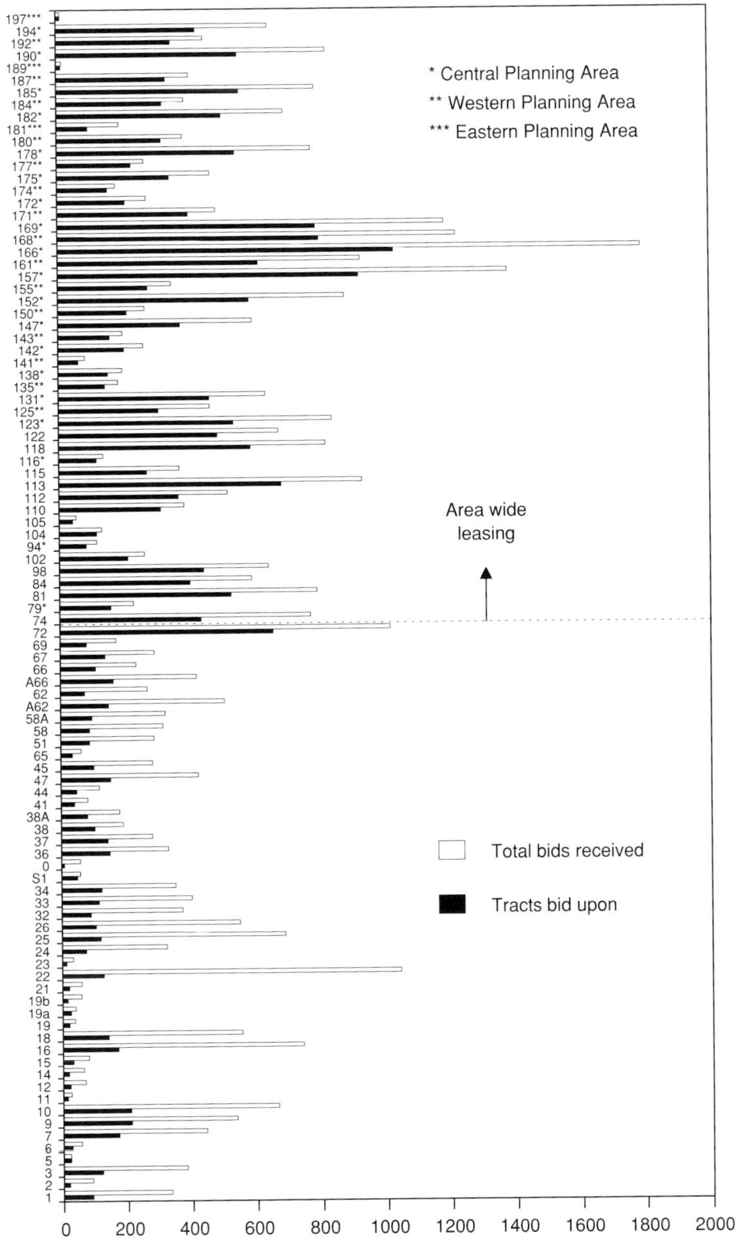

Source: MMS

Figure 10.1: Behaviour of Key Indicators in GOM Acreage Auctions, by Sale Number, 1954–2004

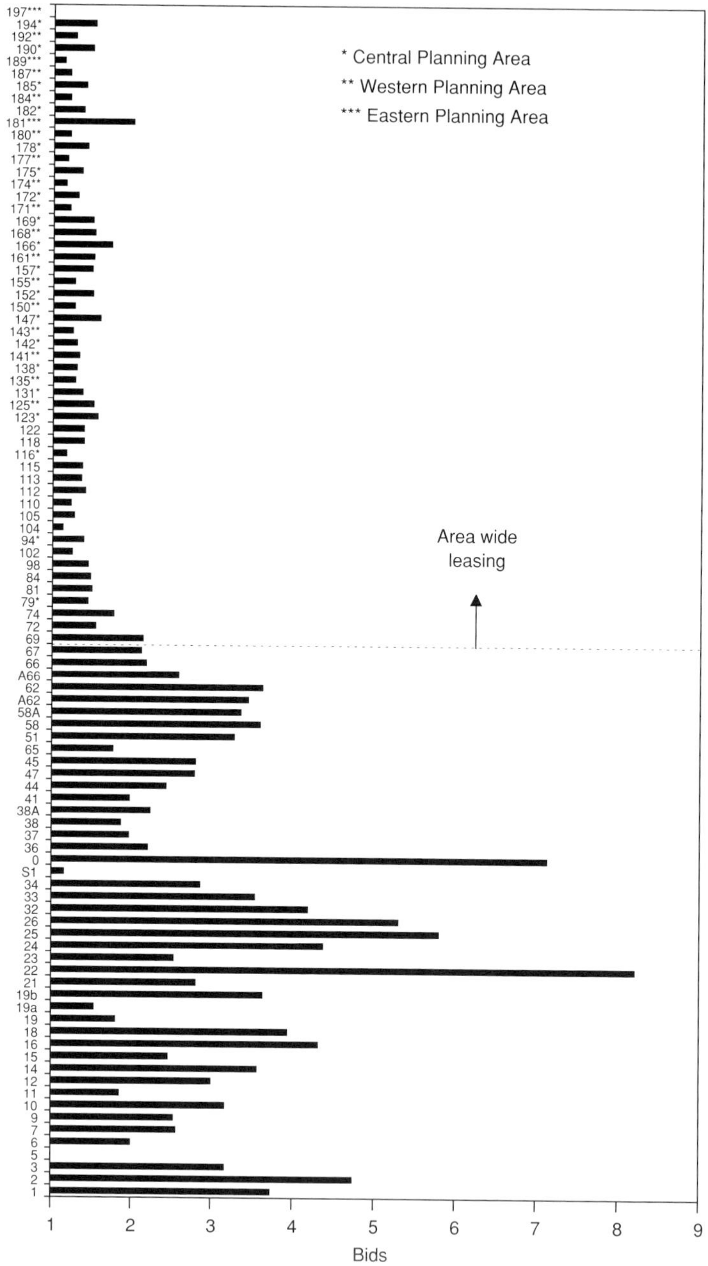

Source: MMS

Figure 10.2: Average Number of Bids per Tract Leased in GOM Federal OCS, by Sale Number, 1954–2004

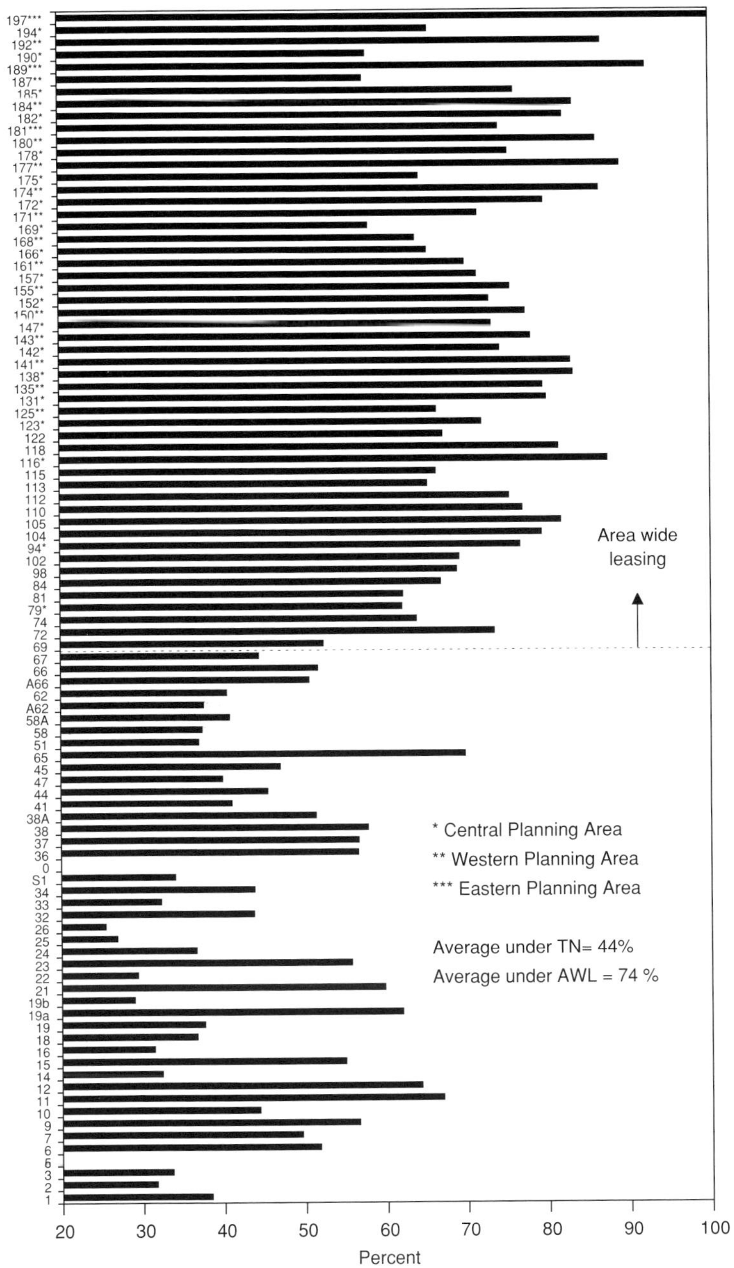

Source: MMS

Figure 10.3: Ratio of Winning Bids Against Total Monies Exposed in GOM Acreage Auctions, by Sale Number, 1954–2004

Table 10.1: Number of Bids Received per Leased GOM Block, by Sale Number, 1993–2005

Sale number	*Total blocks leased*	*Bids received* *1*	*2*	*3*	*4 or more*	*Average bids per tract*
157	924	647	174	64	39	1.49
161	617	444	89	56	28	1.51
166	1,032	637	209	95	91	1.73
168	804	563	137	66	38	1.52
169	794	596	120	36	42	1.50
171	402	347	40	5	10	1.21
172	207	166	27	7	7	1.31
174	153	135	15	1	2	1.16
175	344	262	59	16	7	1.36
177	226	197	21	6	2	1.18
178	547	398	93	35	21	1.43
180	320	272	38	6	4	1.21
181*	95	48	19	14	14	2.00
182	506	565	93	30	9	1.38
184	323	343	36	9	3	1.21
185	561	646	97	33	17	1.41
187	335	356	42	6	3	1.21
189*	14	15	0	1	0	1.14
190	557	667	97	42	22	1.49
192	346	352	49	14	6	1.20
194	403	301	75	26	26	1.52
197*	10	10	0	0	0	1.00

* Eastern Gulf Planning Area

Source: MMS

early-1970s vintage lease sales (Table 10.2), when DOI itself aired complaints in the sense that the rapid acceleration of leasing initiated by the Nixon administration had meant 'the fraction of tracts bid on that received six or more bids … declined rapidly and consistently from 38.5 percent in the 19 June 1973 sale to 9.6 percent in the 16 October 1974 sale'.[3] These complaints were echoed and expanded upon in a Congressional assessment of the premises and shortfalls of Nixon's accelerated leasing programme:

> while the area offered for bids nearly doubled, the average number of bids per tract (*a good measure of overall competition*) declined sharply from 5.3 bids per tract in the first sale of 1973 to 2.2 bids per tract on the last sale of 1974. The decline was accompanied by a considerable increase in the proportion of tracts leased on the basis of only one or two bids,

the level of competition identified by a Department of Interior analysis as being low enough to jeopardise the receipt of fair market value by the public. In the first sale of 1973, 37% of the tracts leased, representing only 9.3% of the bonus money accepted, received no more than two bids. But by the last sale of 1974, the fraction leased on the basis of only one or two bids had risen to 66.9%; more importantly, these facts represented 39.4% of the bonus money accepted.[4]

Table 10.2: Effects of Increased OCS Acreage Offerings on Aggregate Measures of Competition, 1973–4

Sale number	*Date*	*Blocks leased*	*Average bids per tract*	*Bonus per acre*	*Bids received (%) 1*	*Bids received (%) 6 or more*	*Bonus paid for tracts receiving 1 bid as % of total*
26	19/16/1974	100	5.2	2,908	24	38.5	3.8
32	20/12/73	87	4.2	2,804	23	31.5	1.8
33	28/3/74	91	3.5	4,968	20.9	22.7	5.7
34	29/5/74	102	2.9	2,605	40.2	14.6	17.8
36*	16/10/74	136	2.2	2,248	34.6	9	17.0

* Data exclude 10 tracts involved in a royalty bidding experiment

Source: OTA 1975b

If all of this constituted such a problem back in the days of the First Oil Shock, it seems legitimate to ask why, nowadays, nobody within MMS (or the American organs of government, for that matter) seems even remotely concerned by the fact that the number of blocks receiving six or more bids in Sales 157–192 has never consistently exceeded even one half of 1 percent, especially bearing in mind that blocks leased as a result of these sales account for 85 percent of the cumulative acreage assigned since the beginning of the offshore leasing programme!

In light of the above, it is no surprise to see that the vast majority of deepwater tracts assigned under AWL have attracted bids of USD 100 per acre or less, even after the province came of age during the 1990s and officially became the place to be in the petroleum world. Indeed, after 1983, there has never been a repeat performance of a situation like the one recorded in the lease sale of March 1974 (to cite but one example), when a block that received a record bid of USD 168.9 million also drew a further nine bids, three of which exceeded the USD 100 million (USD 34,000 per acre) mark! Out of the top ten blocks receiving the highest number of bids throughout the history of the offshore leasing programme, eight were put up for auction before the advent of AWL (Table 10.3). The post-AWL blocks that made it into this list are both shallow water blocks, and the winning bids that they attracted are but a fraction of the value of the high bids for the other blocks on this list. Interestingly, commercial accumulations of hydrocarbons have been found in only one of the blocks on this list (EC371), and then only in modest volumes (cumulative production in this block as of 2004 is slightly more than 6 MMBOE).

Table 10.3: Top Ten Highest Number of Bids on a Single Block for All Lease Sales in the GOM Federal OCS

Rank	*Number of bids*	*Block number*	*Sale Area*	*Date (MMUSD)*	*Planning*	*High Bid*
1	16	PL5	45	4/25/1978	Central	55.19
2	16	HIA555	31	6/19/1973	Western	68.58
3	16	HIA317	31	6/19/1973	Western	77.71
4	15	GA313	122	8/23/1989	Western	6.11
5	15	WC248	113	3/30/1988	Central	18.22
6	15	HIA322	31	6/19/1973	Western	31.28
7	15	HIA273	31	6/19/1973	Western	44.62
8	15	HIA572	31	6/19/1973	Western	45.03
9	15	SA13	51	12/19/1978	Western	93.89
10	14	EC371	30	12/19/1972	Central	30.17

Source: MMS

In the opinion of James Watt, the Secretary of the Interior responsible for AWL, the way in which DOI handled access to acreage up until 1982 (predicated as it was on short-term budget needs) certainly served to prop up acreage prices. But Watt claimed that, at the same time, it stunted competition and unnecessarily curbed OCS production (thereby making the USA's economic prosperity a hostage to the intolerable whims of OPEC). Clearly, since the aim of AWL was to remedy some of the negative consequences arising from the DOI's alleged fixation with acreage prices, one cannot condemn that policy on the basis of the negative effects that it had on this particular variable. Rather, AWL has to be evaluated on Watt's own terms, by looking at other indicators of the vitality of competitive forces in the market for GOM offshore leases, before and after its introduction.

10.2 The Market for OCS Acreage: from Functional to Genuinely Cursed?

James Watt relentlessly pilloried his predecessors at DOI for forgetting the cardinal rule of the energy game in America; namely, that 'if you are interested in consumers, you want to deliver energy to them. If you are interested in national security, you want energy, and energy comes about through competition, not through restricting supply'.[5] Watt was not averse to taking personal credit for having supposedly re-introduced the rigours and discipline of competition to the offshore leasing game. Unfortunately for his grandiloquent claims, any meaningful indicator that one cares to examine strongly suggests that competition in the auction market for GOM offshore leases in fact declined grievously after 1983 (Figure 10.4).

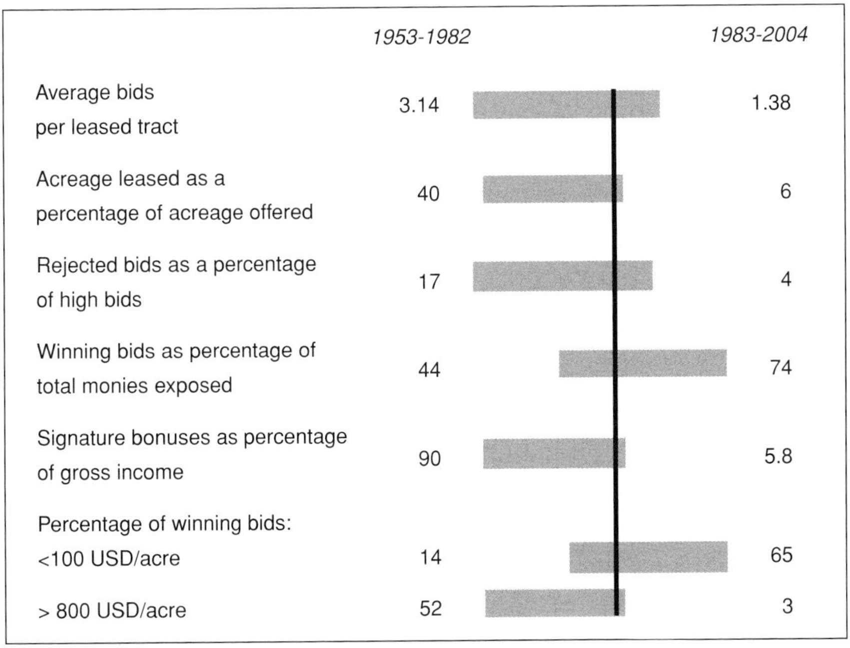

Source: MMS

Figure 10.4: Indicators of the Intensity of Competition in the Market for Leases in the GOM Federal OCS

As early as 1985, GAO ventured the suggestion that 'the increased pace of offshore leasing through the area-wide programme decreased competition (in terms of the number of bids received for each tract) and reduced government revenues (in terms of the amount of high bids for individual tracts)'.[6] Its statistical analysis established that, throughout the life of the OCS leasing programme, each additional bid that MMS received for a given tract on average translated into an increase of USD 1,082 per acre in the amount of the high bid submitted for that tract (i.e. USD 6.2 million for a typical tract).[7] Unfortunately, during the period covered by the first ten areawide sales, the overall number of bids per tract declined from 2.44 to 1.65; in the GOM region, the average number of bids received per tract fell even more steeply, from 2.67 to 1.56.

Not a great deal has changed in this regard over the lifetime of AWL, as confirmed by a recent study undertaken by the Centre for Energy Studies at the Louisiana State University on commission from MMS. This study (whose broad objective was to ascertain whether 'consolidation of ownership and control in the petroleum industry [has] reduced the extent of competition for or lowered the value of oil and gas leases in the Gulf of Mexico OCS'[8]) states that, 'on average, the proportion of leases with just one bid from 1983 to 1999 was 75 percent. The proportion of leases receiving at least three bids was less than 10 percent during this period ... [while] nearly 70 percent of leases had three or more bids during the period 1969–73.'[9] Indeed, the average 'number of bids per lease was 2.72 for 1954–1966 and 3.90 for 1966–77 ... [while] the average number of bids per tract over the period 1954–1973 for the entire US OCS [was] 3.56'.[10]

The remarkable thing, though, is that the overall decline in competition in the market for offshore leases described in the GAO and LSU studies coincided with the lowering of the entry barrier constituted by steep acreage prices. This result flies in the face of the well-established notion in auction theory and design literature that even modest bidding costs constitute a serious deterrent for entry in capital-intensive industries (oil, telecommunications) underpinned by competitive auction processes.[11]

Ever since the start of E&P activities in GOM, the much higher costs of offshore operations had dampened the wildcat spirit characteristic of non-major players in US onshore basins, making them very 'cautious about expensive and unproductive experimentation'.[12] This caution was readily detectable in the empirically ascertained fact that wealthier firms ascribed higher *ex ante* values to OCS leases, and also required smaller rates of return than less well-off firms.[13] Despite this, before the coming of AWL, small and medium-sized companies used to participate enthusiastically in offshore acreage auctions. Indeed, the theoretical deterrent effect of high bonuses notwithstanding, by the early 1970s companies from outside the ranks of the majors (ranging from small E&P outfits to very large independents) were contributing upwards of 50 percent of total bonus payments, a figure significantly in excess of their share in the total market capitalisation of the US oil sector. This degree of participation is all the more remarkable when one considers that, in the days before AWL, the market for GOM Federal

OCS leases was seen as a prime example of a market in thrall to a condition known as the winner's curse.

Although the earliest formal description of this phenomenon saw the light of day in 1969 (in a paper that coincidentally presented one of the first published treatments of what is now called a common value bidding model), the term 'winner's curse' itself is indissolubly associated to a seminal 1971 article penned by three engineers working for Arco. The direct motivation for the article came from the 'rather careful look' that several major oil companies had had into 'their record and that of the industry in areas where sealed competitive bidding is the method of acquiring leases ... [with] the Gulf of Mexico ... [being] the most notable of these areas, and perhaps the most interesting'.[14] In the particular case of Arco, the event that prompted the 'careful look' had been an acreage sale where Atlantic Refining (one of the two companies that merged in 1966 to create Arco) 'bought virtually everything it had bid on at a Gulf lease sale', whereupon the company found itself not only 'in a budgetary bind for a couple of years' but also in the market for 'some operations research-type solution'[15] that would hopefully prevent a recurrence of this unfortunate experience.

In auction markets like that for OCS blocks, the winner's curse 'reflects the danger that the winner of an auction is likely to be the party who has most greatly overestimated the value of the prize', a danger that looms more ominously in circumstances 'when bidders have the same, or close to the same value for a prize, but they have different information about that actual value'.[16] It is important to note that the winner's curse preys on the minds of large and small bidders alike. However, small bidders are especially wary of it, out of recognition of the fact that 'they are only likely to win when they have overestimated the value by more than usual'. In contrast, larger bidders can to a certain extent afford to be less cautious, 'since beating very cautious opponents need not imply one has overestimated the prize's value'.[17] In practice, and other things being equal, 'the winner's curse affects weak firms much more than strong ones', and *in extremis*, it can conceivably deter all entry by weaker bidders, thereby becoming self-reinforcing: 'the advantaged bidder wins most of the time ... and because its rivals bid extremely cautiously [or not at all], it also generally pays a low price when it does win'.[18]

The winner's curse can manifest itself in two distinct guises that need not occur in unison, but can and often do. The first one is an absolute measure of dead economic loss (and hence can be called the strong version of the curse): the negative difference between whatever sum was offered for a tract, and the real (lower) value of that tract (i.e. the market value of the production eventually obtained from it, minus exploration, development and lifting costs, as well as royalties and taxes). The second one is a relative measure of foregone profitability (and hence can be called the mild version of the curse): the amount that a winner offered over and above the lowest possible winning bid. This is usually referred to as 'money left on the table' (henceforth MLOT) and is often expressed in ratio form as the absolute difference between the highest and second highest bids, divided into the former. Clearly, the worst possible situation for a company involves the submission of a bid where both aspects of the winner's curse put up an appearance.

It should be noted, in passing, that the winner's curse is usually seen as arising in situations where there are multiple bids. As Capen, Clapp and Campbell expressed it in their original treatment of the issue, 'if one wins a tract against two or three others he may feel fine about his good fortune. But how should he feel if he won against 50 others? Ill.'[19] Having said that, the more malign version of the winner's curse – overbidding relative to the value of the prize – can just as easily occur in the absence of any competition whatsoever (and, if there is anything worse than beating fifty others to a prize, that is surely overpaying for an item that nobody else thinks is worth having).

The generalised (but by no means universal) opinion was that, in the days before AWL, the winner's curse affected GOM acreage auctions in a manner that was both pervasive and intense. However, its effects were seen as being felt primarily at the level of MLOT and lower than expected profitability, rather than at the more ruinous level of excessive payments for insufficiently valuable tracts. The attention vouchsafed to this phenomenon (especially from within the ranks of oil companies), was traceable to the impression that even though there was 'a lot of oil and gas in the region, the *industry ... [was] not making as much on its investments as it intended*',[20] rather than to a feeling that it was actually *losing* money through its E&P activities in the region. In fact, the only E&P proposition seen as a consistent money loser was the purchase of blocks attracting a really high number of bids (Hendricks, Porter and Boudreau, for example, found that net profits for blocks receiving seven bids or more were negative, even though the tracts themselves proved quite valuable[21]).

In 1995, Porter calculated the incidence of the winner's curse for 2510 OCS blocks auctioned between 1954 and 1979 (Table 10.4). Porter's analysis indicated that while both winning bids and MLOT behaved as increasing functions of the number of bidders, the ratio of money left on the table to the winning bids was a decreasing function of the number of bidders. According to his calculations, MLOT amounted to an average value of 44 percent of the winning bid on the 1608 tracts that received two or more bids and, at 30 percent, MLOT was quite high even for the 180 tracts that received ten or more bids.

Porter's calculations have been replicated in Table 10.4 for tracts leased under AWL rules in sales for which data on all bids submitted – as opposed to just the winning bid and the sum of all bids – is available online (Sales 157–192). At first sight, the results of this exercise seem to imply that AWL sales have seen a marked recrudescence of the winner's curse, as reflected in MLOT. In actual fact, while MLOT did go up in *relative* terms in all bid categories bar the last one, it is clear that the winner's curse has been less strongly felt under AWL rules. For one thing, the difference between the weighted averages for the highest and second highest bids was 90 percent lower for AWL sales and that in *nominal* – not even real – terms. Even more importantly, the percentage of blocks receiving only one bid, for which MLOT was

Table 10.4: Incidence of the Winner's Curse in GOM Federal OCS Acreage Auctions, by Period

	Number of bidders per tract						
	1	*2*	*3*	*4*	*5-6*	*7-9*	*10-18*
1954–1979 period							
No. of tracts receiving bids	902	463	255	212	264	234	180
Average maximum bid (MMUSD)	1.283	2.667	4.07	5.523	7.871	14.103	21.778
Average second highest bid (MMUSD)	-	1.2028	2.0757	2.9824	4.8328	9.3644	15.2881
(Highest bid – Second highest bid)/ Highest bid (%)	-	0.549	0.49	0.46	0.386	0.336	0.298
1996–2005 period							
No. of tracts receiving bids	6,760	1,450	542	190	125	38	8
Average maximum bid (MMUSD)	0.51	0.998	1.523	2.974	4.515	6.090	13.690
Average second highest bid (MMUSD)	-	0.335	0.588	1.297	2.614	3.502	10.661
(Highest bid – second highest bid)/ Highest bid (%)	-	0.66	0.61	0.56	0.42	0.43	0.22

Sources: MMS, Porter 1995: 6

not even a consideration, increased from 36 percent in the sample for 1954–1979 to 74 percent in the 1996–2004 sample.

These results are in accordance with the thrust of Saidibaghgandomi's research on the effect of uncertainty on OCS bidding, which established that, 'generally, when for the first time an area was introduced for auction, the average bid per acre was lower than the average bid per acre for the second sale in the same area'.[22] They also tally with the findings of a paper that compared bidding behaviour across two OCS sales held *after* Nixon accelerated leasing and *after* Capen, Clapp and Campbell published their paper, but *before* AWL came on the scene: the May 1974 Texas Gulf sale, and the August 1976 Mid-Atlantic OCS (Baltimore Canyon) sale.[23] Both sales offered acreage that had never come into play before, but whereas the former featured an area with whose geology the industry was well acquainted, the latter involved a very challenging frontier play. In the face of extreme uncertainty, bidders in the Baltimore Canyon auction adjusted their bids downwards, in a manner that appeared consistent with recognition of the possibility that the winner's curse might strike. In contrast, no such correction was detectable for the GOM acreage, which was much less of an unknown quantity for the industry.

In a more wide-ranging survey of GOM lease sales held between 1974 and 1976, Smith confirmed that industry-wide knowledge of the winner's curse phenomenon had nonetheless not given rise to 'widespread use of non-aggressive bidding strategies of the type advocated in an article by Capen, Clapp and Campbell'.[24] Smith's characterisation of their article as 'advocacy' is entirely apposite, as its publication was the outcome of a deliberate corporate decision taken at the end of a process of discussion and introspection. A recent tribute to Capen's lifetime achievements in the oil industry recounts that, initially, 'Atlantic was careful in releasing information about its system ... [with] the lawyers and some in management ... reluctant to share the company's strategy.' However, the process of adjusting bids downwards at Arco, as Capen himself recalled, had on occasion led to 'bids ... so low we didn't buy anything'. The prospect of ongoing suboptimal behaviour by other participants (i.e. overbidding through ignorance) led Arco's research and development personnel to grasp the 'obvious advantages of telling the whole world' about the winner's curse.[25]

Although Arco's vice-president for R&D had to acknowledge that the publication of the theoretical intuitions underlying the company's valuation strategy amounted to 'legalised collusion',[26] management thought that the potential prize was well worth it: 'if everyone lowered their bids to protect from the curse, the entire industry would be better off … The sellers would suffer a reduction in bonuses … [but they] were doing far better than they should have.'[27] In the end, this pragmatic position carried the day, and Capen was initially 'allowed to give oral presentations' (including one to an overflow audience at the 1970 Society of Petroleum Engineers annual meeting) and was eventually authorised to submit the paper to a trade journal with a wide circulation.[28]

Arco's line of reasoning in deciding to go public with the results of the Capen, Clapp and Campbell model seems impeccable. As Richard Thaler puts it:

> suppose you are Capen and his colleagues and you have figured out the winner's curse. You now have an advantage over other oil firms. How can you exploit your new competitive advantage? If you react by optimally reducing your bids, then you will avoid paying too much for leases, but you will also win very few auctions … Unless you want to switch business, this solution is obviously unsatisfactory … A better solution may be to share your new knowledge with your competitors, urging them to reduce their bids as well. If they believe your analysis, then the game can be profitable for the bidders. This, of course, is exactly what Capen, Clapp and Campbell did.[29]

They did it, moreover, in a highly competent manner, aiming for and obtaining maximum exposure for their findings. In a very short space of time, the entire industry duly became acquainted with the potential savings from Arco's bidding strategies, the winner's curse became a household word, and Capen's model spawned a host of imitations at other oil companies.[30] And yet, remarkably, the market for GOM offshore leases proceeded to confound everybody's expectations, *and continued to behave pretty much as it had done during the previous twenty years*.[31]

According to Saidibaghgandomi, Arco's efforts at legalised collusion actually did pay off: 'despite the increase in the real oil prices during the 1970s and the deregulation of the oil and gas industry … the average bid per acre declined during this period … This finding suggests that bidders may have responded to the suggestions made by Capen, Clapp and Campbell that the bidders should adjust downward their bids in order to avoid the possibility of "winner's curse"'.[32] The validity of this conclusion, however, is undermined by the fact that the econometric analysis underlying it failed to control properly for the total extension of acreage offered up for sale in post-1973 lease sales. Actually, the decline in acreage prices on a unit of surface basis over the period mentioned above, while real enough, was *entirely* attributable to the accelerated leasing programme (between 1964 and 1973, the average rate of leasing was about 0.5 million acres per year, but by 1974 the rate had gone up to 1.8 million acres[33]). In other words, despite the increasing theoretical sophistication behind the bidding strategies of prospective OCS lessees, what reduction in unit acreage prices they managed to obtain in the years after the First Oil Shock came entirely from the significant increase in the extension of acreage offered (a fact not lost on those who were to lobby in favour of AWL at a later stage).

As has been stated before, only a few of the analyses carried out on the acreage auctions held from 1959 until the late 1970s, inclusive, saw the strong version of the winner's curse as having affected such sales to any great extent. Among them, however, one can cite a study led by the influential Walter Mead. This particular study calculated the rate of return on 1233 GOM leases assigned between 1954 and 1969, and found that lessees had suffered an average present value loss of close to USD 200,000 per lease, using a 12.5 percent discount rate (a rather high figure, and above the weighted average cost of capital for the oil industry as a whole at that time).[34] However, Hendricks, Porter and Boudreau called these findings into question. These authors scrutinised the same data set used by the Mead group, albeit on the basis of different assumptions, and concluded that the leases had been profitable, albeit not inordinately so.[35] In their view, the worst that could be said about the leases was that they had not proved nearly as profitable as winning bidders had expected them to be (or, to put it another way, that the auctions where they had been sold had been affected by the mild – rather than the strong – version of the winner's curse).

The marked differences between the positions above beg one obvious question: which of them is more likely to be correct? Choosing between these divergent viewpoints on a strictly quantitative basis is difficult, though, because of the various alternative ways in which obscure factors that were of great relevance to the profitability of the oil industry in those days (the relationship between posted and transfer prices, the oil depletion allowance, to name but two) can be incorporated into the calculations. In theoretical terms, though, the choice is simpler. To the extent that it is possible to leave aside the crucial question of whether the discount rate that Mead chose was the right one, accepting his position that the US oil industry lost money on every single tract leased over this period is tantamount to saying that the management and the shareholders of US oil firms consistently and systematically overlooked the fact that their companies were throwing billions of dollars down the drain, for the space of a decade. It also implies accepting that deep and liquid financial, products and labour markets went along with this,[36] and that US oil firms were making so much money elsewhere that they were able to compensate for their apparent profligacy in GOM auctions (recall that the 1960s is a decade remembered as a golden age of sorts for the US oil industry)

In contrast, the implications of the conclusions of Hendricks et al. are less problematic: the returns the industry made were good enough to continue attracting capital into offshore E&P activities, while the US Federal government managed to collect nearly all the available economic rent through signature bonuses. In a nutshell, what Hendricks et al. are obliquely saying is that during the period under consideration, the market for offshore GOM leases *worked*.[37]

This last assertion is not immediately reconcilable with the widely held notion that the US government was 'doing far better than it should have' as far as the leasing of GOM acreage went. So how was it, then, that the industry came to believe so strongly that the Federal government was fleecing it? The answer is surprisingly straightforward: even though during the 1954–69 period offshore finds were on average much larger than onshore finds, offshore rates of return tended to be lower than onshore rates of return (after due allowances for differences in costs). What the industry failed to appreciate is that this anomaly was due to the fact that its onshore lessors were for the most part private landowners[38] who had to transact in the context of extreme informational constraints, thereby giving oil companies plenty of scope to realise windfall gains (H.L. Hunt, for instance, managed to end up as the world's richest man chiefly by virtue of his dealings with many hundreds of East Texas landowners).

Up until 1983, then, the price outcomes generated by the market for GOM offshore leases were less subject to the winner's curse than is commonly assumed. Since it has also been shown that MLOT shrank drastically after this date, it might seem reasonable to conclude that the adoption of AWL led to the virtual disappearance of the winner's curse from this market. In fact, the exact opposite is true: the winner's curse has become more deeply ingrained and, to make matters worse, it has done so in its more virulent form, chiefly thanks to the desperate efforts of some second tier majors or first tier independents to gain a belated foothold in the GOM deepwater.

These conclusions are derived from the findings of Pulsipher, Iledare and Mesyanzhinov, who looked at the normalised standard deviation as a percent of the mean for high bids submitted in lease sales held over the 1983–1999 period, calculated for five different – albeit not mutually exclusive – categories: shallow water bids, deepwater bids, competitive bids (i.e. two or more bidders), non-competitive bids, and all bids.[39] They calculated an overall score in each category for every individual participant (with a score of zero meaning that the average high bid for the firm is the same as the average of all the firms in the distribution, a score of +1 indicating that the average bid is one standard deviation above the mean, a score of −1 indicating that it was one standard deviation below the mean, and so on). A high positive score was interpreted as being indicative of a tendency to overbid (i.e. to overestimate the value of the leases on offer relative to what everybody else was prepared to pay at a sale), while a low score was taken as a knack for identifying bargain acreage.

Pulsipher, Iledare and Mesyanzhinov found that the number of firms bidding higher or lower than the norm was quite small. Only one firm, Zilkha Energy, could be called a consistently successful bargain-seeker, posting low scores in the shallow water, competitive, non-competitive and all bids categories. Zilkha shareholders could be excused for seeing this as very good news. In reality, though, the apparently canny Zilkha did no more than succumb to the strong version of the winner's curse, paying admittedly bottom-dollar prices but only for bottom-drawer acreage: although the company obtained 348 leases over the 1983–1999 period (the sixth highest total in GOM), it came in at a disappointing 37th place in the production department, which is where any value creation in E&P activities ultimately has to take place. It is quite interesting to note that, despite Zilkha's underwhelming record at translating cheap leases into output, in January 1998, Sonat purchased Zilkha for USD 1 billion, in an all-stock transaction. Almost two years later (October 1999), Sonat itself was purchased by El Paso in a deal valued at USD 6 billion. Subsequently, El Paso managed to avoid bankruptcy only by the thinnest of margins, and it had to take hefty write-downs on the Sonat deal. This is as good an illustration as any that, from the 1980s onwards, the activities of US-centred (and hence opportunity-constrained) E&P companies seem to destroy shareholder value on a consistent basis, and only manage to recoup some value when these companies are sold off to an over-optimistic buyer. The way in which AWL hamstrung their capabilities to compete on an equal footing with larger companies played a not unimportant role in this process.

On the other side of the bidding spectrum, the panorama looks even grimmer (Table 10.5). The study found six firms which bid significantly more than the average acreage value, on a consistent basis, in more than one category. Tellingly, these firms (Marathon, Anadarko, Statoil, Elf, Occidental and Kerr-McGee) are largish concerns, and all of them latecomers to the deepwater party. In their desperation to shore up flagging E&P portfolios and plug the holes left in them as a result of

Table 10.5: Bidding Performance for GOM OCS Blocks by Selected Firms, 1983–1999

	Standard Deviations as a Percentage of the Mean, by Category				
Company	*All Bids*	*Shallow Water*	*Deep Water*	*Non–competitive*	*Competitive*
Overbidders					
Marathon	3.20	1.15	1.32	4.29	1.79
Anadarko	2.96	5.55	-0.60	0.66	4.14
Statoil	2.84	–	0.79	4.08	1.44
Elf	1.94	0.14	0.75	1.77	1.66
Occidental	1.91	-1.37	0.65	0.22	1.58
Kerr–McGee	1.52	2.35	0.10	2.27	1.05
Unocal	0.63	-0.79	0.30	0.39	2.00
Pogo	-0.13	-0.46	1.78	-0.36	-0.16
Seagull	-0.45	-0.77	5.63	-0.19	-0.21
Underbidders					
Mobil	-1.01	-1.17	-0.64	-0.74	-0.76
Coastal	-1.09	-0.76	-0.74	-0.95	-0.77
Pennzoil	-1.09	-0.84	-0.66	-1.17	-0.84
Newfield	-1.13	-0.81	–	-1.38	-0.85
CXY	-1.22	-0.99	-0.29	-1.14	-1.18
Zilkha	-1.64	-1.34	-0.94	-1.87	-1.29

Source: Pulsipher, Iledare and Mesyanzhinov, 2003

their lack of deepwater exposure, these companies clearly succumbed to the temptation of bidding themselves out of trouble, a risky and costly strategy that most of them could ill afford.

The most deluded in this regard seems to be Marathon, whose all bid score is 3.2 standard deviations over the average. An even better measure of Marathon's counterproductive desperation to make up for lost ground is the fact that its bid score in the non-competitive category was a dismal 4.29 standard deviations over the average. Given that Marathon's exploration deepwater programme has been a costly disappointment to date it is obvious that one is looking at a very acute case of the winner's curse in action. Hot on Marathon's heels in the lousy bidding department are Anadarko and Statoil, with all bid scores of 2.96 and 2.85 standard deviations over the average, respectively. Anadarko's score is attributable mainly to its exuberant bidding in the shallow water category, itself the result of its pursuit of shallow subsalt prospects during the early 1990s. Kerr-McGee also demonstrated a certain knack for submitting relatively high bids for shallow water acreage. Both this company and Statoil, in turn, showed themselves to be almost as good as Marathon in the unenviable art of submitting high bids for acreage that nobody else particularly coveted, with Elf only slightly behind them. In the deepwater category, the most notable over bidder, by some distance, was Seagull Energy (at 5.63 standard deviations over the average), distantly followed by Pogo. Finally, Anadarko, Unocal, Marathon, Elf, Occidental and Statoil, in that order, were the firms that submitted the most overvalued bids for blocks on which more than one party made a move.

Up until 1983, 'all measures used for money left on table … had a positive association with the productivity of plots. In other words, money left on table … was significantly higher on productive than on non-productive plots'.[40] However, the rather unfortunate bidding and production record of the firms named above strongly suggests that, at least as far as the deepwater is concerned, the reverse is now probably true. In other words, after 1983, the market for GOM acreage has metamorphosed into a market that is well and truly cursed. In the process, American independent producers have been relegated to tidying up the scraps falling off the majors' table, not necessarily because they were less technically competent and efficient than the majors, but rather on account of their limited capabilities to warehouse and manage offshore geological risk as part of a flawed bidding process.

Nowadays, many observers insist 'independents are slowly gaining more of a foothold as primary leaseholders, particularly in the shallower sections of the deepwater. Increased activity from the independents in deepwater is expected in the future as a result of the recent mega-mergers … [as the merging companies] decide which assets to keep and which to let go. When decisions have been made and some of the properties are dropped it seems likely that more independents will move in.'[41] It certainly cannot be denied that some of the larger independents are doing reasonably well in the deepwater, and the risks of deepwater E&P are much better understood. But for every independent that has managed to establish a position in the deepwater, there are many more whose only hope for participating in the bonanza involves paying large bonuses that they can ill afford, as a first step to mount increasingly desperate exploration gambles. Indeed, despite the fact that bonuses are still very low overall (especially if measured on a surface basis), the generalised perception among industry analysts is that less advantaged players on the whole tend to 'overpay for poor quality acreage, and create balance sheet issues at a group scale'.[42]

The following section shows that responsibility for the plight of the independent sector should be laid squarely on the doorstep of the designers of AWL, and their blatant disregard for some of the key insights of the economics of imperfect information. This disregard is made all the more incomprehensible by the fact that, when the policy was formulated, this theoretical furrow had already been very ably ploughed (George Akerlof's seminal article on the economics of imperfect information came out 12 years before the launching of AWL).[43] And it is made all the more inexcusable given the logical incompatibility of AWL's stated twin goals on the competition front: reducing the magnitude of bonus bids, on the one hand, and encouraging the submission of more bids by more companies, on the other. After all, as Moody observed: 'if the winning bid increases with the number of bids, then increasing the number of bids by encouraging more companies to bid will cause the winning bid to rise'.[44] Therefore, the only way in which AWL could have worked as advertised would have involved quite the balancing act: reducing the winning bid while holding the number of bids constant *and at the same time* increasing the total number of bidders.

10.3 The Relationship between Signalling Devices and High Upfront Costs

The abiding merit of the pre-AWL leasing procedures resides in how effective they were in counteracting the risk aversion and the vulnerability to the winner's curse of small- and medium-sized companies. Often, this allowed such relatively disadvantaged players to bid on tracts for which they had no independent data, in the expectation rather than the hope 'that resources would be found later'.[45] Alternatively, it offered them the possibility of being farmed into more daunting projects by larger companies wishing to offload some of their upfront costs ('water depth was a strongly significant inducement to smaller shares' in bidding consortia in pre-AWL days[46]). The following passages discuss the finer points of how exactly this difficult trick was achieved.

Under the leasing procedures in use up to 1983, DOI used to issue a call for nominations, in which it requested oil companies to identify promising tracts within an OCS region (this is the reason why the procedures were known under the name of Tract Nomination, henceforth TN). After evaluating the nominations that industry submitted, DOI would unilaterally decide which tracts to offer, on the basis of 'the past leasing history of the area, economic and environmental considerations, multiple-use conflicts, and the estimated potential of the sale area'.[47] Importantly, the mere fact that a tract had been nominated did not oblige DOI to offer it in a tract sale. Indeed, DOI had the right to withhold from offer any tracts lying in areas about which the department felt that it had insufficient knowledge, *a faculty that gave oil companies interested in obtaining such tracts a strong incentive to remedy DOI's ignorance*. In doing so, however, they could not avoid remedying, at least in equal measure, the ignorance of potential rivals for tracts.

This particular feature of the OCS leasing system rose to prominence after the quasi-mythical sale of 1962. As Priest points out, this sale 'brought an end to lease sales where most tracts nominated, with a few exceptions, were offered'.[48] But despite the fact that the BLM had proven a relatively 'soft touch' in the 1962 sale, and regardless of the key role that Shell played in energising the offshore leasing programme thanks to its involvement in the 1960 sale, this company (which enjoyed 'a special position in the otherwise young and still rather unstructured offshore sector ... [as] the main contributor to the development of offshore technology'[49]) failed to secure some of the acreage it coveted the most in the 1962 sale, simply because this acreage was found in blocks 'lying in 300 feet of water *where Shell [had been] the only bidder*'.[50]

According to a top engineer in Shell's deepwater programme, the company grasped then and there 'that the only way we could ever have access to those frontier areas was to share our knowledge with the rest of the industry, to give them a base of technology from which they could expand'.[51] And it was clearly this realisation that prompted Shell to hold an unprecedented three-week course on offshore technology for its contractors, its competitors and leasing authorities, a course that was to go down in oil lore as 'the million dollar school of offshore technology'.[52] The diffusion of drilling and production technology that this seminar put in motion culminated, in 1969, in the celebration of the first Offshore Technology Conference (OTC), an event that has since become an important annual fixture in the international oil industry calendar.

After 1962, 'the BLM and the USGS Conservation Division ... [became] more rigorous and scientific in its [*sic.*] approach to evaluating and leasing tracts',[53] with the result that these agencies' independent estimate of the value of each and every tract (an estimate that was never made public) turned into the *primary criterion* for the acceptance or rejection of individual bids that DOI received. In other words, this estimate had precedence over both the magnitude of the highest bid and the amount by which the winning bid exceeded the second highest bid. Importantly, DOI was supposed to arrive at this estimate on the basis of proprietary geological, geophysical, engineering and economic data that companies submitted to it. Their motive to do so, in turn, was to increase the likelihood that the blocks they were most interested in would actually feature in future lease sales. Thus, the mere fact that DOI was offering a tract constituted a signal to less advantaged players not only that some other company had seen something sufficiently interesting in the tract to submit valuable proprietary data to DOI, but also that the agency itself had formed a reasonably positive impression about the tract's prospectivity.

DOI's independent estimates of tract value served to neutralise the incentive that players who were informationally advantaged had to submit low bids for acreage. DOI's estimates effectively functioned as reservation prices, so companies attempting to lowball the leasing authorities faced a tangible risk of not obtaining the acreage they desired: between 1954 and 1982, DOI rejected about 17 percent of all high bids for GOM tracts as unsatisfactory (Figure 10.5).[54] Moreover, according to the Small Business Committee of the US Congress, on those occasions when DOI rejected a high bid for a block, it obtained an average of 13 times more for it the next time it put it up for auction.[55] Of course, the risk that companies faced in not obtaining coveted tracts could be significantly lowered the more closely company bids reflected both their true estimate of the tract's worth and their cost advantages. For instance, this explains why, in the record-breaking GOM sale of March 1974, and *despite the fact that it faced no competition for South Timbalier Block ST26* – a drainage tract adjacent to an area where it had substantial production – Shell decided to put in a bid of USD 65.8 million for it (the only other bid submitted was a lowly USD 834,000). As the *Oil and Gas Journal* correctly pointed out at the time, Shell submitted what turned out to be the second highest per-acre bid of the whole sale (USD 26.3 million) in order to meet what it considered would be DOI's estimate of the value of the block.[56]

Independent tract valuation also functioned as an effective deterrent against collusion. After all, even if bids for different tracts were the outcome of inter-firm agreements, the colluding bidders would at the very least have to exceed DOI's independent estimates of the value of tracts, if they were to obtain the desired leases. In addition, the fact that these estimates of value were unknown to bidders introduced an element of uncertainty that, again, encouraged companies to submit bids that genuinely reflected their ideas about the prospects of tracts. What is more, members of a bidding ring could never be sure that companies that were not part of it would not submit higher bids; indeed, they could not even be sure that companies who were part of the ring would not try to convince the others to submit a lowball bid, only to submit – on their own – a higher bid reflecting information

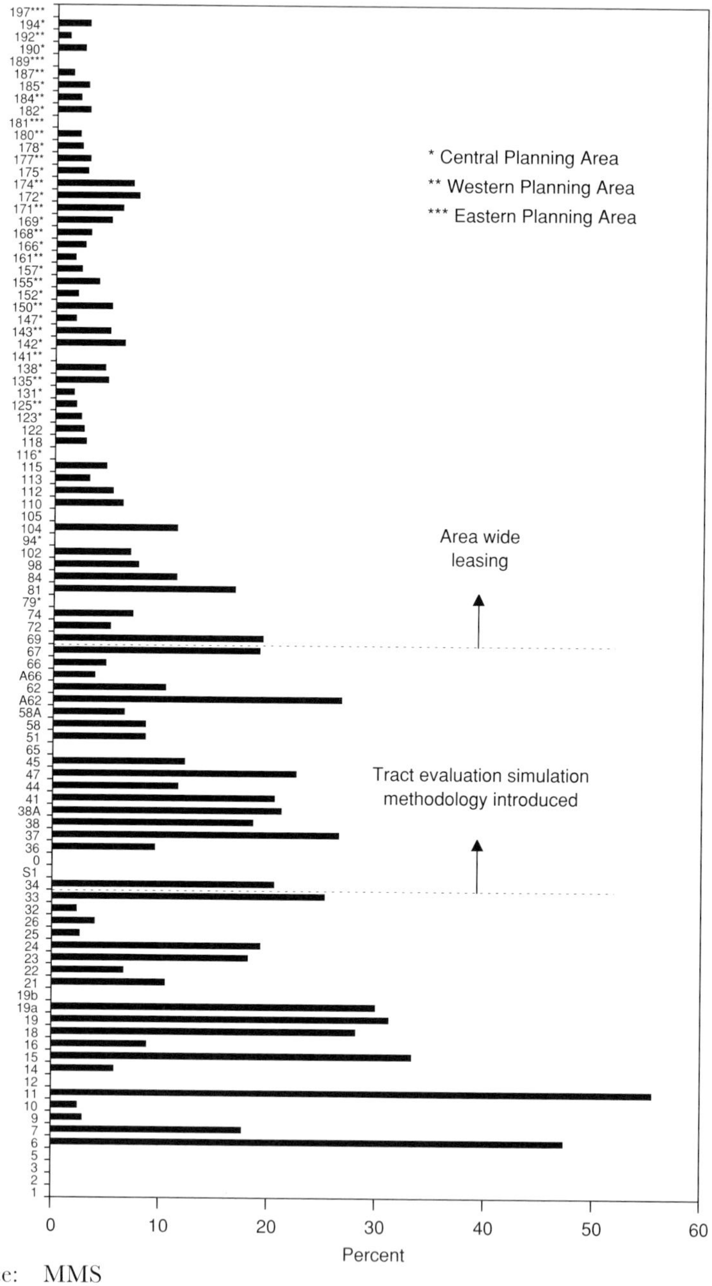

Source: MMS

Figure 10.5: High Bids Rejected or Withdrawn as a Proportion of High Bids Accepted for GOM Federal OCS Blocks, by Sale Number, 1954–2004

obtained on occasions when a collusive bidding strategy might have been agreed upon. And because DOI was fully within its rights to refuse to lease any acreage at all if it was dissatisfied with the bids received, the TN system offered reasonable protection even against the remote possibility that all prospective lessees might become part of a bidding ring. Finally, the TN procedures also made it difficult for advantaged players to nominate dud blocks as a way of distracting the attention of potential competitors for prime acreage (if it did not have information on dud blocks, DOI simply would not include them in a sale, and *ditto* if it did have information showing that the blocks were likely to be duds).

The relatively high bids obtained for drainage tracts, once again, constitute eloquent proof of just how well the tract valuation mechanism worked. An investigation by Hendricks, Porter and Wilson uncovered evidence that whenever drainage tracts came up for auction, only one of the firms producing in blocks adjacent to the tract tended to put in a bid. But even though these particular tracts were leased in the absence of rivalry among potential lessees (indeed, amidst 'evidence of collusion among the neighbour firms' [57]), the acreage still managed to attract large bids (as the Shell case cited above eloquently shows). In other words, the low number of firms bidding on these drainage tracts had a limited effect on the magnitude of the average high bid, which still ended up being far higher than the average bid for wildcat acreage (although not high enough to compensate entirely for the greater profitability of drainage tracts).

It is important to note that, for all of its effectiveness, DOI's tract valuation methodology did not translate into extortionate reserve prices for acreage. GAO noted that, under the TN system 'high bids usually substantially exceeded Interior's estimates of tract value',[58] and a 1980 study carried out by the Los Alamos National Laboratory (in which DOI's estimated values were considered under the light of the number and magnitude of bids received and the levels of production achieved after lease) found that the simulation methodology on the whole rendered quite conservative estimates of the fair market value of tracts. Indeed, so conservative were these estimates that DOI was on occasion accused of 'radically underappraising the land' in order to clear the way 'for accepting just about any bid that came along'.[59] In support of this allegation, for instance, Sherrill mentioned the case of 35 tracts whose value was appraised at USD 146 million, but which were leased for USD 1.49 billion (with two individual tracts valued at USD 144,000 going for USD 91.6 million and 76.9 million, respectively). What Sherrill never explained, though, was why bidders found it necessary to pay such sums if DOI was indeed ready to accept any offer that came its way.

The conservatism of DOI's estimates of tract value is hardly surprising. For one thing, the uncertainty of the exploration business is such that the ability of even the better informed companies to predict accurately the volume of recoverable resources was subject to very significant constraints.[60] For another, the staffing levels at the USGS (entrusted with providing DOI with estimates of resources in place) were such that the agency could not even 'calculate precisely the actual proved reserves on ... producing OCS tracts'. Thus, especially during the early years of the offshore leasing programme, the USGS was 'clearly not able to evaluate all the tracts offered ... in as much detail as the bidders ... [were] able to evaluate the relatively smaller number of tracts in which they are most interested'.[61] Indeed, such was the scarcity of qualified personnel that, according to John Rankin (former regional director of the BLM OCS office in New Orleans), most of the preliminary work in terms of the drawing up of OCS leasing procedures and regulations was done by George Schoenberg, a New Orleans attorney in the employ of Shell Oil.[62] Even as the offshore leasing programme gathered speed, the number of people in the Department of Interior with OCS responsibilities was only about 35, and the regional BLM office in New Orleans consisted of the regional director, the assistant regional director, and two support staff.[63]

Up until the mid-1960s, due to their lack of personnel, money and information, neither BLM nor USGS could even pretend to analyse bids rigorously and scientifically, and specific bids were accepted or rejected merely by comparing them to bids received for neighbouring tracts. From the mid-1960s onwards, the valuation system managed to generate much better 'ballpark figures', as Table 10.6 shows. Moreover, with the adoption of a more sophisticated simulation methodology from

Table 10.6: Assessed Results of DOI's Pre-Sale Tract Valuation System, 1973–4

Sale number	*Date*	*Rejection rate (%)*	*Ratio of total high bids to total pre-sale value*
26	19/16/1974	4.0	NA
32	20/12/73	2.3	10.21
33	28/3/74	25.2	
34	29/5/74	20.5	4.24
36	16/10/74	9.6	5.02
38	5/2/75	18.6	1.93

Source: OTA 1975b

Sale 33 onwards, the accuracy of the pre-sale evaluations increased again. As a consequence of this, the rejection rate for high bids went sharply up, while companies grew accustomed to the harsh realities of dealing with a better-informed counterpart.

The TN procedures gave players with limited resources a free ride of sorts on the ample financial and technological coat-tails of more capable players, allowing the former to concentrate and focus their limited resources more effectively (and depriving advantaged players of the *full* fruits of their geological and geophysical *recherche* and expertise, as a price for being able to win *some*, but by no means *all*, the tracts they coveted). In other words, despite 'the relative advantage on the part of the oil and gas companies in their ability to evaluate and interpret geological and seismic data',[64] the TN procedures induced the more advantaged players into revealing some of their ideas about the prospects of different areas, thereby rendering both the leasing authority and rival prospective bidders less ignorant than they would otherwise have been. The benefits accruing to smaller players from DOI's externalised information can be readily gauged from Mead's finding that majors were actually more prone than smaller firms to acquire dry leases (probably because the latter had to be more careful with their money) and that, conversely, small firms earned higher rates of return on their lease investments than did the majors.[65]

In marked contrast to TN, the way in which AWL worked gave more capable players the scope to minimise the informational trickle from their bids, and this biased the whole auction process in their favour. Under AWL, assignation of acreage still takes place following a competitive bidding process (thereby maintaining the tradition whereby the USA is one of the few countries where exploration acreage is granted solely on the basis of cash bonus payments). However, because of the vast acreage involved in AWL offerings, the primary assignation parameter could no longer be an estimate of tract value, as MMS lacked the resources to subject many hundreds (even thousands) of individual tracts to a detailed examination.[66] Instead, high bids would automatically be accepted for *any* tract receiving three or more bids, regardless of how low the highest bid might appear. Likewise, a high bid would automatically be accepted for any tract that MMS had reasons to believe was non-viable. These reasons did not even have to be particularly *good ones*, though: the mere 'lack of Interior maps on [a] tract',[67] for instance, was justification enough to write it off as non-viable. For those tracts that could not be assigned automatically under the criteria above, MMS would still estimate fair market value figures, in order to compare them with high bids received. However, if a tract received at least two bids but the high bid was lower than the MMS estimate, the agency would have to combine the bids with its estimate of fair market value and then compute a geometric average for the tract. If the high bid exceeded this average value, it would be accepted.

This assignation procedure, still in use today, is highly questionable, not least because in geometrically averaging multiple company bids with a single MMS estimate, greater weight is being given to the former. Even more critically, this procedure is effectively a signal to bidders that MMS is prepared to assign leases in exchange for less than its very conservative estimates of their fair market value. Obviously, the fact that this procedure grants 'more importance to a relatively few bids ... than to ... good supporting data and estimates of tract value'[68] has also given advantaged players a powerful inducement to use strategic bidding practices aimed at generating 'the right combination of ... bids ... [that can] guarantee the high bidder the lease to any tract'.[69] So it comes as no surprise to see that, since 1983, the likelihood that MMS will reject a high bid for a tract has declined to less than 5 percent, from a pre-AWL average of 17 percent. Admittedly, to compensate for the loss of rigour in tract value evaluation that AWL entailed, after 1983 MMS increased the minimum acceptable bid from USD 25 to USD 150 per acre. However, due to the 1986 price collapse (which ravaged the balance sheets of small and medium players), MMS reinstated the previous minimum figure in 1987 and this number has remained essentially unchanged since (although the minimum bid for blocks located in 2600 feet or more of water has now been raised to USD 37.50 per acre).[70]

As has already been mentioned, the key characteristic of AWL was that entire OCS planning areas were offered to the industry. During the early 1970s, the wisdom of offering acreage on this basis had been intensely debated. However, even though the desperate times in the run-up to the First Oil Shock seemed to call for desperate measures, this radical course of action was not adopted because it was feared that the required capital outlays that would be generated by 'offering the entire OCS on the presumption that efficient production is the goal ... [would] move the industry into a less than perfectly elastic range of its supply curve of capital funds. This would increase discount rates, tending to cause too rapid depletion in profitable areas, and making unprofitable some areas that could be efficiently exploited at discount rates implied by a slower pace of leasing'.[71]

The adoption of AWL did indeed lead to a marked increase in discount rates, albeit in the form of greater risk aversion on the part of small- and medium-sized companies. The problem for these players was that their limited resources were literally swamped by the extension of acreage on offer. An elegant piece of analysis by Ed Capen on the importance of technology in the process of identification of oil prospects provides an excellent measure of the over-abundance of acreage. Capen focused on eight GOM lease sales held between 1986 and 1989. A total of 3600 bids were submitted in these sales, and ARCO and Amoco, coincidentally, submitted 240 bids each. Capen posited the following hypothesis: 'if each company [had] used its best technology and technology is able to find the best prospects, then the two companies should have competed many times because technology would draw them to the same best prospects ... at least 50 percent of the time'.[72] However, far from finding a 50 percent degree of overlap between the bids of both companies, Capen discovered that they had put bids for the same block on only 17 occasions (which, in statistical terms, was merely a chance overlap). Capen saw his findings as vindicating the commonly held view that companies 'working at the edge of technology ... seldom identify the same blips on the screen'. However, he neglected to add that this was chiefly a reflection of how much the screen itself had grown (especially in relation to the size of the blips) as a consequence of the adoption of AWL.

Keefer has characterised bidding on OCS leases as,

> a major corporate resource allocation problem, involving enormous uncertainties and significant probabilistic dependencies. Moreover, problem characteristics such as continuous decision variables, initial binary uncertainties, subsequent multiple long-term uncertainties and the option of sharing risks with partners extend into other problem domains. Problems of this type present significant challenges for future research in optimisation methodology as well as in decision analysis and probabilistic modelling.[73]

Obviously, resource allocation conundrums that leave academics – even with the benefit of hindsight – stumped, are magnified greatly for company managers that have to take decisions in the heat of the moment, with what they hope will prove to be reasonable foresight. That is why the difficulties inherent in looking for small blips in an enormous (and unforgiving) screen accentuated the risk aversion of less informationally-advantaged companies, often to the point of paralysis.

When the Reagan administration announced the details of AWL, Kosmo correctly predicted that the new leasing method was 'likely to favour the major and [medium-sized] companies. Its anticompetitive implications ... [would be] restricted to its effects on independent access, since the results do not suggest that the [medium-sized companies] will be hurt. Concentration should not increase at the top of the industry'.[74] In fact, concentration at the top did increase significantly because of AWL, as was made clear by a former Unocal official who had to handle the transition process in that company to the AWL environment: 'Unocal was ideally suited to compete with the larger integrated oil companies only within the pre-1983 framework, but was totally unprepared for the competitive rigours created by the AWL framework. That transition staggered us a step that we never regained.'[75] This statement is highly revealing, not least because Unocal was far from being a corporate minnow: in 1982 (with its problems with T. Boone Pickens still beyond the horizon), Unocal was a fully integrated concern supplying 3.25 percent of the US gasoline market (making it the 9th largest supplier, just below Chevron), and ranking as the 13th largest US oil company by gross revenues, the 11th largest by net income, the 14th largest by total assets, as well as the 28th largest company by net sales among the Fortune 500! Given the effect that AWL had on a company like Unocal, it is not difficult to visualise how much harder it must have hit players located lower down in the oil pecking order.

In view of the above, it is very surprising to see that smaller players have been as unstinting in their praise and support for AWL as have the largest of the majors.[76] Fixated on the idea that low acreage prices cannot be anything but a boon, small- and medium-sized companies have been unable to appreciate as a group that, had leasing continued to take place after 1983 under TN rules, the first movers in the deep-water province would have had to tip their hands. Granted, the first few blocks to be leased in deepwater areas would still have attracted only large companies and modest bids, due to the lack of knowledge regarding geological conditions, and the extra element of technological risk involved (after all, even in TN auctions, the average bid per acre tended to be quite conservative whenever completely new areas were offered).[77] Under such conditions, 'large firms, which are capable of bearing risk and which have sufficient capital to carry out exploration, win the first round ... [and] pay nominal sums to the government per acre explored and ... selected for development'.[78] However, the interest of potentially opportunistic parties in subsequent bidding rounds would have forced parties that had obtained encouraging information from proprietary surveying or drilling to pay top value for tracts adjacent to hot prospects or discovery wells (even if the latter were successfully kept tight). Thus, although 'in effect, the government [would have paid] for the initial broad exploration out of revenues it could have received from the first development tract if its existence had been known and it had been leased directly',[79] its total revenues per acre leased in subsequent rounds would have been much higher (and more than compensate for low receipts in the first round), because the exploration information and reduced risk would have attracted many medium- and small-sized firms into the fray.

Under leasing procedures akin to those embodied in TN, 'claims on a portion of the economic rent associated with nearby tracts' are prevented from being sold bundled together 'with tracts in underdeveloped areas, characterised by high presale uncertainty and low government rent capture'.[80] Indeed, the essence of the TN system, to paraphrase an academic who decried it as being too lax, was as follows: 'require disclosure for tract development data, eliminate the information monopoly rights and later lease the nearby tracts in an environment of lower uncertainty and higher expected government rent capture'.[81] Had the larger oil companies operated under such conditions throughout the 1980s and early 1990s, their E&P activities would have doubtlessly attracted the unwelcome attention of medium-sized players like Unocal, who would have tried to get into the frontier deepwater action earlier, seeing it as much less of a long shot. Indeed, so long as uncertainty was kept reasonably in check, these medium-sized players would have stood a reasonable chance of getting their hands on some of the more prospective acreage, even if they became entangled in the occasional bidding war with larger players.

It is not difficult to find evidence that small- and medium-sized companies are quite capable of punching above their weight as far as the payment of bonuses for OCS acreage is concerned whenever informational asymmetries are reduced (even though this pushes up prices per acre). This applies, for instance, in the case of attractive blocks that are easier to identify and survey. As has been discussed before, after the Mahogany subsalt discovery (1992), the block updip from the strike (which was obviously very easy to identify and survey) received nine bids, and was ultimately leased to Anadarko for USD 40 million (USD 7000/acre). This was until 2006 inclusive the highest bonus figure recorded in any lease sale held after 1985,[82] but even in nominal terms it is still a far cry from the (pre-AWL) USD 165 million (USD 52,457/acre) that a Superior/Pennzoil/Sohio consortium paid in 1980 for block EB304.

Likewise, post-2000 OCS lease sales have witnessed some really spirited bidding for prospective deep gas acreage. In Lease Sale 185, for instance, four of the ten highest bids received were placed on shallow water blocks, and two blocks attracted 18 bids between them, a phenomenon pretty much unseen since 1983. In Lease Sale 187, High Island block HI170 drew a total of 13 bids whose combined value (USD

111.5 million) came to 43 percent of the total monies exposed in the 407 winning and losing bids that were placed on the 335 blocks on offer. The winning bid (submitted by LLOG Exploration Offshore, a privately held E&P independent) amounted to USD 22.6 million. The fact that the magnitude of LLOG's bid could be a cause of general amazement is, in itself, a sobering indication of just how much the market for offshore oil acreage has changed since the adoption of AWL. After all, back in 1981, Exxon had paid USD 68.2 million for exactly the same tract.[83] Indeed, the bidding for the rest of the blocks on offer in Sale 187 is far more typical of the intensity of competitive forces in the AWL age: out of the 335 blocks that attracted bids, only one (the aforementioned HI170) received more than four bids, while two blocks received four bids, six blocks received three bids, 42 received two bids, and the rest went on uncontested single bids.[84]

Perhaps the best example of the ability of small- and medium-sized companies to compete with their larger peers in a level playing field comes from Lease Sale 181, held during December 2001. Political factors (namely political opposition to drilling anywhere near Florida's coastline) meant that only 1.3 MM acres in the Eastern Planning Region could be offered to the industry, by far the lowest figure ever offered in an AWL sale. Total bonuses paid in this auction came to USD 340 million, equivalent to USD 622 per acre, the third highest figure in the history of AWL, and the highest figure recorded since 1985. Non-major oil companies submitted 47 percent of the winning bids, regardless of the fact that all the offered blocks were located in 5000 feet of water or deeper. According to MMS, this level of interest was due to the fact that this lease sale 'was the first opportunity in 16 years for companies to bid in an area immediately adjacent to discoveries in the Central GOM area'.[85] This explanation rings very hollow, though, not least because in Lease Sale 182 (held nine months before), 23.4 million acres within the Central Planning Region itself were offered, and total bonuses paid came to only USD 355 million (USD 144 per acre). Rather, Sale 181 confirms Kosmo's fears in the sense that 'sales that offer more than one million acres [would] impede OCS access for independent oil companies'.[86]

In sum, the marginalisation of small and medium players from the highly profitable deepwater action can be said to stem from *low* acreage prices and the unsound leasing policy that begat them (by making it so easy for larger companies to cover their tracks). That is why, throughout the formative years of the deepwater province, one repeatedly comes across stories like that of two Garden Banks blocks located in 2900 feet of water, which Shell won in a bidding round held in early 1984 (for a bonus payment of USD 8.9 million). After having shot a proprietary grid of seismic over all its blocks around the zone, Shell decided that the prospect with the greatest potential was a large salt dome that it had located in the aforementioned blocks. Having acquired the adjacent blocks in 1985 in the face of no competing bids (for an even more modest bonus payment of only USD 2.4 million), Shell drew up a drilling programme and sank its first exploratory well in 1987 and discovered a field estimated to contain about 300 MMBOE. In 1989, the company announced its decision to develop this field (by then christened Auger) by means of a tension leg platform, thereby launching the deepwater boom in the process.

It is thoroughly typical of the confusion surrounding the whole issue of GOM offshore leasing that the acreage positions that small- and medium-sized firms were able to piece together from the mid-1970s up to 1982, inclusive, were perceived at the time as being woefully inadequate, not to say symptomatic of a structural inability on the part of these companies to compete *tête à tête* with the majors. Consider, for instance, how Sherrill summed up the auction at which the Cognac blocks were leased: 'look at the winning bidders…Gulf Mobil and Texaco teaming up for one tract, Exxon bidding alone on another. It was the same old crowd. Where were the smaller companies, the independents? Shut out by the prices.'[87] Had Sherrill examined carefully the list of winning bids in this auction,[88] he would have seen that hundreds of independents were able to acquire a good number of blocks (as part of consortia when not for their own accounts), and actually ended up with around a quarter of the net acreage leased (indeed, their percentage participation in the – admittedly small – number of deepwater blocks offered was marginally *higher* than the average). Such participation may have seemed like small beer to Sherrill, but for most of the small- and medium-sized companies active in GOM, it compares very favourably with the acreage positions that they have been able to build in lease sales held since 1983 under the auspices of AWL.

10.4 Effects of AWL on Lease Inventory Turnover

One of the many achievements that AWL has been credited with was that it 'made more prospects drillable … [and hence] encouraged better and faster evaluation of the GOM'.[89] This claim is made on the strength of the vast OCS extensions that AWL made available to the oil industry, despite the adverse effects of moratoria. AWL also led to a significant increase in acreage leased, although this was not remotely proportional to the increase in acreage offered (indeed, by 1992–3, the annual amount of acreage leased had returned to pre-AWL levels, although it was to pick up again later).

Back in the early 1970s, serious doubts had been expressed in a congressional report on offshore leasing about the capability of the oil industry to drill enough wells to explore and develop the vast extension (at least 10 million OCS acres) that was supposed to be leased between 1975 and 1980. As the authors of the report saw matters, the futility of offering up areas way in excess of the industry's exploration capacity lay in that 'only an amount that the industry believes can be explored in five years would receive bids, and an even smaller amount would be leased'. In support of this assertion, they cited the fact that, in 1973, 1.5 million acres had been offered for lease, with 1 million acres (68 percent) ultimately being leased, whereas in 1974 an acreage offering of 3.7 million acres had led to only 1.7 million acres (47 percent) being leased. By analogy, they concluded that 'an offering of 19 million acres [would] ... result in no more than 3 to 5 million acres being leased'.[90]

As things turned out, these estimates of acreage take-up undershot the figures achieved after the introduction of AWL by a significant margin (during the peak years for leasing during the 1980s, the annual figures were as high as 6–8 million acres). Having said that, these much higher take-up rates reflected the fact that the primary leasing terms for tracts lying at depths between 1300 and 3000 feet, and depths greater than 3000 feet, were also extended after 1983, from five years to eight years and ten years, respectively.[91] In absolute terms, drilling also exceeded the expectations of the report, partly because of the far larger number of tracts available but mainly because record oil prices induced an exploration frenzy as oil companies burned billions of shareholders' funds in a futile attempt to prove Hubbert wrong. In this sense, it is worth commenting that this same report visualised OCS crude oil production rates reaching 4.5 MMBD by 1985 (assuming an oil price of only USD 11/B[92]), a figure 4.2 times higher than the 1.066 MMBD production recorded in that year!

For obvious reasons, AWL heralded a marked slowdown in the turnover of the lease inventory. According to statistics compiled by Hendricks, Porter and Boudreau, during the TN years around 'twenty seven percent of all leases were allowed to expire without any wells being drilled'.[93] In terms of deepwater acreage in particular, the selectivity fostered by high acreage prices meant that more than 90 percent of the small number of leases assigned over the 1974–9 period were drilled (70 percent of the leases acquired over 1974–5 contained producible hydrocarbons, and about half came into production).[94] In contrast, by the late 1980s, less than 10 and 5 percent of issued deepwater leases were being drilled and produced, respectively. Furthermore, the onset of the deepwater boom did not translate into an improvement in this situation. For instance, out of the approximately 3200 deepwater leases issued from 1996 through 2000, less than 7 percent have been drilled (and there are over 2400 leases from these sales still in their primary term, with more than 750 of those lying in water depths greater than 7000 feet, where barely forty wells have been drilled).[95]

From 1983 onwards, the combination of huge deepwater lease inventories and limited rig availability has meant that only a few GOM leases are tested by the time their primary terms expire (Figure 10.6 shows the number of all active leases by water depth as of the end of 2004, and sets them against the leases that are set to expire in the coming years).[96] Nevertheless, the annual figures for leases drilled have, on the whole, been higher than those registered during the last years of TN. It is widely believed that this higher drilling intensity has meant a greater number of discoveries and producing wells. In this sense, even the imminent expiry of the vast number of deepwater blocks leased between 1996 and 1998 has been presented by MMS as a windfall of sorts, because it will supposedly 'pressure leaseholders to drill and evaluate their holdings and will provide opportunities for other companies to enter an active play by acquiring leases as they expire or by obtaining "farm-outs" from companies with untested acreage'.[97]

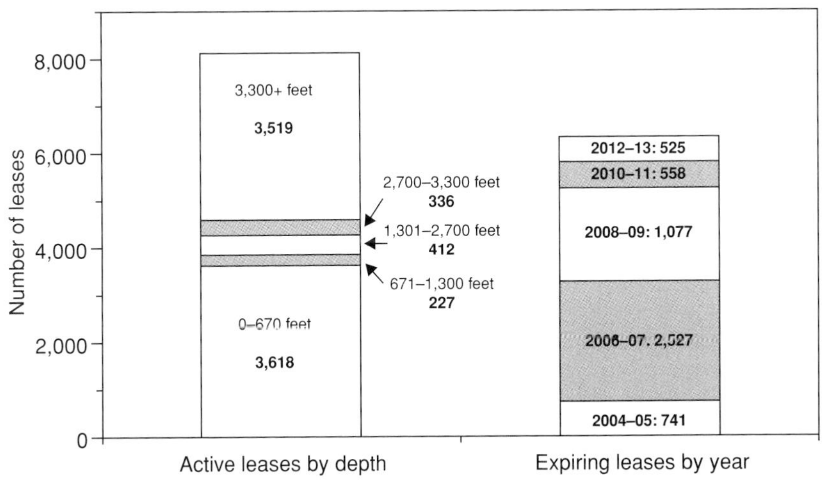

Source: MMS

Figure 10.6: Lease Status in the GOM Federal OCS (as of 2004)

MMS seems to think that, although most of the leased deepwater acreage in GOM has actually been lying fallow over a period of time when American dependence on imported crude oil has been growing steadily, the impending expiry of these leases may allow '*potentially*, a more rapid exploration and development of the acreage'.[98] In fact, what happened was that, by enabling a handful of oil companies to amass lease inventories of an extension that vastly exceeded their capability to explore them, AWL effectively put this acreage beyond the reach and the ken of other companies that would have been both capable of and eager to explore it.

During the early 1970s, the risk that accelerated leasing under the TN system might lead to a situation precisely like this one had been foreseen, and explicit warnings had been sounded that in such an eventuality, accelerated leasing could very well 'fail to increase production faster than would a lower leasing rate'.[99] AWL essentially amounted to Nixon-style accelerated leasing but on an even grander scale,[100] so it could hardly be expected to do better in this regard, and it certainly did not. But to make matters worse, there is also abundant albeit not entirely compelling evidence pointing towards a more disturbing possibility: as a result of the pernicious effects of AWL on the auction market for offshore acreage, GOM production might have been lower than it would have been in the context of a less concentrated deepwater lease ownership structure.

Throughout much of the history of the offshore leasing programme, the average winning bid submitted for a tract always was a 'decreasing function of the time until the first well [in that tract was] drilled'.[101] In other words, for a very long time, OCS leases receiving the highest bids were not only more likely to be drilled a short time after assignation, but also the most likely to be productive.[102] In contrast, 'tracts with zero gross profits receive[d] lower [and fewer] bids on average than tracts with negative or positive results',[103] and also tended not to be drilled.[104] Drilling decisions for these low value tracts, furthermore, depended entirely on 'drilling outcomes on the high value tracts'.[105] As a result of the patterns described above, most discoveries in the GOM Federal OCS tended to occur within the first 18 months after the assignation date of the block where a strike was made. Recently, though, there has been a significant departure from this historical precedent.

It is still the case that, as Richardson et al. point out, 'for any given lease-sale year, almost 50 percent of tested leases were first drilled within three years of lease acquisition ... [while only] 23 percent were drilled in year eight or later'.[106] What is no longer true is that the majority of hydrocarbons will also be found during the first half of the primary term of any given lease. According to the latest MMS deepwater data, 'twenty-nine percent of the hydrocarbon volumes were discovered during the first three years of [the] lease term, but 44 percent of the hydrocarbon volumes were discovered in year eight or later'.[107] The latter figure includes the volumes of such important fields as Thunder Horse, Thunder Horse North and Mad Dog, all of which were found at the very tail end of their respective lease terms.

MMS analysts consider that 'the ... surprising ... amount of major discoveries found in the later years of some leases' terms' is a reflection of 'the difficulty in recognizing the best prospects at the beginning of a lease's term'.[108] This difficulty is undeniable. After all, the interpretation of the 3D seismic data for deepwater subsalt reservoirs like these is a task that may literally take years. Having said that, the foremost reason why such a long time had to elapse for BP to discover the three fields named above was that the company was busy elsewhere in the deepwater GOM. In essence, BP focused on tackling the development of more amenable prospects while keeping more challenging prospects on a low burner. And it was able to do this because the 10-year primary leasing term gives companies a long enough period of time to evaluate their lease inventories at a leisurely pace, and then to drill the most promising deepwater structures that they manage to identify (even if these turn out to be quite challenging). Unfortunately, the long primary leasing term also means that any leases that are not tested in the allotted time will still be denied to other players until expiry, or until the leaseholder finds it convenient to farm part of this acreage out. Because of this, large deepwater lease portfolios constitute a way for their holders to steal a march on potential competitors, firstly by keeping acreage out of the latter's hands and, secondly, by ensuring that potential developers of expired leases that they used to hold will have to use infrastructure put in place by the early movers. In other words, large deepwater lease inventories work as an entry foreclosure mechanism, whose effective operation is related to the very low acreage costs. In this situation, signature bonuses function as low cost call options to drill and develop, rather than as a mechanism to capture windfall profits.

Figure 10.7 seeks to illustrate this point by making reference to the three large, late, finds already mentioned above. Most of the Thunder Horse blocks were leased back in 1988 to Conoco, for altogether unimposing sums. BP acquired them in April 1994 and was happy to sit on them while it attended to more pressing matters (like the development of Pompano, Troika and Marlin, not to mention the digestion of Amoco, Arco, Vastar and Burmah). However, BP's near simultaneous

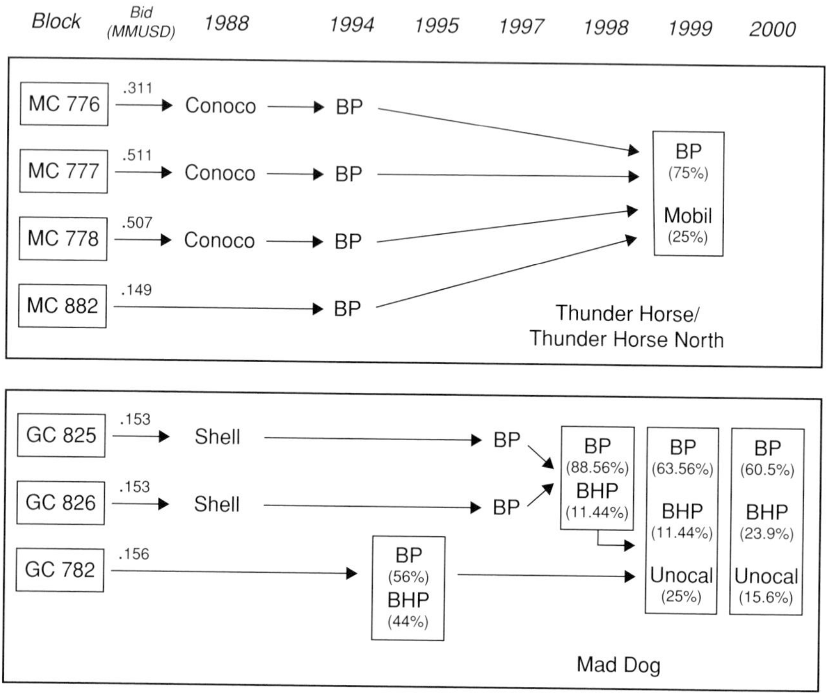

Source: MMS

Figure 10.7: Ownership History of Selected Deepwater Leases

acquisition of block MC822 in May 1994 also shows that, at that point (six years *after* the other three Thunder Horse blocks were leased), it sensed that it was on to something. BP decided to go for broke in late 1997, but this was easier said than done, as proven by the fact that the discovery well was only completed in July 1999 (more than one year after the tenth anniversary of lease assignation). Almost simultaneously with the Thunder Horse discovery well reaching its target depth, BP sold Mobil a 25 percent share in the blocks (including the one where the Thunder Horse North field would be found by a well completed almost two years later, in February 2001).

The story of the Mad Dog blocks is slightly more convoluted, but it also illustrates how large companies like BP might find large idle lease holdings useful, even when these are in the hands of other major oil companies. The GC825/826 leases had originally been sold to Shell in 1988, but BP bought them in 1997, by which time it was following a hot trail in the vicinity, at blocks GC699 and GC782 (both of which had been acquired by BP and BHP Billiton in 1995, with the former block becoming the site of the Atlantis development). At that stage, and once again clearly believing that it was on to something important, BP also bought a 50 percent share in Shell's nearby block GC644 (now the site of the Holstein development), where BP was to drill a discovery well in 1999. In May 1998, BP also spudded a well in block GC826, but this was not immediately tested. Instead, it brought BHP Billiton on board the two Mad Dog leases, with an 11.44 percent share. In 1999, on the eve of the Mad Dog well being announced as a discovery, the partnership was widened with the inclusion of Unocal as a 25 percent shareholder. Further rearrangements in 2000, undertaken in the light of likely development costs and field unitisation, saw the consortium taking its present shareholding structure: BP with 60.5 percent, BHP Billiton with 23.9 percent and Unocal with 15.6 percent.

So what is the lesson of these development sagas? It is clear that the development of the Mad Dog and Thunder Horse blocks was compromised by virtue of their figuring in very long lists of idle leases. BP picked up the blocks and eventually developed them, but only a company with its enormous resources and know-how could have hoped to pull off such a stunt so near to the expiration date of the leases. In other words, had BP not developed an interest in these leases (a tad fortuitously), current US domestic oil supply would have been denied some sorely needed 200 MBD of Southern Green Canyon blend, not to mention the even greater volume that Thunder Horse is expected to produce when it reaches its peak.

Supporters of AWL may argue that this line of reasoning is undermined by the fact that BP *did* develop an interest in the leases, however belated. Point granted, but one is impelled to ask after the fate of the many hundreds of potentially productive but idle leases that will have failed to excite anybody's interest and which, therefore, will fall through the cracks in the majors' lease inventories upon expiry (as is scheduled to happen in 2006–7). This outcome is irrelevant as long as one believes that, in the words of a BP officer, 'only a leading energy company … can muster the organisational capabilities and capital to take on projects … [the] size [of Mad Dog, Holstein, Thunder Horse]'. If only large majors can develop this sort of acreage, it follows that in the fullness of time they will eventually get round to doing so. Moreover, the waiting period need not even be inordinately long. After all, as the same officer put it, 'we at BP did five such projects at the same time. That is unprecedented even for a major oil company.'[109] However grating BP's degree of self-congratulation might appear, it is indisputable that only a company of its size and expertise could confidently take a five-project bite in the GOM deepwater without choking in the process. By the same token, it is clearly untrue that smaller fish do not have anything to look for in the deepwater pond. The development record of projects like Red Hawk, Medusa, Boomvang, Nansen suggests that smaller players are perfectly capable of developing challenging fields, if given sufficient time, access to leases and above all, a reasonable chance to form an idea of potential tract profitability before bidding takes place.

Critics of the supposedly 'out-dated'[110] TN system suggest that this mechanism would have acted as an obstacle to the discovery of prolific GOM deepwater fields because fiscal revenue considerations would have prompted DOI to continue offering exhausted shallow-water acreage instead of more prospective deepwater tracts (as this course of action would have held the promise of attracting higher bids). This criticism rings true to the extent that, especially during the late 1960s, DOI made TN somewhat more restrictive than it need have been by its penchant for often failing to include in lease sales many of the tracts nominated by the oil industry. However, a TN system in which industry nominations were invariably respected would certainly have seen GOM deepwater tracts come into play and bid upon as soon as the 2000 feet depth threshold had been comprehensively pierced in some other oil province.

Such a system would have offered a reasonable compromise between fiscal revenues and the timely and opportune access to prospective acreage for the industry, so long as one elementary rule was scrupulously observed. In the words of Logue, Sweeney and Willett, nominated tracts would have to be 'bid upon by all interested parties; however, in the event that only one bid [were] received, the tract would be withdrawn to be offered again at a later date. Only when two or more ... bids by independent parties [were] received would the tract actually be leased'.[111] Such a solution would have been elegance and simplicity personified not least because, as Moody and Kruvant pointed out, 'it [would not have] restrict[ed] the supply of leases artificially and it [would have] *incorporat[ed] the information from the nomination process into the lease price*'.[112] Unfortunately, driven by ideological prejudice and a well-attested preference for regressive tax policies, the Reagan administration instead opted to adopt a clumsy and wasteful form of corporate welfare that transferred billions of taxpayers' dollars into the coffers of the largest members of the Brotherhood of Oil.

The long-term effects of this policy choice are still being felt today on many planes, even though technological entry barriers to the deepwater have come down in recent years. As explained below, although entry into the deepwater via the exploration route has risen, those entrants fortunate enough to discover a field have also found themselves having to turn over a significant chunk of the value generated to incumbents, in exchange for access to the extant transportation and processing systems of the latter.

10.5 Effects of the Ownership of Pipelines and Processing Hubs on Access to Infrastructure and on Ultimate Recovery

The economics of developing deepwater projects are highly sensitive to the presence of infrastructure in their immediate vicinity, and the ease and cost of tie-ins to existing pipelines, on the one hand, and to processing facilities, on the other. The price that such access commands, as explained in the section on lifting costs, is very high. To a certain extent, this situation is a consequence of the myriad difficulties inherent to operating in the deepwater. However, it also reflects the market power that incumbents have derived from their commanding logistical positions. These positions, in turn, are the product of two related factors: first, effective leveraging of their deepwater infrastructure with extensive assets lying in shallower parts of GOM; second, their earlier start to the leasing and development race in the deepwater. The scope that deepwater incumbents have had to augment their production revenues significantly by processing and transporting hydrocarbons for third parties has given them an obvious competitive advantage over new entrants. Aside from this, the competitive vulnerability of new entrants is heightened by the possibility that incumbents may choose to ration, via prices, the access to their transportation, storage and processing facilities.

BP's aforementioned challenge to the HOOPS tariff confirms that there are indeed situations in which a transportation company affiliated to a producer may have little need, want or desire to reduce tariffs to increase the throughputs of third-party volumes, notwithstanding the high fixed costs of deepwater infrastructure. However, the HOOPS hearing ended with FERC decreeing the suspension of EMPCO's proposed tariff, and enjoining the parties to negotiate alternative rates (which they now have done to their mutual satisfaction). Does not this positive outcome demonstrate that existing regulation is adequate to prevent anticompetitive outcomes?

The answer to this question is: not necessarily. For one thing, it is legitimate to wonder what would have happened in the HOOPS case had the aggrieved party not been the formidable BP. After all, as a recent US Senate report on the workings of the American oil market warns, 'the laws and regulations governing access and control to ...

[pipelines] are complicated and often not well understood – even by the parties most affected by them'.[113] For another thing, FERC's observance in matters such as those that arose during the HOOPS controversy has been known to slip and its enforcement record is not exactly stellar.[114] Indeed, MMS' highly controversial royalty assessment rule of June 2000 expressly disallowed the use of 'FERC tariffs … as a substitute for actual costs in non-arm's-length situations', due to this agency's belief 'that FERC tariffs often exceed the transporter's actual costs'.[115] But the main reason why existing regulation may not be enough to safeguard anticompetitive outcomes involving access to deepwater infrastructure is that, as has already been mentioned, FERC lacks jurisdiction to enforce both the Interstate Commerce Act (ICA) and the Natural Gas Act (NGA) with respect to lines located *wholly* on the OCS.

In 1992, FERC reached the conclusion that intra-OCS pipelines did not engage in interstate commerce, which meant that they were bound not to contravene the access provisions of OCSLA, but did not need to comply 'with any of the requirements of the ICA with respect to their facilities on or across the [OCS]'.[116] This legalistic distinction is of enormous significance, because ICA expressly stipulates that rates must be '*just and reasonable*' as well as non-discriminatory, and it is the former two provisions that are the key safeguards that shippers have against price gouging. In contrast, aside from not making it obligatory for pipelines to file any tariffs specifying the terms and conditions of transportation services, OCSLA only requires that access to OCS pipeline facilities be '*open and non-discriminatory*'.

In the case of the HOOPS tariff, FERC was able to invoke its attributions under the ICA to decree a downward adjustment in a tariff that appeared unreasonable in the context of likely throughput scenarios, but only because this particular pipeline is one of the very few deepwater lines that crosses a state boundary (a consequence of the fact that the Hoover and Diana fields are located offshore Texas, where infrastructure is much thinner on the ground than offshore Louisiana). Crucially, had HOOPS been more like one of the lines that terminate at one of the many hubs that lie beyond the Louisiana state boundary, FERC would have had no legal grounds for rejecting the tariff so long as there existed one shipper willing to pay it (as ExxonMobil would certainly have been). In such an eventuality, BP would have had no option but to cough up, and the same would have been true for a player seeking to bring on stream a much smaller field than Hoover or Diana, and who might have been prevented from doing so by the steepness of the tariff.

Many of the trunklines catering to deepwater transportation

requirements do not cross any state lines, and terminate in areas exclusively under Federal jurisdiction: the Mars/Amberjack system terminates at the West Delta WD143 hub facility, the Caesar pipeline will terminate at the Ship Shoal block SS332 facility, while the Auger and Bonito pipelines terminate where they meet the Ship Shoal system). Thus, FERC's disavowal of ICA jurisdiction for pipelines lying entirely within the OCS makes shippers on such lines 'vulnerable to the risk that [their] owners will impose transportation rates *that are unreasonable under the ICA but non-discriminatory under the OCSLA*'.[117] Shell, for instance, was so convinced that this risk was both tangible and unacceptable that it took the highly unusual (and ultimately fruitless) step of appealing *against* the verdict that FERC had reached *in Shell's favour* regarding Pennzoil's obligation under § 5(f) of the OCSLA to connect the Bonito pipeline system with the pipeline transporting crude from the Auger production platform.[118]

Given the above, the ruling determining that the ICA does not apply to many OCS oil pipelines may make a negative contribution towards the ultimate recovery of oil in the GOM deepwater, chiefly because of the highly concentrated industry structure that prevails there. The potential fate of small pools in the deepwater GOM is paradoxical, given that the whole point behind the fiscal sacrifices underlying AWL was to make sure that even the most marginal of oil and gas accumulations were brought into production. However, it is not entirely surprising, because a similar problem regarding ultimate recovery appears to be currently unfolding in the UK sector of the North Sea, where it is alleged that many small oil pools (10 million barrels or so) are in danger of being left stranded as production facilities located nearby are abandoned and decommissioned by some large companies who nonetheless refuse to mark down their tariffs in a way that might allow small operators to tap such pools.[119]

According to the UK Department of Trade and Industry (DTI), over 8 billion barrels of oil reserves remain to be developed in the UKCS (this figure does not include as yet undiscovered pools). These fallow reserves are not uniformly distributed however. In the Northern North Sea, for instance, where several competing pipeline systems coexist, just 23 fields remain to be developed. In contrast, in the area served exclusively by the BP-operated Forties pipeline, there remain 79 fields and ten probable developments, which between them contain around 70 percent of the undeveloped total. Ullage in the Forties pipeline system is ample (350 MBD plus and growing), and yet potential developers of small fields have stressed repeatedly that 'where infrastructure owners are the obvious and only option, the levels of tariff and commercial terms are always far from acceptable. This can lead to a protracted period of negotiations and the resultant potential delay to developments'.[120]

When one looks at the Forties tariff structure, the complaints of small oil companies seem well-founded: not only is the base tariff high (45 pence per barrel) for what is after all a fully depreciated line, it also escalates with higher oil prices, which have nothing to do with the cost of providing transportation services (the base tariff is applicable at prices up to USD 15/B, but for prices above USD 25/B the tariff is an extortionate 150 pence per barrel). As Arnott observes,

> BP would argue that it put up its own equity to fund a pipeline that carried a risk that it would not be filled ... [but in] reality it was the UK taxpayer who carried most of the risk as the effective rate of taxation meant that BP paid just 20 pence in the pound of the capital cost of the original pipeline ... When the pipeline was rebuilt and increased in size after 1991, there was minimal risk that the pipeline would not pay for itself, as significant exploration activity had already proved up substantial new reserves in the Central North Sea ...[and] BP still obtained the maximum level of tax shelter for the redevelopment of the pipeline ... [The] fact that throughput doubled between 1991 and 1994 proves the point.[121]

Indeed, returns on BP's Forties investment have been around 14 percentage points higher than the company's cost of capital, and yet no one in a position of authority has seen fit to question how this is to be reconciled with the 1996 Offshore Infrastructure Code of Practice, which clearly stipulates that allowance for capital recovery in pipeline tariffs should be set at a level that will earn the owner an '*appropriate*' return on his investment.[122]

North Sea pipeline operators have successfully defused controversies regarding access by making reference to the ullage available in their transportation systems (even though this argument is a *petitio principii*) and, more convincingly, by pointing out that frustrated potential entrants have always had the right to refer alleged access irregularities to the Secretary of State for Trade and Industry for adjudication. The fact that no parties have ever availed themselves of this right is presented as proof that no abuse of market power is taking place. Given the readiness of some companies to air grievances in consultation papers, their reluctance to file formal complaints is indeed puzzling. But then again, so is the British government's unwillingness to regulate tariffs for lines like Forties, despite its stated objective of encouraging 'new entrants ... to extend the life of the region, to maximise recovery and to bring on stream small undeveloped fields'.[123]

In view of the formidable lobbying powers of the US oil industry, the governmental quietism regarding the status quo in the Central North Sea sets a disheartening precedent for GOM. Nevertheless, if the fact that intra-OCS pipelines are not bound by the rate reasonableness, non-discrimination, or tariff filing provisions of the ICA were to give rise to major problems, these could probably be defused through legislation. Indeed, the fallout from the 1995 *Shell vs. FERC* and the 2003 *Williams Cos. v. FERC* cases, coupled with the worsening natural gas supply situation, has already led MMS to conduct a public enquiry regarding 'the scope, magnitude, and seriousness of any instances where access or discrimination problems were encountered by service providers or shippers of natural gas, both for lines that do not operate under the jurisdiction of the NGA and those that do'. The ultimate objective of this exercise was to 'help the MMS to gain a better perspective on the need for a regulatory framework to ensure open and non-discriminatory pipeline access'.[124]

If this consultation process were to crystallise eventually into amendments of the MMS regulations regarding pipelines transporting oil or gas under permits, licenses, easements, or rights-of-way on or across the OCS, the GOM deepwater province would not necessarily become a paradise for competition as a result. This is because high tariffs do not necessarily constitute the most serious threat to competition in the offshore upstream. A more tangible threat, and more difficult to counteract due to its insidious nature, is the reliance of the GOM deepwater transportation system on hubs controlled by very few companies providing integrated transportation and processing services.

This logistical model lends itself particularly well for such players to circumvent common carriage duties through the imaginative use of a mix of regulated (i.e. pipeline) and non-regulated (i.e. storage, processing and gathering) facilities. As the US Senate has warned,

> control of critical transportation and storage facilities are a less visible and very effective way to influence cost, supplies and market competition ... Although on the surface common carriage appears to be a neutral means of transporting supplies ... [the] parties who control the transportation and storage facilities *can* take advantage of the complexity of the laws and regulations to circumvent the requirements of the law and limit competition in the market.[125]

The Senate unearthed plenty of evidence about how transporters very often do take advantage of these factors in order to limit competition. For instance, one products pipeline in Michigan (Wolverine) was found to have withheld common carrier service from shippers by claiming that it did not have control over tankage owned by a party affiliated to the line, which was 'essential for transportation movements into [the]

… line'.[126] In this sense, it is worth recalling that, quite apart from its objections regarding the magnitude of the proposed HOOPS tariff, BP also objected to the conditions under which HOOPS shippers were to be granted access to ExxonMobil tankage facilities at the pipeline's landfall in Quintana (Tx.), on the one hand, and to EMPCO's agreements covering the lease of certain SPR lines through which crude shipped in HOOPS was to be taken to markets in the Gulf Coast and beyond, on the other.[127] Both of these complaints bring some of the most disquieting findings of the Wolverine case to mind.

10.6 No Outright Collusion. The One Bright Spot for AWL?

The AWL experience confirms that a leasing authority operating in conditions of pronounced adverse selection, and still expecting to capture excess profits through signature bonuses, really has only two choices open to it in terms of the procedures to allocate acreage to investors. The first consists in allowing 'allocations [to be] distorted away from first-best allocations and toward low-powered schemes'.[128] For the most part, such distortions will affect the contract of the *least* efficient firm, which will therefore receive breaks that it does not apparently deserve. But, in contraposition, the contract will be made 'less attractive to the efficient firm, thereby reducing … [its] informational income'.[129] The trade-off makes sense because the magnitude of the informational income that would otherwise accrue to the more efficient player is much greater than the value of the breaks given to the inefficient player. This regulatory response to asymmetry of information can take many forms. They range all the way from bureaucratic mechanisms like the discretionary assignation of acreage (which on the whole provides a poor antidote against creeping oligopolisation, due to the pervasiveness of regulatory capture), at one extreme, to the elegant market-centred mechanisms underlying the TN leasing system, at the other.

The second approach to the problem consists in the leasing agency pretending that informational problems are either of no consequence or else easily remedied by a 'natural' rivalry that is heroically assumed to exist between all firms, regardless of their respective sizes. This self-deluding option amounts to a form of regulatory capture by default, as it were. Nevertheless, it has proved to be very popular with leasing authorities operating in jurisdictions where the concept of regulation is invested with all sorts of negative (obstructionist, bureaucratic, smothering, and so on) connotations. The quid of this approach, which lies at the very heart of AWL, is simply to ignore the key insight that an economy in the real world necessarily 'has to choose between two different imperfections: imperfections of information or imperfections of competition'.[130] It also presupposes that not even the simultaneous presence of both forms of imperfection will be problematic for the operation of a market, although there might be 'no particular reason [to think] that these imperfections will be 'balanced' optimally'.[131]

Proving that, as a policy, AWL is steeped to the quick in wishful thinking requires nothing more than examining the MMS definition of fair market value (a definition that enjoys full judicial backing, having been accepted by a Federal district court in *California v. Watt (II)* and then upheld by a US Court of Appeals judgement in lawsuits brought against the Federal government by the Texas and Louisiana state governments after the adoption of AWL). The definition is as follows: fair market value is 'the amount in cash … for which in all probability the property would be sold by a knowledgeable owner willing but not obligated to sell to a knowledgeable buyer who desired but is not obligated to buy … *This market value which is sought is not merely theoretical or hypothetical but it represents, insofar as it is possible to estimate it, the actual selling price*'.[132] This is tantamount to saying that, as long as compulsion is absent, the mere fact that an economic actor succeeded in selling something to an arm's-length buyer proves by definition that the price at which the deal was done realised the fair market value of the good. Such a facile assertion would be disputed by anyone even superficially acquainted with the way in which auction markets work, of course.[133]

Back in the days before AWL gave MMS an acute case of cavalier torpor, this assertion would also have been vigorously challenged by the DOI itself, as is patently demonstrated by the following extract from a 1970s-vintage memorandum in which the department gave air to its misgivings regarding the sudden acceleration of the offshore leasing programme:

> If OCS leasing is accelerated merely by offering more tracts … there will probably be a decrease in the average number of bids received on each tract. Furthermore there are strong indications that the lower the number of firms bidding on a tract, the lower the level of the winning bid … Thus, the government may not be receiving fair market value for those tracts receiving only one or two bids.[134]

In light of the complacency that underlies the current MMS definition of fair market value, it is hard to believe that there was a time when the receipt of two bids (let alone one) for an OCS block was seen as implying a 'level of competition identified by a Department of Interior analysis as being low enough to jeopardise the receipt of fair market value by the public'.[135] Of course, the tautological formulation of fair market value that MMS adopted *après* Watt does nothing to support this Secretary of the Interior's boast in the sense that his policies 'demonstrated that the marketplace is the right place for decisions to be made regarding the allocation of natural resources'.[136] The way in which offshore leasing took off in the wake of the adoption of AWL, to use Sherrill's phrase, 'only demonstrated that oilmen know a sucker when they see one'.[137]

The only – minor – consolation that MMS can take from the sorry saga of AWL is that the dismal showing of many major oil companies in the deepwater province negates the possibility that the current pattern of lease ownership might be the product of an explicit collusive agreement between bidders. Simply put, it is inconceivable that parties to such a collusive bidding ring would have idly stood by while Shell took them to the cleaners in the manner implied by Figure 10.8.

The conclusion that the industrial structure in the deepwater GOM is not the product of outright collusion is supported by the findings of Pulsipher, Iledare and Mesyanzhinov. According to these authors, 'neither aggregate measures used to analyse concentrated market and industry structures nor patterns of joint bidding among firms active in

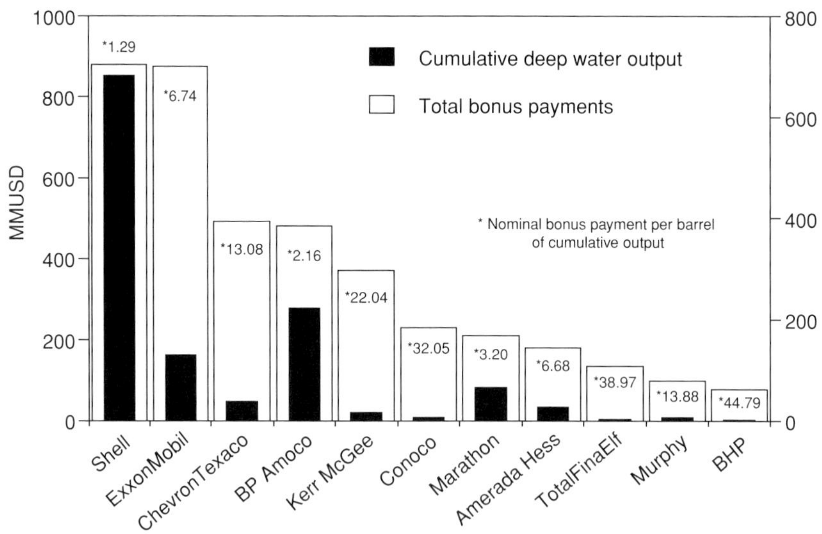

Source: MMS

Figure 10.8: Deepwater Bonus Payments and Cumulative Production by Company (Including Payments and Production by Predecessor Companies), until 2000

the offshore Gulf of Mexico suggest either a decrease or a deficiency in the competitiveness of the US Minerals Management Service.'[138] In actual fact, though, their econometric analysis does not confirm that competition in this market has been healthy. Rather, it indicates that the bidding process has not been vitiated by collusion (collusion and competition are antithetical, of course, but the absence of the former does not presuppose the existence of the latter, due to the strong entry deterrence effect that the winner's curse can exert on disadvantaged participants in an auction market, for instance).

Before the adoption of AWL, it had been incontrovertibly demonstrated that 'joint bidding tend[ed] to raise the number of bids received by the government ... [with the] lowering of entry barriers apparently more than offset[ing] any cooption of rivals'.[139] As Smith observes, joint bidding constituted 'an effective institutional device for diversifying risks that are by nature indivisible ... [with the] direct effect of diversification ... reflected in higher bids [and] the indirect effect ... reflected in the rapid historical growth of consortia formation relative to the practice of solo bidding'.[140] Pulsipher, Iledare and Mesyanzhinov confirmed that cooperation between bidders in lease sales held under AWL rules still did not lead to lower prices, and in fact continued to be associated 'significantly with higher winning bonus bids than is the case for bids by solo ventures'.[141]

These authors observed 'a higher degree of competition for leases in which non-majors were involved either through joint or solo bidding arrangements than those that involved only majors under the same bidding environment from 1983–1999'.[142] About 12 percent of the total leases awarded to larger companies (usually on the Restricted Bidders List) were obtained through joint bidding (in years when at least some of these companies were not included in the restricted list, obviously), with these leases accounting for a disproportionate 33 percent of the total bonus outlay of restricted bidders.[143] In contrast, leases won by large firms bidding on their own 'were subject to fewer bids by fewer bidders and were won by significantly lower bids than leases won by firms that did not include restricted bidders'.[144] Again, both of these findings indicate that while AWL may have been the cause of a host of problems, collusion amongst bidders (expressed through joint lowball bids) has not been one of them.

The leading bidders in GOM acreage auctions (Shell, BP and Exxon) on average resorted to joint bidding less extensively,[145] in part because they appeared in the restricted bidders list more often than other companies, but also because the combination of their size and expertise and the very low acreage acquisition costs at the onset of the AWL era allowed them to put together large deepwater lease inventories while sparing them of the need to bring other companies on board to share the burden. Among Shell's winning bids, 15 percent were made jointly, and these accounted for 22 percent of the company's bonus outlay. For its part, 25 percent of Exxon's high bids were made jointly, and these bids accounted for 19 percent of the company's bonus outlay. As far as BP goes, 12.5 percent of its winning bids were made jointly, but these bids accounted for only 10 percent of the company's bonus outlay.

Other large companies were much slower off the mark in the 'leasing-up' process of the deepwater GOM. As a result, these companies had to resort extensively to joint bidding, and this had the paradoxical effect of driving up their lease acquisition costs. For instance, Mobil was close to the group average in terms of the bids it submitted jointly, but these bids accounted for 66 percent of the company's bonus outlay. The comparable figures for Amoco were 42 and 31 percent, respectively. Texaco, for its part, had the highest proportion of joint bids among companies on the restricted list (56 percent), with these bids representing 60 percent of the company's bonus outlay (putting it second only to Mobil in this respect).[146] Significantly, 28 percent of Texaco's total of 493 joint winning bids involved cooperation with Chevron, and 96 percent of these Chevron-Texaco joint bids were placed during three sales held in 1996 and 1997, years before the merger between these companies. Pulsipher, Iledare and Mesyanzhinov attribute this conduct to an intense desire on the part of these two companies to broaden their exposure to the deepwater GOM quickly, and to catch up with Shell.[147]

As a final point to conclude this section, one should make it clear that Shell deserves great credit for being able to press home to the greatest effect the many advantages that AWL gave it. Indeed, its deepwater success is a testament not so much to the inexcusable economic illiteracy that underlies AWL as to the excellence of this company's applied geosciences, on the one hand, and its unflagging commitment to research and development, on the other. To a far greater extent than other majors (especially Exxon), Shell was able to translate its extensive lease holdings and its R&D efforts into deepwater output, in the process achieving an enviable figure for acreage acquisition cost per barrel of oil equivalent produced.

As far as BP goes, this company succeeded in carving out a position for itself in the deepwater despite its late starter status, but only by taking enormous risks and running bills that very few companies would have had the resolve or the resources to imitate. BP's success can be traced to its bold decision to abandon the proven Shell deepwater exploration strategy of looking for complex stratigraphic traps adjacent to salt bodies. Taking advantage of advances in imaging and drilling technology, BP essentially opened up a new deepwater province, through the very risky and costly exploration strategy of looking for what appeared to be less complex and much larger traps under the mammoth salt sheets that cover vast swathes of the northern GOM. It was in one of those traps that BP found Thunder Horse, still the largest field in the GOM deepwater. None the less, one should not forget that BP underwrote its ambitious deepwater exploration and development programme at least in part with the cash flows generated by its highly profitable natural gas operations in GOM. Output from these operations, which were part of Amoco's dowry, came for the most part from blocks leased under the auspices of AWL. BP's case, therefore, tends to reinforce rather than undermine the case regarding the anti-competitive effects that AWL has had on the structure of the offshore upstream sector.

10.7 By Way of Conclusion

As mentioned before, conventional wisdom has it that the amount of money paid for an offshore lease depends on the level of interest that it manages to arouse in competing bidders, a factor that in its turn 'is more directly related to an area's resource potential than to the method of leasing'.[148] This proposition implies that throughout the 1980s and early 1990s, prices for deepwater blocks were depressed because – particularly in the aftermath of the 1986 crisis – no one wanted to accumulate acreage whose development odds appeared very remote, given the prevailing cost and price expectations. We have argued, though, that bonus payments would have been much higher had MMS bided its time, and offered promising deepwater acreage on the basis of a mechanism similar to tract nomination. In such circumstances even smaller companies – i.e. more risk averse and less capable of handling overseas operations – would have participated in the bidding process almost from the very outset, as this would have given them the chance of securing challenging but highly prospective acreage only a couple of hundred miles away from home. As it was, though, AWL allowed large companies – able to carry the costs of very long-term speculative investments even after the 1986 price crash – to establish a corner on the most prospective deepwater acreage.[149]

In a doctrinaire neoclassical *Weltanschauung*, 'the power to license' is seen as being synonymous with 'the power to exclude', as it is held to be self-evident that regulatory entities cannot resist 'the imposition and administration of restrictions on entry, and on what might otherwise have been independent and competitive price and output decisions' in favour of the firms that they are supposed to oversee.[150] In actual fact, as Laffont and Tirole point out, 'the question of whether agencies favour or discourage entry ... cannot be resolved on purely theoretical grounds ... [since] only a detailed industry study can indicate the relevance of [the factors that determine] ... whether cartelisation or excessive entry occurs ... [namely] the pro- or anti-competition nature of the agency's information ... whether the political principal is active or passive and whether competition destroys industry rents'.[151] The detailed industry study presented in this chapter shows that, under TN rules, DOI was a very active regulatory principal, with the crux of its activity consisting in coaxing advantaged players into revealing their ideas about the prospectivity of different areas, and then signalling the relative attractiveness of the areas to the rest of the industry. This information had a very salutary impact on both entry and competition, as it shifted substantial surveying and exploratory risks and costs onto the shoulders of more affluent players. Rivalry between bidders in acreage auctions, in turn, was highly effective in eliminating industry rents and delivering these into the pockets of the Federal government. The very dear acreage prices paid by lessees up until 1982, *far from being pathological*, constitute the best evidence of the fundamental soundness of the market for offshore oil and gas leases up until the introduction of AWL. Steep bonuses, in a way, were precisely what gave small- and medium-sized players a chance to compete head-on with the majors for tracts.

AWL, in contrast, transformed DOI into a passive principal, by turning the responsibility of deciding which acreage would come into play over to the companies. As a result, DOI was no longer able to convey valuable information to risk averse players (which happened to constitute the majority of the universe of bidders), a development that exacerbated informational asymmetries in the market for offshore leases and hamstrung competition. As Thaler rightly says, 'acting rationally in a common value auction can be difficult. Rational bidding requires first distinguishing between the expected value of the object for sale, conditioned only on the prior information available, and the expected value conditioned on winning the auction'.[152] AWL made it exceedingly difficult for the informationally-disadvantaged players to calculate the expected value of acreage, and forced them into behaviour that smacks of irrationality (withdrawal from lease sales initially, systematic overbidding for low value tracts later).

In loading the dice against its supposed beneficiaries, AWL marginalised these firms from the crucial formative stages of the opening of the GOM deepwater province – the key source of long-term growth for the US domestic oil industry found since Prudhoe Bay – to a far greater extent than appears warranted by the admittedly daunting technological nature of operations in this region. Although it is undeniable that AWL amounted to a massive tax break, to call it an industry-wide break would be both unfair and inaccurate: very few oil companies were in fact in a position to take advantage of the 'great acreage giveaway' and, ultimately, AWL dealt the independent oil producers a crippling blow from which, as a group, they have never really recovered.

Oddly enough, the origins of AWL hark back to a debate held during the early 1970s, in which the pros and cons of the US Federal government going into the oil exploration business were earnestly discussed. The proponents of this option had an eye on worldwide oil depletion – quite a hot topic in those days – and the spectre of exhaustion led them to conclude that it would be 'unwise, perhaps even irresponsible, for the government to sell the rights to resources with great potential value without having a very clear idea of how much they are really worth', not least because of 'the relatively greater ability of the oil companies to estimate the true resource potential of OCS lands, compared to the ability of [DOI]'.[153] By the same token, they were as keen as the rest of the US political establishment to mitigate the price effects of the OPEC revolution through an increase in domestic production. They grasped that time was of the essence in this task, and acknowledged the desirability of mobilising the talent of US oil firms at all levels, in the nationwide effort undertaken in response to foreign pressures famously decried as 'the moral equivalent of war'. However, they also understood the need to avoid squandering scarce resources in a way that would magnify US vulnerability to supply shocks at a later date. Hence, in order to square these conflicting requirements, they proposed that strictly as an *emergency* measure, the US Federal government should seek to reduce uncertainty and at the same time enhance upstream competition through a limited exploration programme initiated and managed by DOI, and jointly planned by BLM and USGS.[154]

The objective of this programme would be to obtain 'exploration data and interpretations on the major traps in the Mid-Atlantic, Southern California, and Gulf of Alaska frontier areas', as well as the deepwater GOM.[155] BLM would then offer the most prospective of the identified traps (instead of tracts), and would hold marginal land for the future. Thanks to the 'identification and evaluation of hydrocarbon deposits prior to leasing ... financing [would be] much easier to obtain even for small firms, since the relatively well-defined value of the resource in the ground would provide substantial security for the investment.'[156] Obviously, this would have a salutary effect on the number of firms participating in bidding rounds, which in turn would enable BLM to exploit to the fullest extent the phenomenal breadth and depth of the talent pool available to the US oil industry. A welcome by-product of all this, but by no means an outright goal, would be to 'increase the competitive pressure on each bidder to offer as a bid all of the expected present value of a tract beyond a normal return to capital'.[157]

Given the political clout of the US oil industry, it is obvious that this plan never stood the slightest chance of getting anywhere. Nevertheless, revisiting its details is worthwhile on account of the clarity with which it identified informational asymmetries as the key obstacle standing in the way of a more intensive development of frontier petroleum provinces. The story of the stillborn 'exploration prior to leasing for production' proposal is also useful to illustrate the analytical rigour that DOI and Congressional bureaucracies used to bring to bear on even the most politically charged issues. However, such rigour was deemed surplus to requirements once James Watt took over as Secretary of the Interior, and evangelical supply-side economics became the order of the day (in public testimony, Watt chillingly summed up DOI's environmental stewardship and conservation roles thus: 'We don't have to protect the environment – the Second Coming is at hand').[158]

According to former MMS director William Bettenberg, 'the notion of areawide leasing ... was really being developed by [offshore leasing] program professionals in '79 and '80', in response both to the perception of 'a large inventory of valuable prospects that were being held off the market' and to industry pressure 'for larger and larger sales'.[159] Initially, DOI approached the issue of how it could best accelerate offshore leasing even further in its typical cautious and nuanced way. This approach was denounced as woefully inadequate by the free-market zealots who were swept into power in 1980 on the crest of the political backlash let loose by long gasoline lines, brownouts and sundry other traumas associated in the minds of US consumers with OPEC's 1970s muscle-flexing. As a result, the policy that eventually emerged as AWL differed in some critical aspects (notably its total lack of concern for the revenue-raising dimension of offshore leasing) from most of the accelerated leasing options discussed at DOI before Ronald Reagan was elected. And just to make sure that the progress of AWL would not be impeded by the good habits and critical faculties that many BLM personnel had developed over decades spent managing offshore leasing, the bureaucracy-hating Reaganauts created a completely new agency

(MMS) to carry out exactly the same tasks that BLM and USGS had been discharging satisfactorily until then (as proven by the fact that most of the original MMS staff consisted 'of the same personnel ... from the Conservation Division [of DOI] and the Bureau of Land Management').[160]

The creation of MMS proved an effective tool to stifle opposition to AWL from within the Federal bureaucracy, but it did not and could not eliminate it entirely elsewhere in the government apparatus. That much became clear in the thorough evaluation on AWL that GAO carried out in 1985. GAO was not opposed in principle to the acceleration of offshore leasing, but its review of the first ten AWL sales reached conclusions that cast the policy in a very unfavourable light. However, GAO warned that its findings – problematic as they appeared – could only be seen as preliminary, as not enough time had elapsed to determine the policy's effects on overall domestic production, imports and prices.[161] The evolution of key financial and output variables after 1985 appeared to substantiate some of GAO's worst misgivings, but the agency's Congressional masters never requested that it reopen the case, most probably because they knew and feared what such an enquiry was bound to reveal. In the event, it took more than twenty years (during which more than 2200 production platforms were installed, and countless wells drilled) before a US government agency – in this case MMS – once again bothered to request an assessment of the impact that AWL had had on upstream competition.[162] Typically, though, the terms of reference for this study restricted 'the analyses ... to 1983–1999 to correspond to the period since the area-wide leasing policy began',[163] which meant that its authors could not address the question of whether or not the programme had in the end lived up to its promises.

When one surveys from above the tangled wreckage that is AWL, it is not difficult to see why governmental actors would want to keep this policy as far from the limelight as possible. But just how were they able to achieve this objective, given the sums of money involved in offshore leasing, the wide availability of data, and the intense academic scrutiny that the OCS programme had traditionally attracted? The answer to this question constitutes one of the sorrier sideshows to offshore leasing, and one from which the American economics profession manages to emerge with almost as little credit as the ideologues who designed AWL.

The urgent question of whether the US government was realising fair market value for OCS mineral rights was indeed taken up during the early 1980s by a number of academic studies that also focused on bidding behaviour in OCS oil and gas lease sales.[164] All of the studies concluded not only that the cash bonus bidding system was the most efficient way of making OCS blocks available to investors, but also that competition in the market for offshore leases was alive and well, and that there was no need of greater government involvement to nurture it. Although the tenor of these studies' findings was at odds with those of the GAO report cited above, this divergence makes sense, for the simple reason that the earlier publication dates of these studies meant that they were based on data covering the period 1954–1977.

Remarkably, though, the passage of time did not see the tenor of these conclusions changing meaningfully, at least if our extensive literature review on the subject is anything to go by. In other words, academic articles published *after* 1983 continued to maintain, tacitly or otherwise, that the market for offshore acreage did not change a great deal despite the radical about-face in leasing policy that AWL represented.[165] How can one square these conclusions with the arguments presented throughout this chapter? The answer is that one cannot. However, that is not a problem in terms of the validity of this chapter's findings because the studies cited above continued to rely entirely on pre-AWL auction data to underpin their analysis,[166] a key flaw that Saidi and Marsden identified in a 1992 paper. In that same paper, the authors went on to claim that their own research confirmed that OCS cash bonus bidding in a post-AWL context was still producing clearly competitive outcomes. In other words, they asserted that the findings of the articles they had criticised on grounds of their reliance on old data nevertheless happened to be correct. This conclusion was not devoid of self-serving implications, though, because Saidi and Marsden themselves only used auction data covering the 1973–1977 period!

When viewed against their valid criticisms regarding the chronological inconsistency of the data sets used in other studies on OCS leasing, Saidi and Marsden's exclusion of AWL auctions from their own analysis seems borderline perverse. Their decision was explained on methodological grounds thus: 'we ended our analysis at the 1977 date to avoid inclusion of sales where, in an apparent attempt to foster competition, the US government banned joint bidding by specific company groups'.[167] Their chief concern, in other words, lay with the internal consistency of long time series for modelling purposes. A similar explanation was put forward by Moody and Kruvant, who chose not to consider tracts leased after April 1983 because, in their eyes, 'substantial procedural changes implemented at that time … [had] resulted in the loss of vital information. For example, some tracts were sold without being evaluated, so that we have no estimate of their value of riskiness.'[168] Once again, then, data consistency is at the forefront. Authors of quite recent papers and articles on OCS leasing have continued to rely on pre-AWL data on similar methodological grounds,[169] even though little profit can be expected from examining current trends from the perspective of 30+ year old data.

From the above, it is clear that the fixation of many authors with pre-AWL data is explainable in terms of a *scholastic* interest in the *data* themselves, and their suitability or lack thereof for building very complex mathematical models of bidding behaviour at auctions. Data from pre-AWL sales are preferred because they are admittedly 'meatier' and cover more aspects worthy of being modelled. Unfortunately, models structured around these data are of little use in understanding how the market for offshore leases *currently* works. This is a rather serious drawback unless one happens to be more interested in proving the reality of models than in building models of reality. Thus, the majority of the more recent academic articles dealing with OCS leasing can be said to constitute tangible proof of the extent to which the priorities of many economists have become distorted through an 'excessive focus on sophisticated theory at the expense of elementary theory ... [with] too little attention [being] paid to the wider economic context'.[170]

Although most of the post-1983 articles dealing with OCS leasing are latter-day equivalents of debates involving angels and pinheads, among them can be found a handful of more worldly articles (i.e. interested in the market *per se*, rather than what it intimates about the behaviour of participants in a certain type of auction). Unfortunately, these articles are not all that useful to understand the state of the market either, because they *also* tend to ignore data from post-AWL lease sales. In this case, though, the determination of the chronological coverage of the data sets used seems to have more to do with the grinding of ideological axes on the part of their authors.

Sadly, the foremost practitioner of this form of data selectivity was probably Walter Mead, widely seen as the doyen of OCS leasing studies on the strength of a couple of exhaustive studies which he undertook in the early 1980s on behalf of the USGS (and which gave birth to the enormous computerised data sets that have been used time and time again by academics who have turned their attention to this subject).[171] In two of his last published pieces on OCS leasing, Mead extolled the rude health of competition in OCS leasing. He also asserted that AWL had, if anything, not gone far enough in making submerged lands more readily available to investors, in the way that privatising the submerged lands outright would have done. [172] But even though Mead's analysis was submitted for publication in 1993, he settled on 1981 as the cut-off point for his statistical data. All too conveniently, this spared him from having to reconcile both his rosy view of the market for offshore acreage and his radical policy proposals with some of the more unsavoury implications arising from post-1981 data.

In a 1998 article, Mead and a co-author went one better: on the basis of the even older auction data that Mead used for the USGS studies, they declared their belief that, over the lifetime of the offshore leasing programme (*including* the AWL years), 'the government [had] collected *at least* the full economic rent available'. They certainly suggested that quite a few aspects of the programme deserved to be reformed, but emphasised that the bidding process was not one of them: 'the experience ... with bonus bidding ... strongly hints at the optimality of this approach, despite a royalty requirement diluting the optimality of a pure bonus approach and other market interventions, such as the five-year rule and diligence requirements'.[173]

Mead's example, in terms of his avoidance of inconvenient data, seems to have been all too widely imitated by policy analysts or industry observers who turned their attention to OCS auctions after 1983. For instance, even though bid rejections declined markedly with the adoption of AWL, Kobrin's recommendation that all remaining rejection procedures be scrapped was based upon pre-AWL data exclusively.[174] Gordon, for his part, took exception even to the residual lip service that Mead paid to the largely toothless bid rejection procedures, and flatly stated that DOI 'should be free to lease for as little as $1 in toto when no competition is evident ... [to avoid] delaying resource use for what are probably reductions in net payoff of leasing due to excessive limits on offers, overly stringent rejection criteria and too elaborate appraisal methods'.[175] The designers of AWL openly confessed that they found this sort of argument 'philosophically compelling',[176] but they also acknowledged that the policy would never be adopted unless it was seen to be 'a mid-course alternative at the conceptual base' that at least incorporated modest safeguards 'to insure receipt of fair market value'.[177]

The extensive literature review conducted for this study managed to unearth only a couple of papers that steered clear of the dubious methodological path blazed by Mead. One of them was an article by Kosmo, who concluded on the strength of his analysis of late 1970s and early 1980s sales that AWL would 'favour the major oil companies ... [and] should [therefore] be reconsidered in light of its anticompetitive implications'.[178] The other was an article co-authored by one of the key contributors to the 1985 GAO report. The central conclusions of this article, unsurprisingly, were diametrically opposed to those espoused by Mead: 'the evidence does not support the Federal government's position that areawide leasing had no effect on lease prices ... We believe that the government has made a serious and costly mistake in its adoption of the areawide programme.'[179]

Moody and Kruvant's conclusions were not really the sort of thing that people in authority within high policymaking circles in the USA wished to hear, and their recommendation that AWL be jettisoned and replaced with a revamped version of TN sank without making a ripple. In fact, during the 1990s, discussions on the future direction of OCS leasing policy moved even further away from the idea of resurrecting the TN procedures, and towards the far more radical proposal of privatising the OCS submerged lands in their entirety.

As a first step in this privatisation process, MMS would be required by law 'to replace royalty-in-value with royalty-in-kind through competitively selected qualified marketing agents (QMAs) ... [that] would buy and dispose of the federal government's royalty oil and gas'.[180] In this way, the royalty entitlements (and associated cash flows) would be severed from the production activities proper, and they could then be sold following the example of the privatisation of BNOC's entitlement oil by the British government.[181] The taking of royalty in kind (RIK) by the US Federal government, therefore, would function 'as a mechanism to facilitate greater reform toward market reliance, namely, privatising the income streams objectified by QMAs'.[182]

This type of RIK provision was actually incorporated in the Royalty Enhancement Act of 1998 (House Resolution 3334), later disembowelled (to President Clinton's great chagrin) by representatives from Texas, Alaska and New Mexico. Despite this setback, in the late 1990s MMS went ahead with three small pilot schemes, which were completely overshadowed from April 2002 onwards by the start-up of the joint DOE/MMS initiative to fill the Strategic Petroleum Reserve (SPR) to capacity with oil exchanged for royalties taken in kind. Thanks to this initiative, RIK activities in MMS have received an enormous boost: in late 2004, around 80 percent of OCS crude oil royalties were being taken in kind (with 50 MBD involved in the small refiners programme, around 3 MBD involved in straight exchanges with commercial purchasers and around 115 MBD going to the SPR).

When the SPR reaches its 700 MMB capacity sometime in the near future, this will throw a spanner in the works of the RIK programme. No specific provisions to cope with this eventuality have been laid out in MMS's recent *Five Year Royalty in Kind Business Plan*, which is supposed to outline the 'business principles, objectives, and specific action items that will guide and evolve the Federal RIK programme from fiscal years 2005 through 2009.[183] Nevertheless, the existence of such a plan – to say nothing of the way in which results achieved thus far by the GOM RIK programme have been trumpeted out of all proportion to their modesty (Table 10.7) – demonstrates MMS' commitment to the wider OCS privatisation agenda (especially when one considers that in the late 1990s, the agency had said that the RIK programme would cost the government a minimum of USD 357 million per year!).[184]

Full OCS privatisation, as Bradley and Mead saw it, would involve the capitalisation of extant leases 'pursuant to a mutually acceptable method to the lessee or lessor for sale and transfer', and the award of new leases to 'the firm(s) or organisation(s) submitting the highest qualifying bid',[185] with such leases being granted in perpetuity albeit 'without "diligence" or "five year" requirements'. Bidding for these perpetual leases would take place under the same conditions that contributed to make AWL such a disaster on both the fiscal and competition fronts,

Table 10.7: MMS. FY 2004 RIK Programme Revenue Performance

	Volume (MMCFD or MBD)	*Gain (loss) USD/MMBTU or USD/B*
Natural gas		
Pipeline system		
ANR Nearshore	56.7	0.003
Columbia	18.3	0.034
Central Texas Gathering System	25.3	0.083
Garden Banks	52.0	0.100
High Island Offshore System	52.4	0.024
Manta Ray	19.2	0.123
Matagorda Offshore Pipeline System	10.9	-0.095
North High Island System	53.8	0.070
Seagull/Blessing	20.2	0.017
Stingray	43.3	-0.005
TGP 500/Viosca Knoll	34.5	0.126
TGP 800	29.2	0.004
Texas Eastern Transmission Co.	14.8	0.252
Mississippi Canyon	42.7	0.031
Viosca Knoll	36.6	0.049
Total natural gas	509.9	0.051
Crude oil		
Small refiner programme	35.4	0.22
Unrestricted RIK programme	2.29	0.42
Total crude oil	37.69	0.23

Source: MMS

except that far larger blocks would be offered. In this way, 'bidders … [would be able to] select contiguous tracts to avoid drainage competition … [hence removing] the need for mandatory unitisation'.[186] By complete coincidence, this feature would also enable informationally-advantaged bidders to buy large extensions of OCS land far more cheaply than would be possible if tract sizes were kept small.

In the paper where they spelled out these details of their proposal for OCS privatisation, Bradley and Mead acknowledged that they had not examined the latest available data for lease sales. However, they were confident enough in the fundamental soundness of the market for offshore leases to issue an open invitation for 'an updated analysis to assess the merits of bonus bidding in the outer continental shelf'. According to them, 'such a study … [would] strengthen the economic rationality findings [on the mode of operation of OCS leasing] … due to technological advances in drilling and production that reduce reservoir uncertainty and bidding error'.[187] This chapter constitutes not only a response to their invitation, but also a vigorous denial that their expectations of optimality have come anywhere near being realised, despite the fact that the advances in drilling and production technologies since they published their paper have actually been far greater than they ever imagined. Indeed, this chapter shows that AWL failed miserably in delivering any of its promised results and, instead, it gave rise to all sorts of undesirable consequences.

In the panic to respond to the OPEC revolution, DOI relaxed its previously unimpeachable leasing standards quite significantly, although the negative effect that this could have had on the going rate for OCS acreage was overshadowed by the tidal wave of high oil prices. However, Nixon's accelerated leasing initiative was in fact a model of prudence compared to Watt's brainchild. However unsatisfactorily, the former still sought to create and maintain leasing conditions that reduced uncertainty and enhanced competition to a large extent. The promoters of the latter, in contrast, saw in these aspirations nothing more than unwanted ballast and summarily jettisoned them by the wayside. In so doing, they ended up by providing yet another confirmation that, as Joseph Stiglitz observed in his Nobel Prize for Economics address, 'policies based on models that depart as far from reality … often lead to failure'.[188]

In view of the above, it is surprising to see that AWL is everywhere hailed as a major success despite the ready availability of data indicating otherwise. Indeed, in some quarters, AWL is seen merely as a stepping stone to even more far-reaching policies, which are not only based on equally spurious premises but would also be largely irreversible if they were ever to be adopted. So how is it that spectacular failure has nonetheless failed to put a dent in the image of AWL?

One plausible explanation for this may lie in the counterproductive fixation that many current practitioners of the dismal science have regarding the concept of 'optimality', and the way in which this has made them unable to distinguish between viable and ideal modes of organisation in the regulatory and business arenas. As another Nobel laureate in economics once put it,

> contemplation of an optimal system may provide techniques of analysis that would otherwise have been missed out and, in certain cases, it may go far to providing a solution. But in general the influence has been pernicious. It has directed economists' attention away from the main question, *which is how alternative arrangements will actually work in practice*. It has led economists to derive conclusions for economic policy from a study of an abstract of a market situation.[189]

This type of unrealistic conclusion has inevitably begotten unrealistic policies, which is why one hears so much in oil policy making circles about *ex ante* variables like optimal rates of taxation or resource extraction, optimal allocation of risk, perfect foresight in the calculation of excess profits and *expected* fiscal revenues,[190] and so little on *ex-post* outcomes (like those detailed in this chapter). And the fact that so few industry observers seem troubled by this is a testimony to the way in which US energy policy has fallen prey to the condition so appropriately designated by Stiglitz as market fundamentalism, a 'simplistic view of competitive markets with perfect information, inappropriate even for developed countries ... [is] particularly inappropriate for developing countries'.[191] The next chapter discusses the unfortunate consequences that have followed from the spread of market fundamentalism to the exceptionally inappropriate milieu represented by developing countries that depend on oil exports for their subsistence.

NOTES

1 Riddle, Snyder and George 2001: 4.
2 Mead 1993 and 1994.
3 OTA 1975b: 20.
4 *Ibid.*: 4; italics ours. GAO (1975: 4) described the Nixon programme as having been 'hastily conceived ... under pressures exerted by the presence of the energy crisis ... developed with little input by the operating levels of BLM and USGS and based on overly optimistic assumptions and inadequate data, adopted by Interior policy officials despite opposition from

programme personnel … and developed and adopted without considering environmental impacts … or alternatives to large-scale expansion of Shelf leasing'. In particular, GAO expressed amazement at the fact that 'at most, 2 weeks was spent drafting the proposal before it was submitted' (*ibid.*: 8).

5 Bradley 1996, v.1: 306.
6 GAO 1985a: 18.
7 *Ibid.*
8 Pulsipher, Iledare and Mesyanzhinov 2003: v.
9 *Ibid.*: 32.
10 *Ibid.*: 31.
11 Klemperer 2002: 186.
12 Kreidler 1997: 197.
13 Millsaps and Ott 1985: 383.
14 Capen, Clapp and Campbell 1971: 641. The previous paper was Rothkopf 1969. However, according to the paper's author (then in the employ of Shell in the USA), a major source of inspiration for it was a letter sent to him by none other than Capen, suggesting the incorporation of uncertainty in his bidding models. This letter was illuminating for Shell, among other things, because 'it implied that ARCO was using models for oil tract bidding' (Rothkopf 2000: 10). In the opinion of Rothkopf (*ibid.*), the Capen, Clapp and Campbell paper 'is more cited in the economics literature than any other paper from the petroleum engineering literature'. For his part, Thaler (1988: 197) calls the Capen, Clapp and Campbell article 'marvellous'.
15 *AAPG Explorer*, December 2004: 24.
16 Klemperer 2002: 173.
17 *Ibid.*
18 *Ibid.*
19 Capen, Clapp and Campbell 1971: 645, italics ours.
20 *Ibid.*: 641.
21 Hendricks, Porter and Boudreau 1987: 538.
22 Saidibaghgandomi 1987:173.
23 Gilley, Karels and Leone 1986.
24 Smith 1982: 256.
25 *AAPG Explorer*, December 2004: 25.
26 *Ibid.*
27 *Ibid.*
28 *Ibid.*
29 Thaler 1988: 200.
30 Those of Shell and Mobil are described in Rothkopf 2000, and that of Gulf in Keefer 1991.
31 Five years after the publication of the Capen, Clapp and Campbell article, in the same journal, Dougherty and Lorenz (1976) concluded that, if one compared 1950s and 1970s vintage auctions, neither MLOT indicators nor bidding patterns changed a great deal, despite the fact that during

the former period there had been no shortage of oil, and prices had been low.

32 Saidibaghgandomi 1987: 9.

33 McDonald 1979: 175. The effect that this had on acreage prices led McDonald to conclude that the 'sharp increase in the rate of leasing of OCS land seems not well calculated … to increase unambiguously the present value of expected economic rent' (*ibid.*).

34 Mead, Moseidjord and Sorensen 1983: 42.

35 These conclusions broadly parallel the findings of a previous study that Mead himself had carried out on behalf of the USGS (Mead and Sorensen 1980a and 1980b). This study concluded that the IRR for all GOM leases was 11.43 per cent (and 19.40 per cent for profitable leases). Both of these figures were significantly in excess of the rate of return on total assets that US oil companies earned over this period.

36 As Roll points out, if one takes the efficient market hypothesis seriously, a scenario like the one underlying Mead's analysis would imply that either 'financial markets are ignorant of the relevant information possessed by bidding firms, or product markets are inefficiently organised so that potential synergies, monopolies or tax savings are being inefficiently exploited … or labour markets are inefficient because gains could be obtained by replacement of inferior managers' (1986: 201).

37 This is a conclusion that Mead advanced in other studies: 'the close correspondence between after-tax rates of return for OCS lessees and for manufacturing industries [10.74 per cent and 11.7 per cent, respectively] generally implies that lessees have not been able to profit at the expense of the Federal government by acquiring offshore leases at bargain prices' (Mead *et al.* 1985: 54).

38 It is estimated that there are a total of 4.5 million royalty owners in the USA at present (Rutledge 2003: 5), with 1 million of them linked to production in Texas alone.

39 The standard deviations for each category were as follows: 54 per cent for shallow water bids, 86 per cent for deepwater bids, 51 per cent for non-competitive bids, 41 per cent for competitive bids and 46 per cent for all bids (Pulsipher, Iledare and Mesyanzhinov 2003: 37).

40 Saidibaghandomi 1987: 324.

41 Riddle, Snyder and George 2001: 13.

42 Traynor *et al.* 2002: 27.

43 Akerlof 1971.

44 Moody 1994: 346.

45 Gramling 1995: 160.

46 Millsaps and Ott 1985: 382.

47 GAO 1986: 9.

48 Priest 2004: 50.

49 Veldman and Lagers 1997: 84–5.

50 Priest 2004: 49, italics ours.

51 *Ibid.*: 50.

52 Veldman and Lagers 1997: 84–5. See also Priest, *ibid.* The million dollar moniker was coined in reference not so much to the profits that some companies were able to reap thanks to the three-week-long course, as to the astronomical fees that they had to pay for the privilege of attending: the fee per delegate was a staggering USD 100,000 (about USD 460,000 in money of 2000). Nevertheless, the SONJ (now Exxon) affiliate Humble sent ten delegates and SOCONY (later Mobil) twelve. The US Geological Survey also signed up to the course.

53 Priest 2004: 50.

54 The rejection rate for drainage tracts was somewhat higher: between 1959 and 1979 the government rejected the highest bids on 58 out of 295 drainage tracts offered for lease (Hendricks, Porter and Wilson 1994: 1416). Bids over 5 million dollars for drainage tracts were generally accepted, leading these authors to conclude that the purpose of the rejection policy was 'to reduce the incentive that firms might have had to bid the preannounced minimum price on tracts that, on the basis of public information, were regarded as low value' (*ibid.*: 1417).

55 Sherrill 1983: 238.

56 *O&GJ*, 8 April 1974: 40.

57 Hendricks, Porter and Wilson 1994: 1416.

58 GAO 1985a: 30.

59 Sherrill 1983: 238.

60 Uman, James and Tomlinson (1979: 490) looked at 14 lease sales held between 1969 and 1975, and reached the sobering conclusion that 'the ability to predict the volume of recoverable resources … appears to be limited to within a factor of 10'. Davis and Harbaugh (1980: 1047) highlighted 'a tendency to overestimate resources in advance of drilling when the presale estimates [were] … less than around 10 [MMBOE] … and to underestimate resources when the presale estimate … [was] greater than 10 MMBOE'.

61 OTA 1975b: 23. With the passage of the OCSLA, the responsibilities of the Bureau of Land Management and the Conservation Division of the US Geological Survey expanded to include offshore mineral leasing, but neither was given additional resources to discharge their new functions. Thus, the BLM and USGS held a consultation exercise with the oil industry and simply adopted most of the industry's suggestions. The position of the agencies in terms of the practicalities of handling lease sales was not significantly better. For instance, even though the first call for tract nominations was put out during June 1954 – for lease sales set for October and November – the USGS only opened its new regional Oil and Gas Leasing Branch office in New Orleans in July, with a staff of ten, while the BLM would wait a further year before opening its own regional OCS office in the city.

62 Priest 2004: 36.

63 *Ibid.*: 44.

64 OTA 1975b: 22.

65 Mead 1993: 226.
66 DOI's Monte Carlo simulations to determine the value of a tract required at least 1000 iterations per tract.
67 GAO 1985a: 94.
68 *Ibid.*: 43.
69 *Ibid.*: 39. For instance, the submission of three bids of USD 1 million, USD 2 million and USD 5 million would guarantee that the high bid would win valued at up to USD 62.5 million.
70 *Federal Register*, 20 July 2000: 45103. In principle, MMS has the faculty of changing the minimum bid level every time it conducts a lease sale, although it must announce the figure well in advance of it. In practice, changes to the minimum bid level have always tended to point downwards. For instance, DOI's 1998 appropriations bill included a rider that prevented MMS from requiring a minimum bid on offshore leases (US Congress 1998: 40).
71 Logue, Sweeney and Willett 1975: 204.
72 *Tulsa Geospectrum*, March 2004: 5.
73 Keefer 1991: 393.
74 Kosmo 1985: 93.
75 Personal communication with John L. Sweezy, LLOG Exploration Co.
76 In interviews reported by Seydlitz, Sutherlin and Smith (1995), different officials at independent oil companies credited AWL with allowing their companies 'to become active in Gulf exploration and production' (7), 'to enter and compete in the Gulf' (31), and – even more implausibly – to 'participate in generation of prospects Gulfwide without having to necessarily compete with companies that nominated blocks, or try and generate prospects only on lands not nominated by others'(74). The puzzling support of non-major oil companies for AWL has a precedent in the Mandatory Oil Import Programme (MOIP), a programme which independent oil companies tenaciously defended even though the majors benefited from it to a far greater extent than they did.
77 Saidibaghgandomi 1987: 176.
78 OTA 1975b: 79.
79 Ibid.
80 Reece 1978: 381.
81 *Ibid.*: 381–2.
82 In Sale 198 (March 2006) Amerada Hess bid USD43 million for Block GC287.
83 Even though the block had produced gas from shallow sands up until 1994, ExxonMobil was forced to relinquish it because the company did not take any positive action after the shutdown of the last productive well within the lease.
84 *PON*, 21 August 2003: 4.
85 Richardson *et al.* 2004: 37.
86 Kosmo 1985: 95.
87 Sherrill 1983: 239.

88 *O&GJ*, 8 April 1974: 38–9.
89 Seydlitz, Sutherlin and Smith 1995: 74.
90 OTA 1975b: 34. The supporters of AWL disingenuously suggested that the ratio of acreage leased to acreage offered (slightly less than 50 per cent) reflected not so much limitations on the physical capabilities of the industry to explore as the inherent inefficiency and incompetence of a governmental bureaucracy: 'The wisdom of the bureaucracy since 1953 served up roughly 50 million of the 1 billion [OCS] acres. The industry actually leased only 22 million of the 50 million acres. Obviously, what the bureaucracy felt was attractive for oil potential was not considered attractive by the industry more than half the time' (MMS 1983: 19–20).
91 In addition, up until 1996, holders of 5- and 8-year leases had to submit either exploration plans or statements of intention to explore by the end of the fourth or the fifth year of the lease term, respectively. Holders of 10-year leases never had a set milestone for submission of exploration plans.
92 *Ibid.*: 32. At the time, the USGS still believed – disarmingly – that OCS production by 1985 could be as high as 20 MMBD, on the strength of a fanciful estimate of 400,000 MMB of undiscovered recoverable oil (*ibid.*: 3).
93 1987: 518.
94 *Hart's Oil and Gas Investor*, Deepwater supplement August 2002: 8.
95 Richardson *et al.*2004: 107.
96 The figures for lease expiration assume that all currently fallow leases expire at the end of their primary lease terms and are not given lease-term extensions.
97 Richardson *et al.* 2004: 42.
98 *Ibid.*
99 OTA 1975b: 6.
100 The accelerated TN programme contemplated increasing the leasing rate to an average of 3 million acres per year over the period 1974–9, about 2 million acres per year for the period 1980–84 and about 1.5 million acres per year for the period 1985–9 (McDonald 1979: 90). In practice, leasing figures under AWL for the 1983–9 period averaged 5.16 million acres.
101 Hendricks, Porter and Boudreau 1987: 533.
102 According to Hendricks, Porter and Boudreau (ibid.: 519), the winners of leases drilled a short time after assignation were able to capture 'a higher percentage of social rents, as much as 37 per cent, versus 26 per cent for all leases'.
103 *Ibid.*
104 *Ibid.*: 525–7.
105 *Ibid.*: 537.
106 Richardson *et al.* 2004: 104.
107 *Ibid.*
108 *Ibid.*: 107.
109 *Hart's E&P*, special supplement April 2005: 49.

110 Riddle, Snyder and George 2001: 1.
111 Logue, Sweeney and Willett 1975: 204.
112 Moody and Kruvant 1988: 38; italics ours.
113 US Senate 2002: 187–9
114 In its investigation on the US gasoline market, for instance, the US Senate reported that it was only as a result of a fortuitous challenge on a rate request of a regulated products pipeline in Michigan (Wolverine) that FERC discovered not only that this pipeline and its affiliates had engaged in practices that 'violated the Interstate Commerce Act [ICA], *some for over twenty years*' but also that 'had not the rate request been challenged, it is likely these discriminatory practices would have continued, and it would have been more difficult for independents to compete' (US Senate ibid.: 16; italics ours). Tariffs can be long-lived, but they are not set in stone (in 1930, the US Supreme Court said that a rate order can never be *res judicata*). FERC itself has made it clear that '[t]he fact that a rate was once found reasonable does not preclude a finding of unreasonableness in a subsequent proceeding' (*OXY USA, Inc. v. FERC*, 64 F.3d at 690, 1995).
115 Barents 2000: 17.
116 Hearing order: 61 FERC 61,051 (1992) for Oxy Pipeline, Inc. In addition, on October 10, 2003, the U.S. Court of Appeals for the District of Columbia Circuit affirmed the district court decision which found that sections 5(e) and (f) of the OCSLA grant FERC only limited authority to enforce open access rules on the OCS. The FERC regulations that the court held invalid are those that require owners of OCS gas pipelines to file information indicating the rates the pipelines charged, the conditions of the service they provided, and whether they were affiliated with any of the shippers using their pipelines. The FERC regulations addressed OCS natural gas facilities that perform production or 'gathering' functions, and do not fall within the FERC's jurisdiction under the Natural Gas Act (NGA) of 1938 (*Williams Cos. v. FERC*, 345 F.3d 910, D.C. Cir. 2003).
117 *Shell Oil Company v. FERC* (1995).
118 Hearing order: 61 FERC 61,050 (1992) for Bonito Pipe Line Company. Shell's objections were dismissed by the Court of Appeals for the District of Columbia Circuit.
119 By 2020, it has been estimated that the Forties line will be the only economically viable system in the UKCS (Arnott 2003: 16).
120 *Ibid.*: 17.
121 *Ibid.*
122 DTI 1996: 2.
123 Arnott 2003: 18.
124 *Federal Register* 69 (70), 12 April 2004: 19138.
125 US Senate 2002: 187–9; italics ours.
126 *Ibid.* As a consequence of this gambit, independent shippers in the line were not in a position to acquire an 'assured commitment … to access the tankage storage [*sic.*]', which meant that they could not make 'use of the through rates posted in the tariff', and also that they were discouraged

'from entering into long term purchases of commodity products which could lower procurement costs' (*ibid.*).

127 EMPCO has leased the lines that connect the SPR site at Bryan Mound (located in the vicinity of Quintana) with refineries in the Texas City area, on the one hand, and to the Jones Creek SPR site (itself connected to a major pipeline running to the Midcontinent), on the other. BP extricated the details of the lease agreement from SPR through a series of Freedom of Information Act (FOIA) requests and appeals lodged at DOE (see Case No. VFA-0503, 27 DOE 80,216 and Case No. VFA-0522, 27 DOE 80,236).

128 Laffont and Tirole 1993: 60.

129 Macho-Stadler and Pérez-Castrillo 1997: 157.

130 Stiglitz 2002: 471.

131 *Ibid.*

132 GAO 1985a: 34; italics ours. The judicial acceptance of this definition went very much against the grain of established precedent. As Cox, Isaac and Smith (1983:48) point out, up until the early 1980s there was 'substantial evidence to suggest that the intent of Congress … [was always] to equate "fair market value" with "greatest possible revenue"', as witnessed by the use of the latter term in the key 1977 House Report on offshore leasing that heralded the introduction of alternative bidding methods in OCS lease sales (for the latter see Boué 2002). MMS underpinned its novel distinction between fair market value, on the one hand, and maximum dollar return, on the other, by alleging that the latter was the product of the government behaving like a monopolist in the assignation of Federal OCS lands (MMS 1983: 25). But although it is undeniable that the government is a monopolist in this regard, there is no evidence whatsoever that it exploited its position to extort from the industry anything more than the Ricardian rents generated in the OCS.

133 See Klemperer 2002.

134 OTA 1975b: 20.

135 *Ibid.*: 4. Back in 1974, there was genuine unease within DOI ranks when the acceleration of offshore leasing took the proportion of tracts receiving two bids or less to 63 per cent of the tracts leased in Sales 34 and 36, while accounting for 34.1 per cent of the accepted high bonuses. In the previous three sales, the proportion of tracts receiving only one or two bids had only been 37.1 per cent, and they had accounted for only 10.9 percent of the bonus money accepted (*ibid.*: 20).

136 Sherrill 1983: 485.

137 *Ibid.*

138 Pulsipher, Iledare and Mesyanzhinov 2003: 15.

139 Moody and Kruvant 1988: 281. In large part, this is due to the fact that most joint ventures involved large and small firms. Hendricks and Porter (1992: 509) discovered that bids by joint ventures involving large firms were only marginally higher than the average winning bids submitted by large firms on their own. According to these authors, 'joint ventures enhanced

competition ... by allowing large firms to compete more vigorously on tracts where expectations of deposit sizes [were] high' (*ibid.*: 506). In addition, joint bidders relatively adjust their bid upward by a greater amount than the solo bidders in response to more competition, suggesting that joint bidders have been more aggressive than the solo bidders (Saidibaghgandomi 1987: 230).

140 Smith 1982: 258.

141 Pulsipher, Iledare and Mesyanzhinov 2003: 13. According to their calculations, 'when leases were won through competitive joint venture bidding, the average value of the high bonus bid per lease was USD 3.29 million, more than twice the value of noncompetitive, joint venture high bonus bids, which was USD 1.18 million. The average high bonus bid for competitive solo venture bids for the period 1983–1999 was USD 1.75 million in comparison to an average high bid value of $0.614 million for noncompetitive solo ventures' (*ibid.* 3–4).

142 *Ibid.*: 54.

143 Between 1983 and 1999, restricted bidders accounted for about 47 per cent of the 13,946 leases awarded (with Shell high bids alone accounting for 14 per cent of the total). Restricted bidders, as defined by the Energy Policy Conservation Act of 1976, are those companies whose global petroleum production is in excess of 1.6 MMBOED and, for that reason, are banned from submitting bids jointly in Federal OCS lease sales. The Restricted Bidder List is compiled on an annual basis, so companies can be in it one year and out of it the next.

144 Pulsipher, Iledare and Mesyanzhinov 2003: 11. This finding totally contradicts Tyler Priest's contention in the sense that, with tract nomination, 'even the largest firms could not afford to bid alone, and had to bring in partners to offload some of the capital risk – and, although they would not admit this, reduce the competition' (in McKay and Nides 2005: 80).

145 Pulsipher, Iledare and Mesyanzhinov 2003: 62.

146 Interestingly, before the advent of AWL, Mobil tended to bid chiefly in partnership with other large firms, and only very occasionally to bid by itself or with smaller companies. Texaco, for its part, enjoyed the dubious reputation of being the company that tended to overestimate most egregiously the value of OCS tracts (Hendricks, Porter and Boudreau 1987: 528).

147 Pulsipher, Iledare and Mesyanzhinov 2003: 64.

148 OTA 1985: 154.

149 For instance, between 1983 and 1993, Shell managed to amass a total of 600 GOM deepwater tracts, which between them reportedly covered 60 per cent of the deepwater area that could be illuminated with traditional 2-D seismic techniques.

150 Laffont and Tirole 1993: 538.

151 *Ibid.*: 537.

152 Thaler 1998: 192.

153 OTA 1975b: 17.

154 *Ibid.*: 28. BLM would contract for seismic and drilling services with the advice of USGS, which would also oversee and regulate the conduct of the exploration effort and interpret the results. BLM would provide both the results and their interpretations to designated federal and state agencies and make both the results and interpretations available to the public at large. The US government would have exercised full management control and had complete control of the data (retaining full control over whether and when to produce any reserves discovered), but it would also have had to pay the full cost of exploration.

155 *Ibid.*: 30.

156 *Ibid.*: 37.

157 *Ibid.*: 45.

158 Watt's testimony to Congress, where he discussed his plans as the steward of the Nation's wildernesses, also included this memorable phrase: 'after the last tree is felled, Christ will come back.'

159 Farrow 1990: 137. The key promoter behind this idea seems to have been John Bookout, then head of Shell USA. Bookout arguably deserves to be called the real father of AWL and, in the light of Shell's subsequent exploration successes in the deepwater, probably made as much money for this firm as any other employee in its history.

160 *Ibid.*: 27.

161 GAO 1985a: 143.

162 Pulsipher, Iledare and Mesyanzhinov 2003.

163 *Ibid.*: 1.

164 Mead and Sorensen, 1980a and 1980b; Teisberg 1980; Gilley and Karels 1981; Rockwood 1983.

165 See in particular Mead *et al.* 1985.

166 Saidi and Marsden 1992.

167 *Ibid.*: 336.

168 Moody and Kruvant 1990: 279. Moody (1994: 348) resorts to the same argument in an investigation on alternative bidding systems. This particular article illustrates well the way in which academics are scholastically fascinated by auction results and not very concerned about their empirical relevance: by the time the article was published, these alternative bidding systems had not been used for nearly 15 years.

169 See Hendricks, Porter and Tan 2000 and 2003.

170 Klemperer 2003: 1.

171 Mead and Sorensen 1980a and 1980b.

172 Mead 1993 and 1994.

173 Bradley and Mead 1998: 213.

174 Kobrin 1986.

175 Gordon 1986: 111.

176 MMS 1983: 28.

177 *Ibid.*

178 Kosmo 1985: 95.

179 Moody and Kruvant 1988: 37–8. The authors quantified the cost of this

mistake at USD 1337 million per acre leased.
180 Bradley and Mead 1998: 228.
181 Vickers and Yarrow 1988: 320–3.
182 Bradley and Mead 1998: 229.
183 MMS 2004a: 1.
184 US Congress 1998: 40.
185 Bradley and Mead 1998: 211.
186 *Ibid.*
187 *Ibid.*: 213.
188 Stiglitz 2002: 485.
189 Coase 1964: 195, italics ours.
190 The reader is referred to any one of the works of Alexander Kemp, for instance.
191 Stiglitz 2002: 485.

CHAPTER 11

CONCLUSIONS: WHAT SHOULD THE WORLD LEARN FROM THE SUCCESS STORY OF THE DEEPWATER GOM?

Prima facie, the current GOM output bonanza is but the latest instalment in a long-running and fruitful relationship between technology and oil entrepreneurship. The stock interpretation given to the development of the deepwater province is that it proves yet again that, regardless of how daunting the technological challenges might appear, the collective ingenuity of the oil industry will in the end win the day provided that the geological as well as the investment conditions are right. In this context, AWL and royalty relief are being constantly put forward as models that other governments should strive to imitate, by rising above pressing short-term financial considerations and scrapping all restrictions on access to their upstream sectors. Only thus, conventional wisdom goes, can the long-term interests (and revenue-generating power) of their respective oil industries be properly safeguarded.

However, if the post-AWL evolution of the GOM upstream sector (as chronicled in this study) shows anything, it is that the likely outcome of simply scrapping all barriers to upstream access at one fell swoop is the emergence of a very concentrated industrial structure, dominated by large players who tend to prosper in uncertain environments that allow them to play to their financial and technological strengths, to the detriment of less advantaged players, who are hamstrung by their inability to manage certain types of risks.

There would be those who would argue that such an outcome does not constitute a calamity. After all, it is reasonable to posit that the development of an oil sector dominated by such players is bound to involve both the highest feasible standards and the best international practices, simply because larger companies tend to be more efficient operators (because of scale effects), can take larger risks and have a much greater staying power, to say nothing of better credit ratings, access to technology and capital markets, more of a corporate reputation to lose if something goes wrong, and so forth.

As Ernst and Steinhubl put it, 'structural considerations continue to bestow significant economic advantages on petroleum companies in a number of countries where there are few competitors, the right to own or access reserves is limited, and capital and risk requirements cannot be met through financial markets'.[1] These arguments are quite compelling, so much so that oil company apologists have not been alone in putting them forward. In this sense, it is worth recalling that one of the founding fathers of OPEC, Juan Pablo Pérez Alfonzo, on one occasion (when he already occupied the Venezuelan oil portfolio) openly stated his preference for seeing the smaller companies active in Venezuela abandoning the country, thereby leaving the government to deal with the majors only. His opinions in this regard are at radical variance with the niche he occupies in the collective psyche of the international oil industry, as the relentless gadfly of the oil majors. Indeed, they are to such an extent unknown even within Venezuela that they deserve to be quoted *in extenso*:

> The [independent] American oil companies that acquired Venezuelan concessions in 1955–6 cannot expect either benevolent understanding or assistance with their problems from the current administration. In fact, if they had to close down or sell their assets on account of finding themselves increasingly hemmed in by the rising costs of Venezuelan taxes [*sic.*], labour and others, and the restricted commercial opportunities in the [international] markets, then this would meet with the approval of the Venezuelan government ... The existence of numerous foreign oil companies is not good for Venezuela. This entails the unnecessary duplication of functions, personnel services, etc. It amounts to a waste of money for us ... *Personally, I would prefer to see only four or five large and efficient companies in Venezuela.*[2]

Quite apart from whether one accepts the economic logic behind arguments such as those of Ernst and Steinhubl or Pérez Alfonzo, recent evolutionary trends in the international oil industry appear to indicate that – wherever their preferences in this regard may lie – governments in oil-exporting countries (especially those sitting atop resources located in challenging plays) will indeed increasingly find themselves vying with only a handful of enormous – and enormously powerful – corporate juggernauts, as ‘smaller integrated petroleum companies, as well as midsize companies (including the smaller majors), with broad product and geographic coverage but not world-class scale or distinctive skills ... [are] left out in the cold’.[3] The objective of this chapter is to show that, while this trend towards an ever more concentrated global industry is undeniable, it has not necessarily come about as a result of an irresistible economic imperative pointing towards ever greater corporate sizes.

It is certainly the case that ‘in petroleum ... the attractiveness of a company depends on its “pipeline”, and the ... megamajors have the broadest portfolios and (potentially) the lowest cost structures, *as well as dominant positions in many of the most promising well-established basins*’.[4] However, we intend to show that one of the key reasons explaining why the megamajors do indeed enjoy the broadest portfolios and dominant positions in the choicest acreage worldwide is not merely a function of their size (although this has never exactly been a hindrance). To a considerable extent, the sheer intensity of their dominance is related to the radical laissez faire approach (epitomised by AWL) that governments in many key petroleum provinces across the globe have taken with regard to the manner and conditions under which they are prepared to grant investors access to their petroleum resources. In other words, the unassailable positions that these companies have built for themselves are as much a reflection of their acknowledged capabilities as they are of the *failure* of their host governments to implement effective and dynamic measures to prevent them from achieving such dominance.

This last assertion begs an obvious question: given the apparent superiority of these dominant companies over their peers in the efficiency league, why would governments wish to put obstacles in their way? The answer is really quite simple. The asymmetry in the relevant information available to governments (in their capacity as lessors/licensors), on the one hand, and the largest oil companies (in their capacity as lessees/licensees), on the other, has always been of considerable magnitude. Thus, the former have had to take active steps in order to reduce some of the cost/knowledge advantages of majors for the benefit of smaller and less efficient players, strictly as a means of loosening informational constraints on themselves, and thereby increasing their capabilities to extract rents from all their present and future lessees/licensees. This was the strategy that imperial Iran used in order to cut a deal with Mattei's ENI and, ultimately, to break the stranglehold that the members of the international oil cartel had on that country's upstream, for instance. However, throughout the 1980s and 1990s, governments in many oil-producing countries (including some members of OPEC) turned their collective backs on the valuable historical precedents accumulated throughout the years (mid-1960s to early 1970s) when OPEC members (in the main) seemed to have the Seven Sisters on the run. As a result of this repudiation of the past, it will be a long time indeed before many of these countries see any tangible fiscal fruits that the foreign investment in the development of their respective oil sectors was supposed to bring.

11.1 Basin Mastery: Adding Value in Global E&P Activities

Throughout the 1980s and much of the 1990s, most private oil companies destroyed shareholder wealth on a colossal scale (to the tune of

USD 400 billion over the 1980s alone, according to Conn and White[5]), with a large chunk of this value destruction taking place at the E&P segment. To the extent that many of these E&P costs were tax-deductible, the activities of these firms destroyed public welfare as well, in many cases with the complicity of governments in thrall to false fears over having to import some of their oil (fears of the sort that prompted the Canadian government during the 1970s to allow 105 percent of intangible exploration expenses to be written off or the British government to allow the deduction of 135 percent of the finding and development costs of a given field from its PRT liabilities, for instance).

Amidst a depressing panorama of wrecked corporate fortunes, there remained standing a few beacons of profitability, whose success (all the more remarkable given its rarity) was put down by management consultants McKinsey to the radical new ways in which their respective managements, on the basis of their sound 'understanding of industry dynamics and their own capabilities', had responded to the 'key strategic questions facing the industry'; to wit: which jurisdictions to operate in, which technologies to focus on and how much to invest in technological leadership, whether to focus on oil or gas, how to manage relationships with host governments, how to magnify key capabilities or restructure the competitive environment by means of alliances, whether or not to maintain a presence across entire value chains and finally, how to use asset rationalisation and restructuring as continuous levers to improve performance, rather than one-off exercises meant to put out corporate fires.[6]

Looking at large companies in particular, McKinsey pointed out that the firms that tended to do worse at E&P value creation were those that followed a traditional exploration-led strategy aimed primarily at replacing reserves, a strategy pithily described as 'lurching from acreage round to acreage round, hoping to secure an "elephant" block'.[7] In contrast, the companies that created the greatest shareholder value were those that followed a disciplined and focused approach, not aimed at being in fewer places *per se* but rather at pre-empting other players and building an advantaged *dominant* position in a few selected low cost basins. McKinsey gave the name 'basin masters' to those companies that managed to build dominant acreage and logistical positions in difficult new plays, partly because of their skills at resource development, technology and integrated project management but chiefly by virtue of their stealing a march on competitors in remote areas where scale and infrastructure were of paramount importance. Shell's operations in the deepwater GOM were hailed as perhaps the quintessential embodiment of this new strategic conception of E&P activities.[8]

According to McKinsey, the basin master value creation model in E&P activities was the manifestation in the oil industry of a global trend whereby 'in industry after industry, spanning both the new and the old economies, a small set of companies is creating almost all of the new shareholder value. This phenomenon has created a "winner takes all" dynamic in which 5 to 10 percent of the companies in a given industry create all of the shareholder value.'[9] This 'winner takes all' model, predicated on companies 'playing to capture disproportionate share of value creation potential', became rather tarnished in the wake of a string of implosions involving firms (like Enron) that had been hailed as supreme practitioners of the new value creation arts. Perhaps justifiably, 'winner takes all' has now come to be seen as a corporate malfeasance model, where 5 to 10 percent of the winners in these companies did indeed take it all (to the detriment of shareholders, creditors and employees alike), proving in the process that the 'new economy' emperor was, if not naked, then at the very least immodestly clad.

Although the earnings that 'winner takes all' companies in many industries reported were the stuff of fantasy, in oil and gas E&P they proved to be real enough. The reason behind this is not hard to fathom. Whereas the value creation levers in 'information age' sectors consisted of somewhat fuzzy (and sometimes downright fictional) 'superior intangibles', in oil and gas E&P they were and are a function of the creation of entry barriers linked to *access and tenure of land*. Such barriers are far more difficult to surmount than any technology, information and capital barriers (hence the famous Texan adage: 'buy land, because God is not making any more').

This process of value creation is schematised in the form of a matrix in Figure 11.1. As can be appreciated, the matrix describes a situation in which the one trait that sets the global majors radically apart from the rest of the pack (their sheer size) is initially used in a diffuse (rather than a focused) way to maintain speculative slow-burn acreage positions in a variety of frontier locations and technology plays, very much as in the traditional exploration-led model. At the next stage, the large majors (relying on the advantage conferred by their size, as well as their access to technology that cannot necessarily be obtained off the shelf) start to go into focus mode, selecting those areas where they see the best chances of success for exploration bets (this, again, is not unlike what happens in the traditional model). As soon as they receive even a fleeting indication that they might be on to something, however, the companies up their acreage ante in a major way and also increase the tempo of their exploration activities (even though any initial discoveries might still be a long way from rudimentary appraisal). Crucially, the

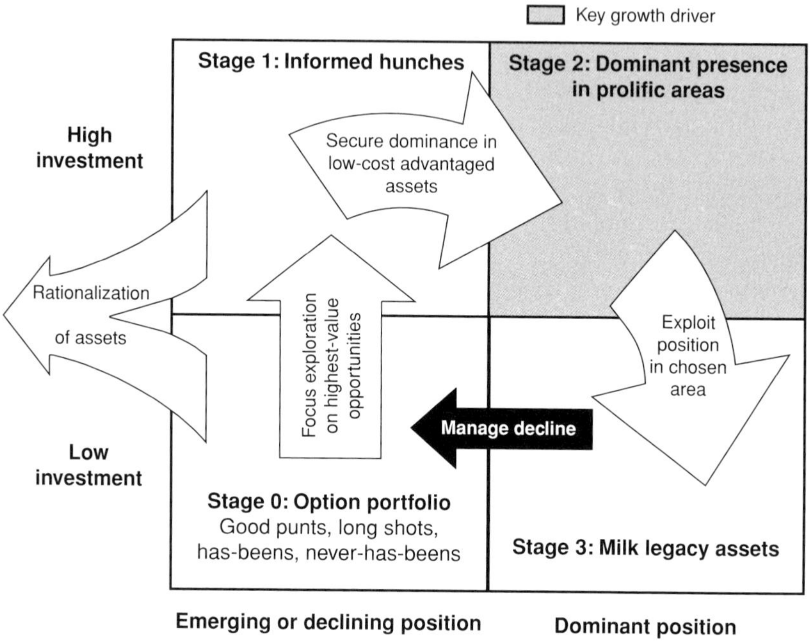

Source: McKinsey and Company

Figure 11.1: The Focused Approach to Creating Value in Global E&P

initial primary objective of this exercise is not the finding, development and marketing of hydrocarbons *per se* (as in the traditional model); rather, it is the foreclosing of future access by potential competitors to the nascent play, through the acquisition of as much meaningful acreage as possible while the going is still cheap, and the installation of early infrastructure corridors.

Once the companies have erected barriers to access that are high enough to ensconce themselves as 'basin masters', they are in a position not only to exploit their finds at their leisure as legacy assets, but also to dictate the pace of basin development, often in a way that *de facto* invests them with the prerogatives of a licensing agency (like, for instance, when they advise governments or NOCs on the specific acreage to be offered in a bidding round, or on what the minimum size of licences should be). The cornerstone of this type of control is their dominant position in early infrastructure corridors (often overbuilt, with a view towards future discoveries), which allows the basin masters to extract rents from other players through access charges to this infrastructure. Frequently, this is bolstered by cosy relationships with governments, licensing agencies and NOC partners (the latter in provinces where PSAs prevail), all of which tend to complicate the lives of potential competitors (through the appearance of bureaucratic and other, less wholesome, types of barriers) while simplifying the lives of the incumbents (cast as they are in the enviable role of 'operator of choice' for the licensing agency). In this way, basin masters can ensure that they will be able to capture the majority of the value in a given province, including that generated from operations in fields not discovered by themselves. As depletion sets in, the basin masters apply portfolio management criteria to exit the province, offloading tired fields to smaller players with lower overheads and materiality thresholds, but sometimes retaining a hold on profitable amortised infrastructure (as BP has done with the Forties pipeline). The basin masters then move on to the next big frontier play.

It is easy to appreciate how this type of first-to-scale foreclosure cycle creates value in global E&P activities. For dominant players, legacy assets like those described above translate into lower operating and drilling costs (on account both of scale and learning curve effects), low acreage acquisition costs, ample cash flows and an enhanced bargaining position against both host governments and competitors. But given that the name of the game is 'winner takes it all' (as opposed to 'winner graciously deigns to share the prize with the natural resource owners'), the assertions above beg a fundamental question: what exactly are host governments supposed to be doing on the sidelines while this cycle is unfolding?

Basin mastery may translate into very comfortable lives for a few bureaucrats and politicians in key positions in the governments of certain countries. However, for these governments as a whole (and even more so for the populations they represent), basin mastery effectively means stunted competition for acreage and consequently lower acreage prices, higher upstream entry barriers, a high degree of fiscal dependence on very few operators and even a negative impact on the eventual recovery of oil and gas in place (*vide* the Central North Sea). Indeed, the example of the GOM Federal OCS shows that imitating AWL is a good way of ensuring, firstly, that certain types of players are discouraged from investing while other players take advantage of their risk-induced paralysis to drive acreage prices and fiscal revenues down; secondly, that future entry into the upstream is compromised by the entrenched position of early movers; thirdly, that fiscal dependence on a reduced number of operators is enhanced; and finally, that more oil may be left in the ground than would have been the case in a more competitive environment. Thus, AWL illustrates clearly that basin mastery is synonymous with a significant downgrading in the capabilities of national governments to collect the excess profits generated by the exploitation of their petroleum resources.

This brings us to the question of why governments should be keen to tax rents in the first place. The arguments usually put forward in response to this question are pragmatic in bent, and draw their inspiration from one of Ricardo's key insights – that rents generate returns over and above those necessary to attract additional capital, so even if investors do not in fact receive these rents, allocative efficiency will not be distorted, nor will the potential output or the rate of production suffer. In other words, pragmatic governments target rents in general because they *can* (and pragmatic *and* desperate oil-exporting governments target petroleum rent in particular because they have to, given that there is generally little else in their countries that is either worth taxing or possible to tax). But there is an additional dimension to the taxation of rent, associated with the name of Henry George. According to George (who proposed levying taxes on rents received by landowners on grounds of equity), rents are unearned income and, as such, contrary to the very *spirit* of capitalism, which is supposed to be animated by toil and entrepreneurship rather than flukes of nature (i.e. the extraordinary fertility of a given tract of land).[10] Thus, regardless of whether rents are easy targets or not, George postulates, governments *ought* to make it their business to tax them. Not doing so is tantamount to their rewarding sloth and the luck of the draw rather than blood, sweat and tears, which is something that can only sap the vigour and drive of a market economy.

George's argument is couched in somewhat Calvinistic terms, and is made weaker for it. However, stripped of its moralistic overtones, it amounts to a robust *normative* proposition: governments *should* tax rents because when they do not, the functioning of free markets suffers and distortions ensue. After all, rents arise from natural conditions of fertility and mineral abundance whose existence is not attributable to any activity or effort on the part of capitalist firms, so players that have access to returns over and above those needed to attract investment and production end up ahead in the competitive race, even though they may not necessarily be the fittest. If rents are retained in the private domain without being subject to taxation, and the magnitude of rents is large enough relative to the size of a country's economy (even if oil is not involved), the development of the forces of production can be significantly hindered and distorted.[11] Thus, capturing these rents is the only option open to a fiscal authority that genuinely wishes to preserve a level playing field for inter-firm competition to take place.

Some would argue that this degree of vigilance is uncalled for, because the market for corporate control can efficiently sort out any imbalances arising from one given firm's access to rent, in a way that still puts its capital to the most productive use in the economy. This line of reasoning is untenable, and not only because it advocates a second-order solution when a simpler, first-order, solution will do. Rather, it is unacceptable because it entirely misses the point that petroleum rent belongs to society at large, and not to the enterprises (be they private *or* state-owned) that exploit the petroleum resources that generate it.

11.2 The Political Economy (and Costs) of Basin Mastery

It would appear, then, that a basin master type of foreclosure may only materialise if host governments, going against their best economic and political interests, allow it. Thus, at least *some* governments must have taken leave of their senses in this way, or the companies identified as successful basin masters would never have made the grade. So how does one account for this apparent anomaly?

Part of the answer lies in the fact that many of the places where basin masters have emerged lie under the jurisdiction of consuming country governments (GOM and the UK North Sea, above all). For such governments, fiscal criteria appear of secondary importance when compared to the imperative of ensuring that as much petroleum as possible is produced within as short a timeframe as possible, in order to keep its market price as low as possible. These priorities were made absolutely clear by the Undersecretary of the Interior entrusted with the implementation of AWL:

> American consumers should not be forced to pay monopoly prices for the offshore oil and gas which they themselves own, simply because the Federal government decides to withhold leases and resources from the marketplace in an effort to get a slightly higher bonus ... The primary goal of any leasing programme ... should be lower energy prices and adequate supplies.[12]

Commendable as this concern for the welfare of consumers might appear, the chronic depression that has gripped the once vibrant US independent sector from the mid-1980s onwards constitutes a powerful lesson about the problems that may arise even when consuming country governments wilfully abdicate the key function of safeguarding competition by taxing rents away.

Basin masters, however, have also prospered under the eye of governments who should have been loath to take any liberties with their oil fiscal income in imitation of the USA or the UK (because to quote a celebrated phrase, the costs of oil production in their countries include the cost of running them).[13] This outcome is a testament to the effectiveness of the well-designed strategy that the governments of developed consuming countries and their international organisations (IEA, IMF, World Bank), as well as oil companies and their consultants, came up with in direct response to the evolution of the oil market during the early 1970s.[14] This strategy was predicated, as Henry Kissinger starkly put it, 'not only [on] economic analysis but – even more – [on the] political, indeed moral, conviction' that it was necessary 'to bring about a reduction in oil prices by breaking the power of OPEC'.[15]

Persuading the governments of most oil-producing countries about the wisdom of reopening their upstream sectors to private investment was one of the central planks of the long-term strategy that embodied Kissinger's convictions.[16] Another one was that this reopening should be done in accordance with a well-defined fiscal prescription, involving the following elements: firstly, net income rather than gross income levies (i.e. income taxes in preference to royalties and severance taxes); secondly, a back- rather than front-loaded character (with acreage being assigned on the basis of bids for the highest marginal tax rate rather than up-front signature bonuses, and full cost recovery in very short spaces of time); and thirdly, streamlined and efficient acreage auction and assignation procedures to facilitate the free and frictionless flow of investment into exploration and, by extension, the highest sustainable rates of production possible.

These policy guidelines amounted to a radical *volte face* in the fiscal and resource management policies that oil producers across the world had pursued, very successfully, for the space of decades (from the 1950s to the time of the OPEC revolution). Thus, they were by no means an easy sell. However, such was the institutional disarray of oil-exporting countries after their catastrophic performance during the 1980s that the message was often well received by its intended audience. A key attraction of the liberal agenda in oil lay in that it offered these countries a way out of the *cul de sac* in which they found themselves after the failure of their development policies, which had been predicated on diversifying their economies away from oil by channelling petroleum rent towards resource-based industries (notably steel, aluminium and petrochemicals).[17] The liberal agenda proposed that, instead of 'sowing the oil' in industries that would never be internationally competitive, governments should see oil as a nucleus of industrial activity in its own right, and pursue a medium-term development strategy whose central objective would be to concentrate policy incentives and support infrastructure on upstream and downstream activities connected to petroleum production itself.

A *sine qua non* requirement of this strategy was that the (mostly nationalised) oil industries in these countries be freed from the many shackles that governments had clamped on them in order to generate the rents that had traditionally been used not only to promote capital accumulation in other sectors of their economies, but also to increase the consumption capacity of their populations. In light of the shocking state of these governments' finances, the assistance of foreign capital and expertise was presented as yet another essential requirement. And it was made abundantly clear that such help would be forthcoming only under certain conditions because, as a high official at a major international oil company bluntly put it, firms like his 'do not like to work for fees ... We want to invest risk capital and get oil in return'.[18]

In the heady days following the OPEC revolution, the governments of oil-producing countries would have had no trouble sourcing investment capital, pretty much under any conditions that they might have cared to impose on desperate oil companies. But by the late 1980s, this situation had apparently undergone a radical change. Governments were informed that oil companies were actually spoilt for choice in terms of possible destinations for their investment capital, thanks to the emergence of a number of new frontier plays (including the GOM deepwater) and the reduction in costs made possible by the rapid advance of technology. Moreover, the fall of the Soviet Union greatly enhanced the potency of the companies' message, as this event introduced into the scramble for oil money, qualified manpower and technology a number of new actors whose needs could only be described as colossal. A former director of exploration at Shell Internationale Petroleum Maatschapij spelled out the implications of the post-Soviet oil order to other would-be recipients of foreign investment capital:

> there are concerns in parts of Africa and South America ... that the potential availability of new areas in the CIS ... will cut the amount of money available for investment. Certainly, the amount of money available for exploration and production is not infinite so, on the face of it, those concerns may be justified because as in any other business, capital and expertise will be attracted by the best opportunities.[19]

In the light of the fratricidal competition for investment capital that pitted producer against producer in the dust storm kicked up by the demise of the Soviet Union, governments who attempted to cling to antiquated fiscal and resource management schemes were warned that they would soon find themselves relegated to the status of also-rans in the investment sweepstakes. However, even though governments supposedly had no choice but to bow down before the new imperatives of a globalised economy, they were reassured that they could do so in the knowledge that this would not amount to a capitulation on grounds of financial desperation (even if it was true that the immediate motivation behind the fiscally-led oil sector reopening in a number of countries was the need on the part of their governments to secure short- to medium-term investment at a critical juncture). Political factors aside, flexibilisation and loosening up of fiscal regimes were the orders of the day *because they made economic common sense*, as they would safeguard the future prosperity of highly productive upstream sectors, partly by limiting the private-sector funds that would be available for upstream investment elsewhere but, more importantly, by ensuring that ultimate oil recovery was not compromised by the regressive effects of royalties and severance taxes.

Now that the post-Soviet dust has settled, the hollowness of the promises implicit in the liberal oil agenda has become painfully apparent. Oil-exporting countries, thanks chiefly to the nature of their socio-political institutions, were the architects of their own misfortunes during a time when the key to a brighter future appeared to be within their grasp. The swing of the pendulum led them into the embrace of peddlers of a Big Idea (privatisation/liberalisation) that would cure all their ills, whereupon the unfortunate inhabitants of these countries once again had the chance to ascertain at first hand that, in Robert Conquest's apt turn of phrase, 'responsibility for Twentieth Century disasters [often] lies not so much in the *problems* as in the *solutions*'.[20]

Currently, the political, institutional, economic and social decay in oil-exporting countries continues unabated (if anything, it has intensified), and nowhere is it more acutely in evidence than in those countries that succumbed most readily to the entreaties of the 'neutral taxation' brigade. The march of events in post-*Apertura* Venezuela, for instance, is a salutary reminder of the crippling price that many oil-exporting countries may end up paying for the privilege of having fiscal regimes that satisfy the investment neutrality requirements as set out in textbooks on optimal natural resource taxation. Suffice it to say that in 2002, in a very healthy international oil price environment, the Venezuelan oil industry paid a paltry USD 470 million in income taxes to the government of the country (on gross revenues of USD 22 billion), a sum equivalent to 38 percent of the income tax liquidated during 1998, a year when the price of the Venezuelan export basket was only 44 percent of the 2002 price! By the same token, during 2004, direct petroleum levies liquidated by the Mexican oil industry came to USD 37.3 billion, compared to USD 16.5 billion in fiscal contributions generated by the Venezuelan oil industry.[21] This difference is nothing short of staggering, when one considers not only that crude production and crude export levels in both countries were broadly comparable (3.84 MMBD and 1.789 MMBD for Mexico versus 3.1 MMBD and 2.279 MMBD for Venezuela, respectively) but also that the average price of the Venezuelan export basket was USD 2/B higher than that of the Mexican basket (USD 33/B versus USD 31/B).

The 1998 catastrophe underscored the fact that the only way in which major oil exporters could limit the funds available for investment was to make the oil price collapse. Furthermore, during subsequent years, the magnitude of oil company share buybacks and cash holdings has made it painfully clear that these companies always had far more funds available for investment than attractive prospects to plough them into (Figure 11.2). The way in which domestic Russian firms succeeded in ensconcing themselves in the driving seat in their country, largely marginalising international oil companies from Western Siberia, has made the dearth of attractive investment opportunities for the latter even more conspicuous.

All these factors notwithstanding, prices for attractive deepwater acreage both in GOM and in other provinces over the period 1996–2004

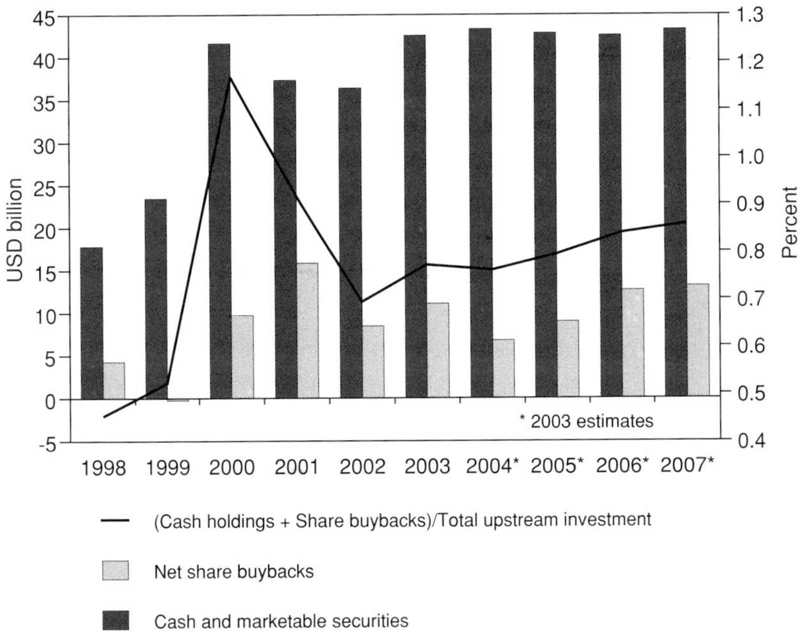

Source: Merrill Lynch

Figure 11.2: Key Financial Indicators for 20 Largest Quoted Oil Companies in the OECD

have been on a declining trend.[22] How can this be explained? Much has been made among industry analysts of the supposedly huge bonuses that oil companies paid in West Africa during the mid-1990s, in the wake of a succession of discoveries in Angola. Indeed, it has become commonplace to say that there is only a slim chance that these bonuses will ever be recouped. However, if one views bonus figures in terms of reserves found, it is clear that they are far from being excessive. Admittedly, at *circa* USD 300 million per block, these bonuses look high when compared to prices registered in GOM auctions (where individual blocks may fetch USD 10 million or more on a good day). However, GOM deepwater blocks are but a fraction of the size of Angolan ones (5670 acres versus 1.2 million acres, respectively). Thus, even though bidding in the 2000–2004 GOM acreage auctions was but a shadow of its former frenzied self, the GOM deepwater continues to boast the highest entry costs of any deepwater province in the world, and by a very wide margin (Figure 11.3). Furthermore, this is true irrespective of whether these costs are tallied on the basis of gross outlays or expenditure per barrel of reserves found.

This last finding sits uneasily with the overall thrust of our discussion regarding the evolution of the GOM deepwater province during the 1980s and 1990s. If it is indeed the case that GOM acreage prices have been artificially depressed by perverse effects on inter-firm competition attributable to AWL, then how come these prices are so very much higher than those observed elsewhere?

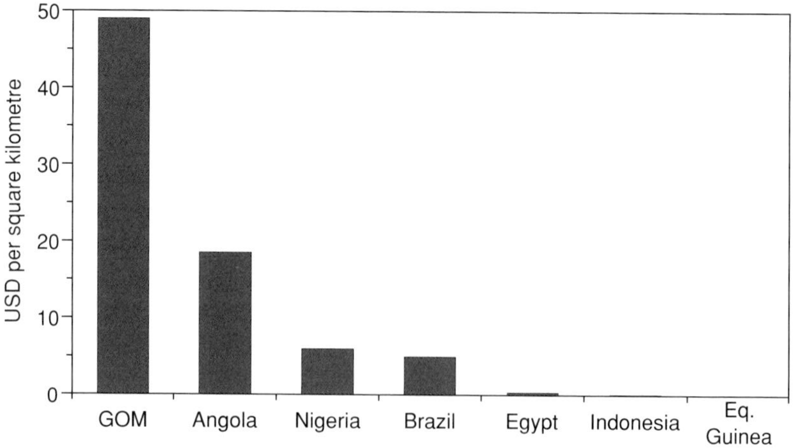

Source: Deutsche Bank

Figure 11.3: Cost of Acreage in Selected Deepwater Provinces

This difference in entry costs between GOM and the West African deepwater provinces would be easy to explain if, for instance, signature bonuses fulfilled different roles within their respective fiscal regimes. In certain provinces (i.e. the British North Sea), bonuses function primarily as tie-breaking devices to aid in the transparent assignation of leases/licences, and are comparatively modest in size because in the end they really are paid merely in exchange for an eventual right to drill. In contrast, when signature bonuses function as the main vehicle for excess profit collection, their magnitude reflects the net present value of expected excess profits net of taxes. Therefore, within such a fiscal framework, signature bonuses can be expected to be relatively large even if (as we contend has happened in GOM since 1983) the leasing agency is not collecting all the economic rent that is potentially available.

This particular explanation does not hold for the case at hand, because signature bonuses are supposed to play a key rent collection role in the West African fiscal schemes, which are structured around PSAs that do not contemplate any additional oil specific levies.[23] Therefore, one is forced to conclude that if acreage prices in GOM since 1983 have been remarkably low, West African acreage prices have been lower still, despite the fact that, in the words of an investment analyst, 'finding multi-million barrel fields in Angola continues to be akin to shooting fish in a barrel'.[24] But then, given the apparent lack of attractive investment alternatives for international oil companies, does this mean that an imminent sharp rise in the price of West African deepwater acreage might be in the offing?

The short answer to this question is no, because there is very little vacant West African acreage left. This is mainly a consequence of the gargantuan proportions of blocks in this region (Angolan blocks, for instance, are twenty times larger than North Sea blocks, and 210 times larger than GOM blocks). Interest in the handful of blocks that have not yet been assigned is subdued, partly on account of the unfavourable location of some of the virgin blocks in comparison to the intense interest attached to relinquished areas (in shallower waters in the Cabinda–Congo border region, for instance) but mainly as a result of a string of very expensive dry wells drilled recently in blocks 33 and 34. These failures have led both oil companies and industry observers alike not only to question the prospectivity of the Angolan ultradeepwater play but also to posit that 'the Angola "boom" may only have a few years left to run and like Nigeria, and most other provinces of West Africa, is unlikely to see a further phase of large discoveries no matter what water depth technology permits'.[25]

In the various countries that make up deepwater West Africa, the extent to which acreage, reserves and infrastructure are concentrated in the hands of a few players is such that it is difficult not to see these countries as case studies in basin mastery. It is also difficult not to conclude that these countries are much less well off as a result of their basin masters than they could otherwise be. Consider an extreme example: Angola's fabled Block 17. The block was leased in the early 1990s, long before deepwater West Africa became the place to be. Since 1996 the owners of this block[26] have discovered some 3000 MMBOE (and counting) of reserves, spread across a number of fields (Girassol, the Dalia complex, Rossa, Lirio, Tulipá, Orquídea, Cravo, Camelia, Jasmím). Indeed, no less than 15 of the first 17 wells drilled in the block were successful. Thus, with the benefit of hindsight, the bonus that the Angolan government received for the block (amounting to a few tens of millions of dollars) looks very modest compared to the USD 1100 billion (plus 100 million in 'social bonus') that Sinopec bid for a 50 percent stake in 10,000 relinquished acres in Blocks 17 and 18 or the USD 902 million that ENI bid for a 40 percent stake in relinquished areas in Block 15.[27]

But what could these countries have done differently to avoid selling their deepwater acreage in exchange for so little money, given that deepwater West Africa was a more or less unknown quantity at the time the first blocks in the province were offered? Offering smaller blocks would certainly have been a good start. Granted, at an early stage in the development of a remote deepwater province, very high sunk costs mean that only very large discoveries offer the chance of achieving positive project economics. Companies need to narrow down the most attractive drilling prospects, and they will be better positioned to do so the larger the area they can explore. It is unquestionable that West African deepwater blocks could not have an extension similar to that of an average North Sea block, but there is no reason to suggest that the Angolan blocks needed to be twenty times larger.

Had the size of West African blocks been kept down, it follows that governments would have had more of them to offer. Would this have made a great difference in terms of their income from signature bonuses? Probably not at an early stage, as bidders would have found their enthusiasm tempered by the extreme uncertainty surrounding the whole play (and the very idea of having to set up shop in Angola). However, discoveries (especially large ones) tend to have a cathartic effect on the propensity of companies to part with their cash, and in the past, licensing agencies were very adept at turning this to their advantage by incorporating reversion clauses into their licensing terms. While these terms gave companies a free hand to explore the blocks they had bought, they also stipulated that once the companies reached a decision as to the locations where they would like to drill, they first had to divide their blocks in a checkerboard fashion using measures drawn up by the licensing agency, and then turn over a set proportion (usually never less than 50 percent) of the acreage back to the agency, again keeping to a set of well-defined rules (for instance, not being able to retain immediately contiguous blocks). In this way, the licensing agency could build up an inventory of acreage that was not completely virgin from the exploration point of view. When drilling led to a discovery somewhere in the checkerboard, the agency was in a position to offer acreage in close vicinity, at a moment when company interest was bound to be at fever pitch.

This type of checkerboard arrangement virtually ensures that no field lies entirely within the confines of one unit, so one of the many positive offshoots that these reversion procedures have is that they allow some of the acreage offered after a strike to be perceived as being low in risk and high in potential rewards. Such acreage can be expected to be more attractive for companies, for obvious reasons, and there is indeed ample evidence showing that bonuses paid for this type of acreage tend to be larger (for instance, a number of studies focusing on OCS lease sales throughout the 1960s and 1970s established that bids for GOM drainage blocks – i.e. blocks adjacent to leases already explored or in production – were considerably higher on average than bids for wildcat tracts).[28]

Such are the advantages of superior information in the E&P game (in this case coming in the form of drilling data providing reasonably precise information about geological strata in adjoining tracts) that this type of reversion procedure does not even compromise the competitive advantage derived from early entry in a play. In their examination of the advantages of incumbency in the pre-AWL GOM, Mead *et al.*, for example, concluded not only that incumbents 'earned a considerably higher after tax internal rate of return on investments [19.8 percent versus 15.9 percent] in drainage leases' than did companies without a position in neighbouring blocks, but also that the former were 7 percent less likely than the latter to buy a dry drainage lease.[29] Other researchers found that the average net profits on all GOM drainage tracts leased over the 1954–1969 period to uninformed firms was zero, [30] chiefly because the average return obtained by such firms was approximately equal to their bids.[31] In contrast, the average return to informed firms amounted to 180 percent of their bids (which were nonetheless much higher than bids for wildcat acreage), in no small part because the strength of their informational advantages had a negative effect on the willingness of less well informed players to get involved in the bidding (the average number of bids on drainage tracts during the period studied by Mead *et al.* was 2.4, significantly less than the GOM-wide average of 3.5).[32]

In this regard, Reece pointed out that the capture of rent by the government was to such an extent 'sensitive to both the level of uncertainty and the number of competing firms', that the bonus bidding system could not, 'even under relatively favourable conditions, *guarantee* that the government will capture substantially all of the economic rent'.[33] Reece also highlighted the likely consequences that would ensue if the Federal government let down its guard: 'if optimal bidding strategies were adopted, [the oil industry could potentially] capture a remarkably large fraction of the economic rent under a wide variety of circumstances'.[34] As we have seen, AWL provided the company with the ideal framework to implement such optimal bidding strategies and a handful of companies did in fact capture a remarkably large fraction of the rent available in the GOM deepwater province.

Even with effective reversion provisions in place, then, licensing agencies wishing to make land available to investors after the incipient stages of development of an exploration play will almost inevitably have to share some of the available economic rent with informed firms (i.e. early entrants). The only way in which this outcome could be prevented would be to ban outright such firms from participating in later licensing rounds (a course of action that is probably unsound on economic grounds, not to mention impractical in political terms and highly suspect in legal terms).[35]

It goes without saying that basin mastery and a licensing regime with aggressive relinquishment provisions are antithetical and, possibly as a result of this, such provisions currently lie very much in a forgotten corner of the panoply of instruments that natural resource owners resort to when making acreage available to investors. Recent vintage licences and PSAs, like those in deepwater West Africa, do make allowances for relinquishment, albeit extremely loose ones: i.e. the forfeiture of a licence if, within a given time, exploratory wells are not drilled or production is not forthcoming. Had these licences/PSAs incorporated more aggressive relinquishment provisions, it is reasonable to assume that the initial bidding rounds would actually have netted the governments involved *less* money. After all, the companies who are now their basin masters in residence would not have been too thrilled at the implications of these clauses, and their acreage bids would probably have been lower than they were (companies, after all, do not like to pay for acreage that they will not, in fact, be able to retain). Crucially, however distasteful companies might have found these clauses they would still have participated in the bidding, as the prospect of marginalising themselves from a potentially lucrative new play would have been too grim for them to contemplate. Moreover, at the first sign of an important discovery, many companies would have flocked to the province to bid for relinquished acreage in competition with incumbent companies, who would quickly see their exploration budgets stretched to breaking point if they attempted to buy up all the blocks on offer.

The end result of such a process would have been akin to what used to happen in the US oil patch every time a prospector struck oil in a novel location: an exploration frenzy, the discovery of a number of fields – most of them with different operators – and very rich pickings by way of bonuses and royalties for the party who owned the mining rights (that is how the University of Texas has managed to end up with an endowment that is second in size only to Harvard's). To return to the case of Block 17, it is clear that had this block been leased piecemeal, bonus payments would have eclipsed the USD 1 billion in bonuses that the Angolan government received during the 1999 bidding round, for instance. As it was, by leasing this gigantic block on a once and for all basis, Angola effectively gave up its chance to extract maximum value from its remarkable fecundity (all the more so since West African deepwater fields are expected to have quite short lifetimes). Furthermore, the exceedingly large size of the blocks (and the length of the exploration periods) has meant that far fewer wells have been drilled than would otherwise have been the case. Of course, parts of the acreage in this and other blocks are to be reoffered in the years up to 2010, but whatever bonuses are collected in the process will never approach the magnitude of bonuses that companies would have been prepared to pay for drainage tracts lying in the vicinity of proven fields. Such are the (paltry) wages that come from allowing an upstream sector to be captured by basin masters.

11.3 The Issue of Tax Breaks

Perhaps even more surprising than the recent trend in deepwater acreage prices is the fact that quite a few oil producers are still busily trying to out-do one another at flexibilising their respective fiscal regimes. After all, oil and gas production are remarkably inelastic with respect to changes in gross income levies like royalties and severance taxes. A landmark study commissioned by the Wyoming legislature (prompted in large part by a fiscal crunch in that US state, whose fiscal dependence towards oil income is second only to Alaska's) found that, *over a forty-year period*, a once-and-for-all drop of 2 percent in the state's severance tax rate would increase total oil recovery by less than 1 percent (50 MMBOE) and employment by 300 persons (i.e. 7.5 jobs *per year*), while causing a 17 percent reduction in the present value of severance tax collections. In contrast, a doubling of the state severance tax rate (from 4 to 8 percent) was found to reduce ultimate recovery by around 6 percent, while increasing tax revenue, in present value terms, by over 90 percent.[36] Likewise, the resurgence of UK North Sea output after what was seen to be its production peak in the mid-1980s is often put forward as a prime example of the power of more flexible taxation schemes to coax higher output from maturing fields.[37] A study focusing on this issue concluded that, out of a total production of 2.676 MMBD in 1995, only about 355 MBD would not have been produced without the modifications to the British fiscal regime introduced from 1983 onwards.[38]

It is also worth recalling that, for a long time, around 75 percent of the oil wells in Texas have produced less than 10 BD (Figure 11.4). It is fair to assume that, twenty years ago, any model of the economics of this type of well would have predicted their abandonment a long time ago. And yet, though stripper wells have never stopped paying royalties, they are still producing (however modestly), because the relentless advance of production technology pushes the real costs of small operators ever

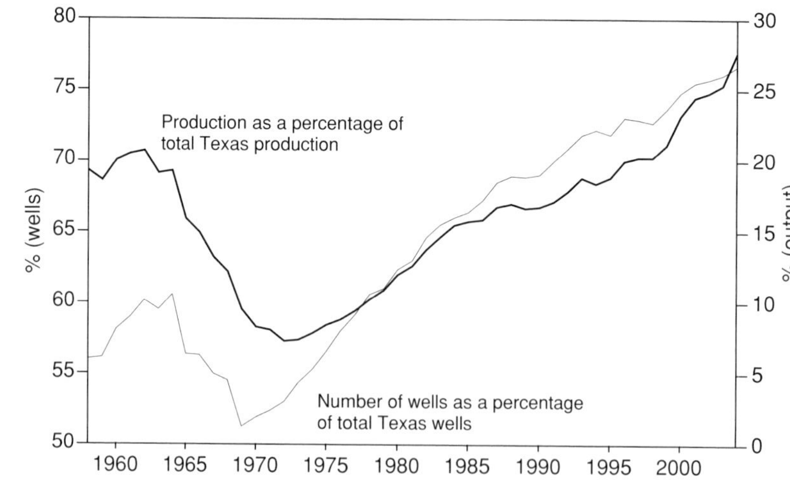

Source: IHS/DOE

Figure 11.4: Texas. Oil Output from Wells With a Daily Production of 10 Barrels or Less, 1958–2002

so slightly down, and allows most of them to make a profit (admittedly minuscule). For some tired wells, of course, not even these cost reductions are enough, and they have to be shut down, thereby adding their small grain of sand to Texas' inexorable output decline. Nevertheless, no politician from that state has ever dared (or will ever dare) to suggest that royalty rates be adjusted downwards, and much less entertain the notion that royalties be abolished altogether, in the interests of greater ultimate recovery. Opinions such as these would saddle him/her with an unhelpful reputation as a radical communist agitator among Texas' million-plus strong royalty owners (who are a key political constituency in that state, of course). This reflection begs an obvious question: if the abolition of royalties is anathema for well-off Texan royalty owners, then why should oil-producing country governments (who are notably less well-off) be any more enthusiastic about it?

The assertion that governments should replace gross income levies with net income levies, all in the interests of ultimate recovery, conveniently overlooks the fact that investors' legitimate claims of obtaining a return on their capital also have a negative effect on ultimate resource recovery: after all, petroleum will certainly be left in the ground when firms no longer find it profitable to extract it. Indeed, petroleum may be left in the ground even though its extraction may still be profitable (such a situation seems to have obtained over the 1999–2004 period, when the pretensions of oil companies to obtain returns significantly above their cost of capital led to the stillbirth of a number of apparently worthwhile projects).[39] So, if ultimate recovery is supposedly at stake, why should it be that natural resource owners are expected to accept royalty holidays without batting an eyelid (thereby sacrificing their own legitimate claims to obtain compensation for every barrel of petroleum severed from their lands), whereas it is never seriously suggested that investors should take profit holidays? One is reminded of the ingenious but ultimately futile objections that a New York Congressman once raised against Senator George H. Bush's impassioned *apologia* in favour of the petroleum depletion allowance on the grounds of the great riskiness of E&P activities: 'Why should we take the risk out of your oil business? You don't want us to limit your profit, George, why do you want us to limit your risk?'[40]

Aside from proving that trying to obtain a higher output through fiscal measures is akin to ploughing the sea (particularly after production decline is underway in earnest), the Wyoming study cited above has also brought to the fore the fact that oil and gas taxes are backward-shifted, a feature which means that the majority of the taxes are effectively exported to foreign consumers, while residents of the jurisdictions where the taxes are levied end up paying cents on the dollar for the public services that these taxes buy. For much the same reason, when these governments lighten the tax burden on companies operating within their countries, their oil industries do not in fact reap the whole benefit of this sacrifice. Due to the interaction between local and foreign fiscal regimes, a reduction in oil production taxes has the unintended and undesirable effect of shifting funds from comparatively poor entities (be it the Wyoming state government or the governments of oil-exporting countries) to entities that are, on the whole, much better off (i.e. the US Federal government or the governments of consuming countries in general).

But, if it is the case that tax breaks are not the panacea that their promoters hold them up to be, then why does the clamour for lower taxation go on *in crescendo*? There is one immediately understandable answer to this question: it goes on because companies will always try to obtain conditions that are as fiscally favourable as possible (capitalist enterprise, after all, exists to minimise costs and in the oil industry, taxes represent the greatest cost item by far). There is more to this than that, though, because many of the companies that complain about being unable to undertake smaller development projects profitably in a number of basins are doing so in (reasonably) good faith. But the root cause of their profitability problems is actually that these low cost operators are trying to do business in places where there are basin masters in residence, and the latter are in a position to target project economics in a way that leaves the former with the thinnest of operating margins, if anything. Thus, even if taxes are lowered in these provinces, the main beneficiaries of the fiscal sacrifices turn out to be the basin masters, rather than the small would-be operators of marginal prospects. In other words, the culprit for the disappointing production response from small fields in places like Oman, or parts of the North Sea is actually the barrier to entry that their respective basin masters represent, rather than excessive taxation. Incidentally, this reflection highlights yet another reason why governments should look askance at the whole idea of allowing basin masters to set up shop in their midst: it is a guaranteed recipe for having to put up with endless gripes about their fiscal regimes, with no prospect of ever being able to give satisfaction to them.

11.4 Final Reflections

From all that we have said thus far in this chapter, it is clear that producer country governments have much to learn from the events and policy transformations leading to the late-1990s resurgence in GOM output. However, if they are to digest the relevant implications of this history, they need to look beyond the technological triumphalism and propaganda that permeates most of the literature available on the subject. In particular, for all of the awe in which they hold the international majors, they need to come to terms with the notion that high technology, on the one hand, and a low degree of upstream concentration on the other, are not necessarily antithetical objectives.

In order to appreciate this point, one need only look at the many offshore upstream projects posing unique and enormous challenges (often at the cutting edge of technology) which medium-sized companies or first-tier independents have managed to tackle successfully none the less. Consider but one case: the Ekofisk field. In the words of a Phillips manager intimately involved with this project, 'it is hard to comprehend the amount of risk, energy, engineering skill, and money involved in ... operating at the limit of knowledge at that time ... in the face of tremendous odds and against the advice of many pessimists'.[41] Total disbursements on Ekofisk (around USD 6.5 billion) have exceeded by a factor of three Phillips' original expenditure estimates.[42] Yet Phillips, a largish but by its own reckoning 'domestically oriented US oil company', was happy to bet the house on Ekofisk, knowing that this discovery alone would be enough to transform it 'into a truly international competitor'.[43] And that Ekofisk certainly did. For all of its outsize challenges, Ekofisk proved to be a company-building elephant (since its discovery, and up until the moment when the intricacies of US anti-trust policy allowed Phillips to buy into Prudhoe Bay, this field was Phillips' most important upstream asset by some distance). And, as things turned out, Phillips was able to master the project, breaking a lot of new ground in the process.[44]

The highly successful development of the Campos basin by Petrobrás (Figure 11.5) also disproves the notion that success in the deepwater frontier is an exclusive preserve of the multinational majors.[45] After all, the Brazilian national oil company has developed a top-drawer deepwater expertise, as proven by its being the record holder for the deepest offshore producing well and, until quite recently, for exploratory drilling in deepwater (9111 feet). In recent years, the company's deepwater reserves have increased at a compounded annual growth rate of 5 percent, making it the most important player in this niche after Shell, a much larger company in market capitalisation terms.

Admittedly, an argument could be made that the successes of Petrobrás are not good indicators of the capabilities of non-major oil companies to prosper in frontier areas, on two grounds. The first one is

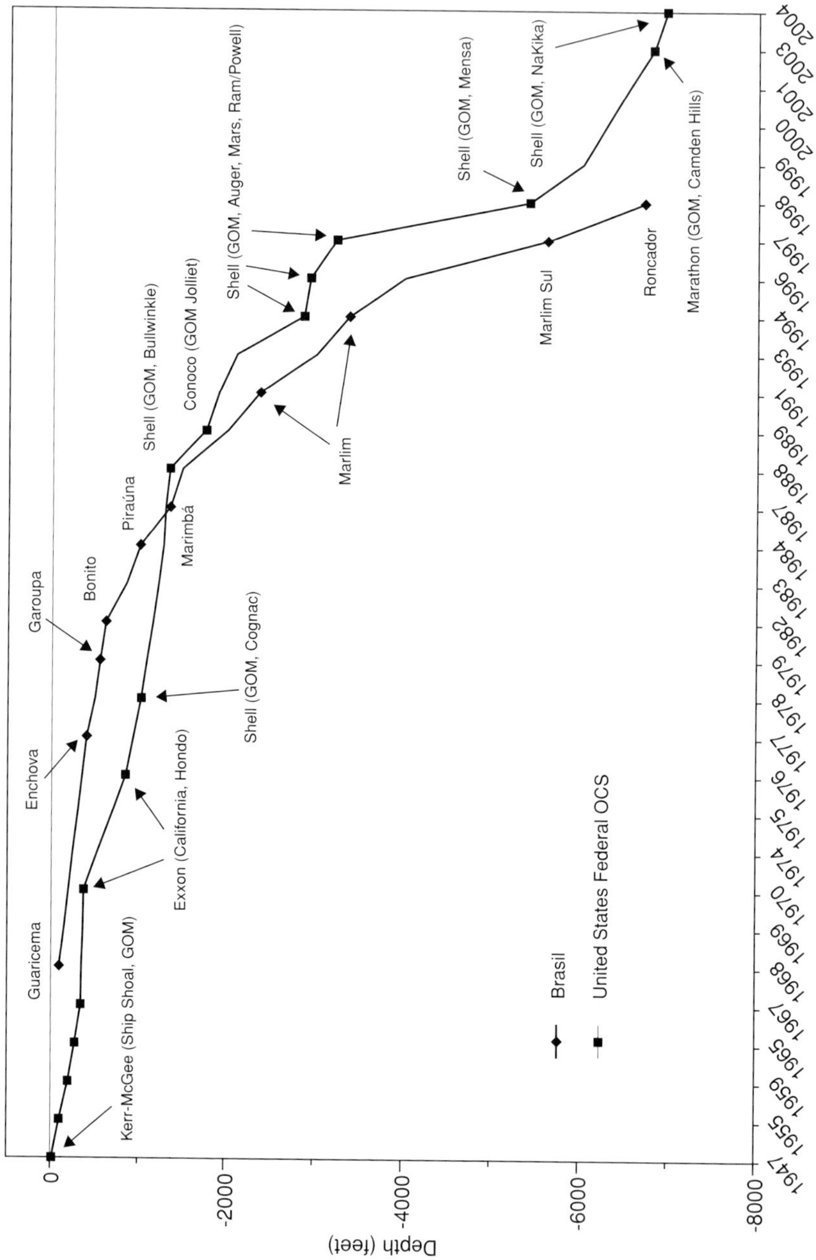

Sources: MMS, OTA 1985, Petrobrás

Figure 11.5: Progression of Deepwater Production Activities in Selected Countries, 1947–2004

that Campos is geologically more tractable than GOM. In strict terms this is true, but that does not mean that Campos is anything less than a thoroughly inhospitable environment in which to operate, by any reasonable standards. The second reason is that the legal monopoly that Petrobrás used to enjoy over all Brazilian upstream activities for a long time protected the company from international competitors, which under equal circumstances might have done the same job for less money. Again, it is undeniable that its cosseted and isolated position was certainly a help rather than a hindrance for Petrobrás, as was the fact that successive Brazilian governments did not balk at bankrolling a distressingly expensive and – for a long time – unsuccessful long-term exploration and development effort (which has cost an estimated USD 35 billion thus far[46]). Still, this does not invalidate the contention that there is nothing in putting together and managing very demanding deepwater projects (like the development of the Marlim field, say) that is *intrinsically* beyond the capabilities of a medium-sized player like Petrobrás. The production infrastructure in Campos, the significant recent increases in Brazilian deepwater output and, perhaps most important of all, the willingness of international investors to underwrite Petrobrás' E&P projects,[47] all bear eloquent witness to this (and to the fact that many of the key elements in deepwater technology and expertise can be bought). Indeed, it is instructive to recall that back in the mid-1980s none other than Shell went to Petrobrás for guidance regarding the complexities of operating in deepwater frontier environments.[48]

Of course, none of the above means that the application in other countries of the very expensive monopolistic solution pursued by the Brazilian government for decades is feasible, even desirable. Neither does it mean that the question of whether relatively smaller companies can beat the large majors on the cost front is irrelevant. But what governments need to understand is that, even though majors do enjoy certain advantages that may make them the obvious choice in terms of the efficient exploitation of their petroleum resources, these companies will not pay them a patrimonial retribution that genuinely reflects the prospectivity and productivity of their basins *unless* their licensing agencies take proactive steps that dissipate and negate some of these scale advantages for the benefit of relatively smaller, slightly less efficient, players. Otherwise, even in a superficially competitive bidding environment, advantaged firms will leverage themselves on asymmetrical information to pretend that they are more inefficient than they really are (or, alternatively, to pretend that the acreage is less prospective than it really is), thereby forcing these countries to share with them the Ricardian rents generated by oil production.[49]

Under adverse selection conditions, laissez-faire of the sort embedded in AWL will inevitably mean '*the most efficient agent receiv[ing] utility greater than his reservation level due to his private information*'.[50] Thus, despite what radical supply-side economists suggest,[51] laissez-faire will not be a way of getting actors with diverse preferences to reveal accurately and honestly their values through a competitive economic process, chiefly because prospective lessees will not be interested in revealing private information if they are in a position to obtain greater gains by keeping it secret (an unavoidable fact that lies behind Stiglitz's lapidary conclusion that 'markets do not provide appropriate incentives for information disclosure'[52]).

In sum, the lessons that policymakers in other oil provinces can draw from the GOM experience is that the fiscal sacrifices that a laissez-faire approach like AWL entails are crippling, and totally out of proportion to the results they can bring about in terms of increased output. That does not mean that the GOM experience is totally bereft of elements that might be worthy of imitation. However, if governments in producing countries want to open up access to their upstream sectors in a manner that is fiscally rewarding but at the same time promotes a competitive industry and efficient market outcomes, then they should focus on trying to devise licensing policies modelled after the tried, tested and enormously successful tract nomination system in use until 1982, while making allowances for their own special circumstances of course.

The story of the evolution and vicissitudes of the broad institutional framework for oil activities in the US OCS also contains valuable lessons for international oil companies (and their champions, the governments of major consuming countries), however. Surely the most important among these is that their pretension to give less developed country governments as little patrimonial retribution as possible in exchange for access to these countries' petroleum resources is a short-sighted and counterproductive policy. It is a policy that will eventually lead to a drying up of investment opportunities in many countries, and all that this entails for companies: declining organic reserve replacement rates, increases in finding and development costs, and excessive reliance on acquisitions for growth. And in order to appreciate this point, one need only look at the long-term consequences of the fact that SLA and OCSLA not only barred US coastal states from having any meaningful say in the development of OCS resources lying off their shores, but also denied them any financial compensation whatsoever.

The rules governing the collection of bonuses and production royalties 'to compensate the general public for the market value of the resources that businesses remove from public lands' enjoin the Federal government to distribute 'a share of those receipts to the states to help state and local governments meet their costs of supporting development activities' on such lands.[53] However, when SLA was signed into law, President Eisenhower made it clear that the submerged lands would be 'administered by the Federal government, and income therefrom should go into the Federal treasury'.[54] Thus, from 1953 onwards, OCS oil leasing has been the *sole* federal programme authorising the leasing, sale or disposal of public resources with no provision for sharing revenues with states affected by the development of mineral resources (coastal states like Louisiana and Texas not only face a substantial part of the risks associated with such activities but also have to foot the bills for providing onshore infrastructure and services in support of them). In contrast to the above, states harbouring Federal leases within their territories (Wyoming, New Mexico, Colorado, Montana and others) have derived substantial benefits from the revenue-sharing arrangements covering royalties from such leases (states receive 50 percent of such royalties from the Federal government, with the exception of Alaska, which gets 90 percent).[55] Thus, in 2004, Wyoming received USD 604.4 million as its share of revenues collected from the production of oil, gas and coal on Federal lands within its borders. New Mexico collected USD 382.8 million, Colorado USD 89.4 million and Utah USD 72.4 million. In contrast, the share of Louisiana and Texas in OCS royalties for that year was USD 41.4 million and USD 14.8 million (out of the approximately USD 8 billion in royalties that the GOM Federal OCS generated that year). It is worth bearing in mind that GOM Federal OCS production accounted for a third of total US domestic oil and gas supply, whereas oil and gas output from onshore Federal lands only contributed around 8 and 5 percent, respectively, to US production.

According to Tyler Priest, this very inequitable arrangement resulted because 'nobody at the time [of the enactment of SLA and OCLSA] could foresee the tremendous revenues that the federal leasing program would take in over the years and the widespread array of petroleum activities leasing would stimulate'.[56] This hypothesis does not hold water, though. Granted, at the time that both acts were passed, their promoters could not remotely imagine that by the end of the twentieth century, cumulative output in the GOM Federal OCS alone would be 40 billion barrels of oil equivalent. However, these legislators had heard testimony that the Federal OCS had a mineral potential enormous enough 'to make its acquisition more important to the nation than the Louisiana Purchase'[57], and yet they still made clear their total opposition 'to the diversion of any money whatsoever from the Federal resources on the Outer Continental Shelf to the abutting states'.[58]

The lack of revenue-sharing provisions in both SLA and OCSLA was always a cause of profound indignation for coastal state governments. When offshore leasing accelerated sharply after the First Oil Shock, DOI's 'lack of awareness of the issues and concerns at the state level ... [once again] served to unite the coastal states on the OCS issue'.[59] With the introduction of AWL, however, these states' historical sense of injustice at their exclusion from the bounty generated just off their coasts finally boiled over into outright hostility, chiefly because AWL vastly increased the magnitude not only of the open-ended liabilities that their support and provision of infrastructure for OCS development would entail, but also of the environmental risks they would face. The 1969 Santa Barbara Channel blowout[60] drove home to these governments, in an awful way, a point that Breeden aptly sums up thus:

> OCS oil development is not really 'offshore' ... [because] OCS oil ... must be transported to land ... to be processed and refined. Consequently, only one end of the independent system of wells, pipelines, oil and gas separation facilities, and refineries will be constructed offshore on the federally owned OCS sea bed. The rest of the system will be located on the sea bottom, beaches, and coastal land area of the neighbouring states ... [where it] will cause major problems.[61]

Given these problems, Breeden concluded with remarkable foresight (seven years before the introduction of AWL) that, 'to the extent that local interests are denied participation in planning the exploitation of offshore oil, state governments can be expected to try to impede federal plans for rapid development of these resources. Hence, failure to reach agreement in the political arena will almost inevitably lead to disputes in the courts'.[62]

James Watt, of all people, echoed Breeden's opinion. When Watt 'unveiled his aggressive offshore lease sale programme ... aimed at infusing the leasing programme with free market economics', he went on record to say that 'he expected the schedule to spawn litigation everywhere'.[63] AWL did indeed go on to provoke major umbrage, even in the quintessentially oil-friendly confines of Texas and Louisiana.[64] Although GOM lease sales were not challenged in court, both the Texas and Louisiana governments initiated legal actions against the Federal government complaining – correctly as the statistics quoted in this study show – that AWL was not conducive to the maximisation of the price for acreage straddling state/federal boundaries. Ultimately, the numerous court actions initiated by coastal states proved fruitless, and the Supreme Court 'did not recognise any meaningful role for the coastal states in the OCS development process'.[65] This provoked severe political fallout in less oil-friendly states like California, Florida and Maine, which led in turn to endless litigation (it can probably be said – and only half in jest – that every barrel of oil produced outside GOM has generated around 10 cents in legal fees[66]) and, more importantly, to a crippling politically-induced paralysis for the offshore leasing programme outside Alaska and the Central and Western Planning Areas of the GOM region.

The intense opposition to AWL by coastal state governments and members of Congress from coastal states crystallised in the enactment of restrictive leasing moratoria on ever more extensive swathes of the OCS. Up to 1982, moratoria covered only 736,000 acres offshore California, but by 1983 they already blanketed 35 million OCS acres.[67] One year later, the DOI appropriation bill had slapped moratoria onto no less than 52.2 million OCS acres (36.6 million in California, 8.2 million in the North Atlantic and 7.4 million in the Eastern Gulf, off Florida), and currently the out-of-bound OCS acreage is 610 million acres.[68] In other words, despite Ronald Reagan's virulent distaste for any sort of disposition that might interfere with hydrocarbons production (and his readiness to deploy his presidential veto against them), the 'antiquated hierarchical approach to decision making'[69] that his administration pursued led to the almost complete breakdown of the offshore leasing programme, within a remarkably short time-span.[70]

The enduring legacy of AWL, even more so than the deepwater bonanza in the GOM region, *is the closing off to exploration and production activities of a vast area of the OCS, potentially very rich in hydrocarbons* (Figure 11.6).[71] It is this non-existence of revenue sharing provisions that, during the early 1980s, prompted the government of California to take the Federal government to court in order to halt Pacific OCS development, while at the same time it was 'issuing offshore drilling permits in the Santa Barbara Channel'.[72] The same motivation impels an energy-importing state like Florida to snarl up E&P activities taking place 100 miles or more beyond its coastline, by invoking the right to decide whether such activities (and the Federal licences and permits underpinning them) are consistent with its own CZMA plan. In a nutshell, it is this that explains why even a governor with the instincts, industry connections and political pedigree of Jeb Bush will simply not countenance drilling for oil and gas off the coasts of his home state.[73] It also suggests that there will be few states that will choose to opt out of leasing moratoria, as contemplated for instance in the Reliable and Affordable Natural Gas Reform Act of 2006.[74]

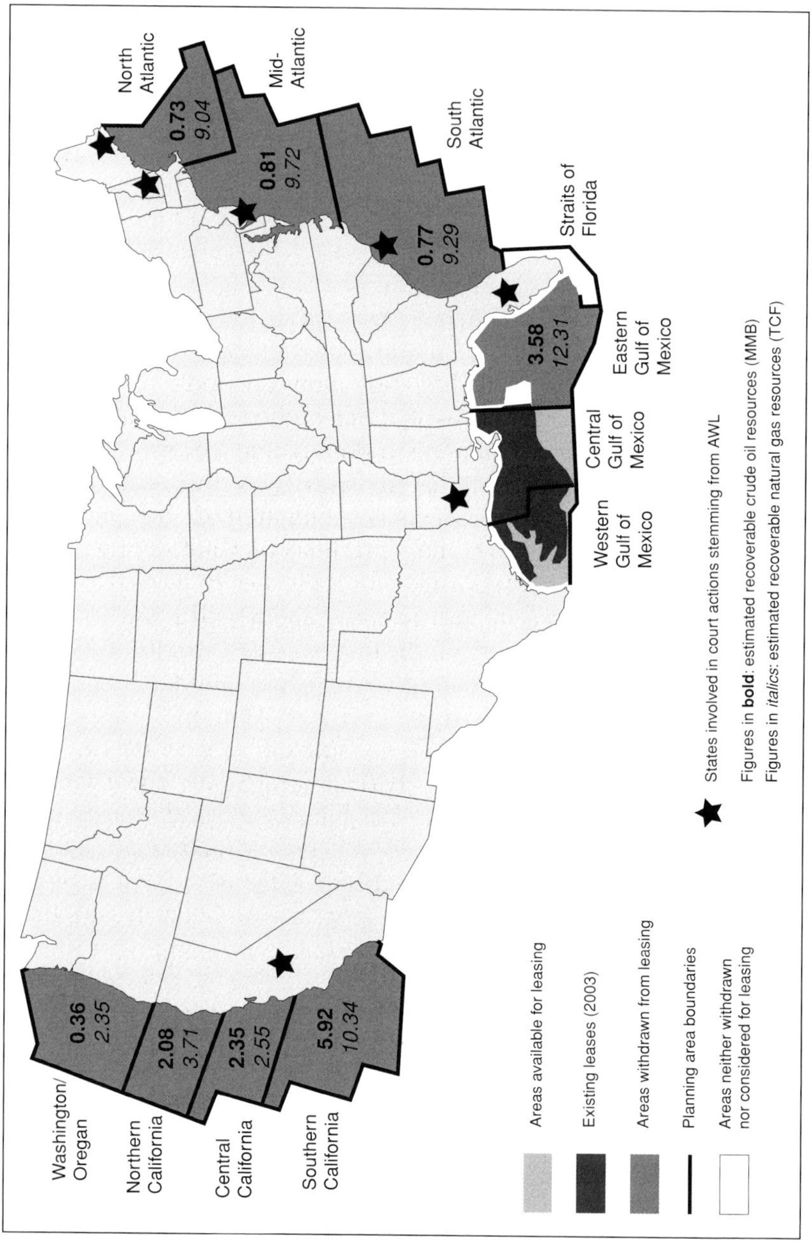

Sourcse: IPAA 2002, MMS

Figure 11.6: Current Status of OCS Leasing Programme in the US Lower 48

As far back as 1985, the DOI had already warned that 'the only apparent solution to reducing the cost of opposition to the OCS programme' would involve providing 'the States and localities with an incentive to support leasing which is perceived by the States and localities as sufficient to counterbalance their perception of the potential harm and risk to which they are subject. OCS revenue sharing is the best incentive to achieve that balance'.[75] When they were penned, these lines were probably a reasonable diagnosis of the Federal/state impasse on offshore oil activities, and revenue sharing might conceivably have broken the deadlock then. Indeed, Fitzgerald is still of the opinion that, with revenue sharing, even 'green' states might support the lifting of the moratoria and the volume of litigation would dwindle to a trickle.[76]

This last assessment is probably over-optimistic, though, not least because the share of the mineral revenues that coastal states would be entitled to receive even under the most generous revenue-sharing arrangements pale into insignificance compared to the amounts of money that states like Florida, the Carolinas or California make from tourism alone.[77] This explains why, in 2005, newly elected *conservative Republican* administrations in both South Carolina and Florida reiterated their opposition to any kind of drilling off their coasts. Hence, one can confidently predict that, regardless of the intense frustration of the oil industry and the US Federal government alike, and however many political manoeuvres are mounted to outflank opposition to offshore drilling in the US Congress,[78] the OCS areas currently under moratoria will likely remain out of bounds until the SLA/OCSLA are radically amended in a way that ensures governments of the coastal states adjoining waters under Federal jurisdiction effectively control access to them and receive the *whole* of the revenues generated by offshore oil activities (i.e. bonuses as well as royalties). And not even such a radical step may be enough to reopen offshore California or Florida.[79] But in any case, it is obvious from the thrust of the discussion regarding the convenience of the MMS taking OCS royalties in kind as a first step leading to the total privatisation of the OCS petroleum resources that, as far as US policymaking circles go, this notion will probably never take hold.

The comprehensive energy bill discussed by the US Senate in April 2005 incorporated a proposal that states with coastlines lying within 200 miles of areas under moratorium should receive 100 percent of the revenues generated in these tracts, in order to put an end to a situation where, in the words of one of the main supporters of the amendment, 'the Treasury ... get[s] all of the revenues and leave[s] the state[s] with the debts'. [80] It also incorporated provisions that would have allowed the governors of states with coastlines adjacent to offshore moratorium areas to petition DOI for these moratoria to be lifted. Predictably, these provisions were subsequently dropped, and the 2005 Energy Policy Act (signed by President George Bush on August 8, 2005) eventually emerged as a hodgepodge of tried and tested (and failed, one might add) policies to 'encourage increased domestic production of oil and natural gas'. At a time when the NYMEX sweet crude prompt month contract was trading at around 60 USD/B, the act preposterously confirmed future additional royalty relief for deepwater and deep gas production in GOM (in the form of higher, albeit unspecified, price thresholds). It also called for a comprehensive inventory of the estimated oil and natural gas resources on the OCS (including areas under moratoria), to be carried out by means of all available technological means including 3-D seismic (but excluding drilling), for submission to the consideration of Congress, in the form of a publicly available report, within six months of the enactment of the legislation (this report would be periodically updated thereafter, at least once every five years).[81] Representatives of key coastal states greeted the inventory proposal with undisguised hostility, and Florida representatives made it clear that they would resist vigorously any attempt by the Bush administration to sidestep the moratorium off the state's coasts by the redrawing of the boundaries of GOM planning areas (which would have the effect of reclassifying blocks that have traditionally been considered as lying offshore Florida as lying offshore Louisiana).

The mobility of OCS operations and their suitability to a remote form of operation means that not even job creation can function as an inducement for recalcitrant coastal state governments to abandon their opposition to offshore oil activities. As Gramling points out in the specific case of Florida, 'few, if any, jobs, or other economic benefits would go to residents of the state. The rigs, support vessels, and crews would come from southern Louisiana, and local purchases would be limited to such items as potable water, which is already in short supply in south Florida, and diesel fuel'.[82] The exact opposite is true in the central and western planning areas of GOM: according to a study commissioned by the Louisiana Mid-Continent Oil and Gas Association (LMOGA), GOM oil and gas operations contribute USD 6 billion per year to this state's economy.[83] This, of course, goes a long way to explaining why OCS development in Texas and Louisiana (and to a lesser extent Mississippi and Alabama) 'is supported as strongly … as it is opposed elsewhere'. [84]

That is not to say, though, that the inequitable distribution of OCS revenues has not kept generating frictions even in these two states, especially Louisiana. In the early 1990s, for instance, a small number of legislators in the Louisiana state senate sought to redress the perceived historical injustice represented by their state's exclusion from the bounty generated just off its coasts through a proposed oil and gas processing levy that would have taxed every barrel of oil and petroleum products and every cubic foot of natural gas processed – i.e. produced, transported or refined) in Louisiana[85] – in exchange for a repeal of the severance tax collected on oil and gas produced within the state (the stillborn processing tax would supposedly have generated more than USD 2.1 billion annually for the state, compared to the USD 400 million raised through severance taxation).[86] At the end of the decade of the 1990s, Senator Mary Landrieu introduced a bill calling for the diversion of OCS royalties to coastal states, but the tacit opposition of President Clinton to the proposal spelled its demise. Then, in late 2004, the Federal–state disagreement on royalty disbursements escalated significantly when the newly inaugurated governor of Louisiana publicly announced that she might consider stopping the issuance of offshore drilling permits, unless more royalties from the GOM Federal OCS came the way of the state ('10 percent of the royalties from offshore Louisiana deepwater gas and oil production [in the Federal OCS] ... to use the money to repair Louisiana's eroding shoreline').[87] The governor also urged the governors of Texas, Alabama and Mississippi to support Louisiana's efforts to force a change in the framework for the disbursement of offshore royalties. These calls went unheeded, but Louisiana politicians in the Federal capital did not give up, and in May 2005 Senator Landrieu put forth a bill (the Stewardship for our Coasts and Opportunities for Reliable Energy Act) that would reserve a significant proportion of total OCS revenues and distribute it among those coastal states that do not have offshore moratoria in place, in proportion to 'their' share in total OCS output.[88] In the end, though, all that the 2005 Energy Policy Act did in this respect was to increase modestly the coastal impact assistance funds available: USD 250 million to be shared annually among the eligible states of Alaska, Alabama, California, Louisiana, Mississippi and Texas from 2007 through 2010, with the annual allocation for each state to be based on the ratio of OCS revenues generated off the state's coastline to total OCS revenues from leases lying beyond three nautical miles of the state/Federal demarcation line, to a distance of 200 nautical miles.

Despite the support (with caveats, at times) of GOM coastal states for offshore leasing, the fact remains that, amidst mounting and widespread concerns within its political system that a rising tide of oil imports may be undermining the foundations of the country's national security, the USA is failing to tap significant domestic energy resources (it has recently been estimated that around 13 percent of the undiscovered natural gas resources in the GOM Federal OCS cannot be accessed because of drilling moratoria).[89] This paradox is a poignant reminder of the failure of the OCS institutional framework and fiscal regime to achieve an equitable Federal/state distribution of the fruits of developing US offshore petroleum resources. Just as importantly, it is a pointer of what is likely to happen in those producer countries which, having invoked their sovereign powers to grant access to their natural resources, now find themselves inhibited of all other meaningful attributions of sovereignty and eminent domain (notably the power to tax extractive industries located within their territories), through a combination of contractual provisions and legal fetters incorporated into BITs and multinational investment agreements (like the Energy Charter Treaty).[90]

Events like the grinding to a halt of the Venezuelan *Apertura* after the electoral success of Hugo Chávez, or the disappearance in recent years of opportunities for direct foreign investment under PSA terms in Russia, suggest that the fundamental asymmetry which lies at the heart of the liberal oil agenda (all profits and rents to the investor; nothing for the resource owner other than the investment itself) is likely to prevent capacity from coming on stream when needed, to a far greater extent than the legitimate claims for compensation on the part of natural resource owners. This is because, sooner or later, access to resources will tend to be compromised if the populations of the territories from which they are extracted feel that they are not getting their fair share of the bounty generated by the liquidation of their mineral assets. And this can be due either to the excessive generosity of fiscal regimes or to mineral revenues being siphoned off by kleptocratic and tyrannical regimes (of the type which oil companies have displayed an unfortunate penchant for always cosying up to), or a noxious combination of both. Just as taxation without representation is a byword for tyranny, access without oil taxes is a recipe for political paralysis and endless litigation. Even in developed countries, Scotland's acquiescence to its marginalisation from North Sea oil revenues is very much the exception that proves the rule (far more typical of the rule is the mess surrounding OCS leasing, or the highly charged response by the Albertan government when Trudeau's National Energy Policy led Premier Peter Lougheed to state that 'the Ottawa government [had], without negotiation, without agreement, simply walked into our home and occupied the living room'[91]).

As in the past, the recalcitrance of natural resource owners to grant access to resources found within their territories could conceivably be overcome by force, as has most recently (and notoriously) been advocated by Deepak Lal. According to this author, an International Natural Resources Fund (INRF) should be set up by amalgamating the World Bank and the International Monetary Fund, with the express objective of levying and administering the rents from the natural resources of failed or failing states. These revenues 'would be put in escrow accounts for use only in the countries in which they were generated ... [and] would be released only on the authority of the INRF for purposes determined by the fund's managers in consultation with the local government – mainly for social and economic infrastructure projects'. In order to make sure that the deployment of these rents was not dictated by political criteria, the aforementioned projects 'would be subject to the international bidding, controls and monitoring procedures of the World Bank'.[92]

So far, so much information-age enlightened despotism. Unfortunately, Lal considers it likely that at least some retrograde and ungrateful members of the populations of these states will obstinately resist the idea that bureaucrats from developed countries should decide not only how much their birthright should be sold for, but also how the proceeds should be spent. Which reflection leads Lal on to a burning question: 'how could predators be prevented from attacking the mines and wells generating the rents'? The manner in which Lal suggests that this issue be addressed is, to put it mildly, bound to raise a few eyebrows. His suggestion is that the international community should follow the example set by 'China during the interwar period', and take the decision to lease to private foreign companies 'territory that they could protect with their own police forces, in return for royalties to the INRF. But even this privatised solution would require the imperial power to maintain "gunboats and Gurkhas" at the ready, in case some local predator decided to mount a challenge to the private controllers of the mines'.[93]

In terms of its breathtaking audacity (not to say preposterousness), Lal's *Modest Proposal* appears to be right up there with Dean Swift's.[94] There is a crucial difference between the two, though, in that Lal's proposal is meant for real (as witnessed by its publication in the pages of the *Financial Times*, no less).[95] However, one could easily be forgiven for thinking that it was actually penned with satirical intent. After all, Kuomintang China may have had its good points (although it is admittedly difficult to think in terms of specifics), but if an unbiased observer were asked to name the one thing about this political regime that is not worth emulating, surely he would single out the pathological role that warlords with private armies played within it (a role which, incidentally, contributed to make Nationalist China a failed state before that term even gained its current notoriety).

As if this were not going far enough, Lal's prescription also requires that a great many assumptions be made regarding such issues as the altruism of the imperial powers, the integrity of the administrators, the unity of purpose of the governments of developed countries, or the extent to which the populations of at least some of them will be willing to countenance this sort of game. These assumptions are so heroic that, even if they were not in the process of being disproved once again by the unfolding developments in Iraq (at an incalculable human cost), they would still tax the credulity of the denizens of the lands beyond the looking glass (some of whom, famously, had no trouble in believing as many as six impossible things before breakfast).

It is obvious that Lal and others of his ilk are not particularly interested in establishing the conditions that would allow resource-rich but otherwise poor and weak countries, on the one hand, and prospective investors in their petroleum resources, on the other hand, to reach a viable and stable *modus vivendi* regarding the timely exploitation of petroleum resources, as well as the distribution of the benefits thus generated. Lal's INRF proposal amounts to merely another attempt to neuter the unalienable prerogative of 'resource-rich countries … to impose conditions on access to their resources and to safeguard their sovereign rights'.[96] Indeed, it is a blatantly transparent and crude bid to subordinate this prerogative to the desire on the part of multinational oil companies and consuming countries to grab – *de facto* if not *de jure* – eminent domain rights away from natural resource owners, pretty much along the lines of what happened during colonial times (albeit with some up-to-date window dressing). This is an exercise in squaring the circle and, even in a unipolar world, is ultimately doomed to end in failure (vide Iraq, again).

The foundations for a lasting agreement can only come through political compromise because, as Mommer asserts, 'short of war, negotiation is the only avenue open to sorting out conflicts over sovereign rights'.[97] True, resource owners have to accept that foreign investment is quintessentially incompatible with the doctrine of 'permanent sovereignty over oil resources', which was embraced by OPEC member countries at the apex of their power (early 1970s), and whose thrust was that the governments of these countries could, if it suited them, renege at will on whatever contractual obligations they had acquired with international oil companies and revise them in their favour. But by the same token, and for all their thirst for abundant and cheap oil and gas and their delusional talk of an endless succession of attractive investment prospects, oil companies and consuming country governments (especially the latter) have to realise that they have got to pay, and pay generously (i.e. not merely with investment), in exchange for access to petroleum resources. Otherwise, even the weakest, most corrupt and/or subservient producer country governments will, sooner or later, refuse to play ball. After all, petroleum rent may not be the future for oil-exporting countries (as the development failures of OPEC countries and Mexico so devastatingly proved back in the 1980s), but the stark fact remains that these countries really have no future without petroleum rent.[98]

NOTES

1 Ernst and Steinhubl 1999: 49–50.
2 Pérez Alfonzo 1960: 143–4; italics ours.
3 Ernst and Steinhubl 1999: 49.
4 *Ibid.*: 50.
5 Conn and White 1994: 10. This figure was calculated on the basis of risk-adjusted market returns. Exxon and Royal/Dutch Shell are not included in the calculation, as they created substantial shareholder value over the interval studied by these authors.
6 *Ibid.*
7 *Ibid.*: 59.
8 *Ibid.*: 64.
9 Campbell and Hulme 2001: 82.
10 George 1888.
11 An excellent study of this problem, focusing on rural Argentina, is Flichman 1977.
12 MMS 1983: 54.
13 Robinson 1989: 6.
14 Mommer 2002: 226.
15 Kissinger 1999: 668–9.
16 See Scott 1994–5.
17 See Auty 1988, 1989 and 1990.
18 *Petroleum Economist*, March 1992: 8.
19 *Shell World*, 7 October 1991: 25.
20 Conquest 1999: ix; italics ours.
21 Income taxes, royalties and dividends from PdVSA came to USD 12.2 billion, which were supplemented by USD 2.4 billion in direct social expenditure by PdVSA, and USD 2 billion channelled to a special development fund (FONDESPA).
22 Traynor *et al.* 2002: 27.

23 In Angolan PSAs, for instance, there are typically no royalties, and the 50 percent tax rate may be paid in lieu by Sonangol. Cost recovery is set at a minimum of 50 percent (exploration costs are expensed), the negotiable uplift on development costs has a 40 percent ceiling, and there is a 3–5 year straight line depreciation schedule.

24 McMahon 2004: 7. Even the ultradeepwater Block 31 has experienced a success rate of 90 percent as of the end of 2005. Exploration success has consistently eluded only Shell, which drilled eight dry wells in Block 16 and has now exited deepwater Angola altogether.

25 *Ibid.*: 8.

26 Shareholdings in Block 17 are as follows: BP Angola 16.67%; TOTAL 40%; ExxonMobil Corporation 20%; Norsk Hydro (Angola) 10%; Statoil (13.33%).

27 Sinopec has bid USD 750 million for 50 per cent of the Block 15 acreage, with Total offering USD 560 million for a 24 per cent stake and Petrobrás USD 531 million for a 50 per cent stake (*PON*, 12 April 2006: 1). Total also offered USD 670 million plus USD 100 million 'social bonus' for the Block 17 acreage, while Petrobrás had come in with USD 690 million plus a USD 17.5 million 'social bonus' for relinquished Block 18 acreage (*PON*, 11 May 2006: 1).

28 Hendricks, Porter and Boudreau 1987: 534; Hendricks, Porter and Tan 1993: 234. Mead, Moseidjord and Sorensen (1984: 506) calculated that the average high bid for GOM drainage tracts over the 1959–1969 period (USD 5.8 million) exceeded the average high bid for non-drainage tracts by nearly 30 percent.

29 Mead, Moseidjord and Sorensen 1983: 506.

30 Hendricks, Porter and Boudreau 1987: 540.

31 Hendricks, Porter and Wilson 1994: 1416.

32 Mead, Moseidjord and Sorensen, *ibid.* Hendricks, Porter and Wilson (1994: 1416) established that 'uninformed buyers are less likely to participate than the informed buyer, but if they participate[d], they bid high rather than low'. Crucially, though, 'uninformed firms won profitable tracts often enough to keep them participating in the drainage auctions' (Hendricks, Porter and Boudreau 1987: 540).

33 Reece 1978: 381.

34 *Ibid.*

35 Hendricks, Porter and Tan 1993: 239.

36 Gerking *et al.* 2000. These conclusions had been foreshadowed by GAO 1990. This study found that some petroleum production incentives actually 'provided incentives to make petroleum production investments that have pretax returns below those of investments in other industries' (p. 5). An even earlier GAO study (1985b: 32) found that a 40 percent reduction in Windfall Profit Tax for EOR projects had led to only one GOM project.

37 Production of liquid hydrocarbons in the UK peaked (for a second time) in 1999 at 2.82 MMBD, after having languished at 1.88 MMBD during 1988–89.

38 Martin 1997: ii–iv.
39 See Antill and Arnott 2002.
40 Quoted in Bryce 2004: 93.
41 Dunn 1995: 47.
42 Indeed, Ekofisk was in the end affected by those 'two immutable laws operating on the Norwegian continental shelf' that became infamous after the Statfjord project ran into major delays and cost overruns; namely that 'regardless of how much has been invested, the amount still to be invested remains constant', on the one hand, and that 'regardless of how far the project has advanced, completion is still six months ahead' (Richardson 1982: 46).
43 Dunn 1995: 47.
44 See Kvendseth 1988.
45 See Nelson de Faria Almeida and Cristina Pinho, 'Marlim Leads Industry into Deepwater', in *Petroleum Economist* 2001: 104–6.
46 Stouffer and Knight 2002: 7.
47 See the prospectuses for the bond issues of Companhía Petrolífera Marlim 1999–2000.
48 When Shell decided to develop its deepwater strikes, it entrusted the engineering aspect of this enormously complex enterprise to Daniel Godfrey, who at the time was working in Brazil (*WSJ*, 4 April 1996: A-1). Godfrey's work was related to a technology exchange agreement that Shell signed with Petrobrás, motivated by the latter company's technological leadership position in deepwater operations (see *FT Energy World*, April/May 1998: 8).
49 Macho-Stadler and Pérez-Castrillo 1997: 157.
50 *Ibid.*: 113; italics in original.
51 See Bradley and Mead 1998: 212.
52 Stiglitz 2002: 472.
53 CBO 2000: 3.
54 Christopher 1953: 57.
55 Similarly, the National Forest Revenue Act allocates 25 percent of the proceeds from timber harvesting in national forests to state and local governments, the Taylor Grazing Act provides states with 12.5 percent of all grazing and stock raising receipts from federal leases, and the Mineral Leasing Act grants the states in whose territory mining takes place 50 percent of all mining receipts.
56 Priest 2004: 34.
57 These were the words that an expert used in the hearings that preceded the passage of the OCSLA (quoted in Christopher 1953: 23).
58 *Ibid.*: 43. Interestingly, in 1949, when the Tidelands controversy was raging with maximal intensity, Federal negotiators proposed a compromise solution whereby Louisiana would receive all royalties and bonuses accruing from production from fields within three miles from shore and, more importantly, 37.5 percent of revenues from all wells located beyond that point. Although the parties were close to an agreement, it was blocked by Leander Perez

(leader of the infamous White Citizens' Council, but also district attorney for the St. Bernard and Plaquemines parishes, and special assistant state attorney general before the US Supreme Court on the Tidelands matter). Perez refused to forfeit Louisiana's territorial claim to the disputed area. With the benefit of hindsight, it is obvious that Perez's decision deserves to go down in economic history in the same bracket as that of the manager of Decca who passed on the chance to sign The Beatles.

59 OTA 1975b: 5.

60 The blowout led to the spill of 71,000 barrels of oil from a Unocal facility.

61 Breeden 1976: 1107.

62 *Ibid.*: 1108.

63 *O&GJ*, 20 June 1983: 57.

64 See *O&GJ*, 18 June 1984 and 25 June 1984; *PON*, 19 April 1984.

65 Fitzgerald 2002: 16.

66 Consider the following passage, where William Bettenberg discusses the manner in which he briefed a recently appointed DOI undersecretary about the legal minefield that is OCS leasing: 'I got to the point where we [MMS] adopted the five-year [leasing] programme, and she said, "Well, then what happens?" and I said, "Well, then we go to court", and she was taken aback. I said, "Well, the Andrus (Secretary of the Interior, 1977–1981) programme was litigated, the Watt (Secretary of the Interior, 1981–83) programme was litigated. We know this will be litigated. We have to take that into account in the way we go about the process – how we write *every* document"' (Farrow 1990: 136; italics ours). The contentiousness of AWL can also be gauged by the fact that, from January 1981 up to September 1982 inclusive, an average of 1.5 DOI witnesses appeared before Congress every legislative day. Preparation of testimony for these hearings consumed more than 37,000 manhours (MMS 1983: 22). Even before the introduction of AWL, BLM officials had pointed out that Nixon's accelerated leasing schedule could lead to problematic legal challenges in both GOM and California (GAO 1975: 5).

67 OTA 1985: 144. Of the 1.5 billion acres of offshore submerged lands under US jurisdiction, only 11 percent is currently available for leasing (NPC 2003, vol. IV: 6–11).

68 NPC 2003, vol. IV: 6–52.

69 Fitzgerald 2001: 274.

70 Gurney (1997: 31) mistakenly says that 'areawide leasing was devised only for the Central and Western Gulf planning areas and the government did not use it in other US OCS regions'. In practice, areawide leasing has not been used much in other areas, but it has certainly not been for lack of trying.

71 In terms of recoverable resources, areas of the OCS currently off limits to leasing activity are estimated to contain about 16 billion barrels of oil and nearly 70 TCF of natural gas. Out of this total, only 230 MMB of oil and 6.79 TCF of natural gas are thought to be found in Alaska.

On 10 May 2006, the House Appropriations Committee partially lifted the Congressional leasing moratorium on offshore gas leasing outside the Western and Central Gulf of Mexico and some parts of offshore Alaska, in force since 1982. The oil leasing ban remained in force as did, more importantly, the presidential ban on any oil or gas leasing in these areas, adopted by the administration of George H. Bush.

72 MMS 1983: 21.

73 On 8 March 2006, Senator Trent Lott from Mississippi introduced a revenue sharing bill that would distribute to four coastal states 50 percent of proceeds from offshore leasing and production activities in acreage included in Sale 181 (i.e. offshore Florida), as originally drawn up (i.e. before Florida's protests led to the withdrawal of most of the acreage on offer). The state of Florida would be explicitly barred from receiving any money (*PON*, 10 March 2006: 2). Lott's initiative came in response to legislation proposed by Florida congressmen that would permanently bar any leasing in the Eastern Planning Area of the Gulf of Mexico within 100 miles of the state's coastline.

74 This bill would allow state governors to petition DOI for waivers of coastal leasing bans for natural gas development beginning in 2007. If oil is eventually found at one of the leases assigned, production would be allowed subject to the consent of the state government. If a state opts out of the moratorium, a neighbouring state would have the right to object to leasing within 20 miles of its coast. As an incentive, the bill would give states opting out of the moratorium half of the bonus bids and royalties on production, and equal revenue sharing also would be extended to states where offshore leasing is currently allowed, but only on production starting five years after enactment of the legislation. For states that want the moratorium on development extended beyond 2012, the measure allows them to seek an extension of up to ten years on areas up to 125 miles off their shores. Up to 2006 inclusive, only Virginia had displayed any sort of enthusiasm for requesting that the ban on drilling off its shores be lifted. A bill passed by the Virginia legislature on 7 April 2006, declares it the policy of the state to encourage members of its congressional delegation and federal executive agencies to take action 'that will provide an exemption to the moratorium' that prevents any offshore activity until 2012 (*PON*, 10 April 2006: 6). The same bill, however, would prohibit drilling within 30 miles of the Virginia shoreline as well as the construction of any onshore natural gas exploration and production facilities on the state's eastern shore.

75 Fitzgerald 2001: 158.

76 *Ibid.*: 276.

77 In 1996, the five western counties in Florida generated a total of USD 8 billion in tourist revenues (Fitzgerald 2002: 49).

78 In February 2005, both the Virginia House of Delegates and State Senate passed legislation seeking to end the moratorium off the state's coast. In April of the same year, this measure was echoed in the US Senate, where a bill was introduced that would give governors of coastal states the power

to override federal drilling bans, and which would also authorise DOI to issue gas-only leases for offshore drilling. A short while later, though, *Platt's* reported that 'the likelihood of significantly expanding offshore leasing to areas now under ... moratoria must be considered remote, given remarks representing three key states at a Senate Energy Committee hearing' held on April 19. The three legislators in question came from South Carolina, Florida (both Republicans) and California (Democrat).

79 In this regard, see US Congress 1998.

80 *PON*, 27 May 2005: 5.

81 The six-month time frame effectively ruled out the possibility of any significant 3-D seismic survey being carried out in areas under moratoria.

82 Gramling 1995: 147.

83 More than USD 1.2 billion of this impact comes in the form of salaries and wages paid to offshore workers. The remainder results from the business the offshore industry does with onshore vendors located in Louisiana. More than 21,000 producing company jobs are said to exist as a direct result of GOM oil and gas activities, of which 16,725 are taken by Louisiana residents. The estimated payroll for them is USD 1.2 billion (an average of USD 74,000 per employee). Nearly USD 6 billion was paid by producing companies to vendors and contractors in support of GOM offshore activities, with more than USD 3.7 billion of that amount (58 percent) spent in Louisiana. Also, approximately 45 percent of vendors in Louisiana derive more than half of their income from OCS activities in the GOM.

84 Gramling 1995: 166. For a rigorous quantification of the spillover effects associated to OCS activities, consult Stiff and Singelmann 2004.

85 The tax would apply to hydrocarbons produced in the state as well as in the GOM Federal OCS, and to imported crude or imported products landed in Louisiana for refining or consumption there, and to all crude oil, natural gas and refined products transported through the state to other destinations.

86 This tax is an offshoot of the First Use Tax, found to be unconstitutional in 1981 by the U.S. Supreme Court (*Maryland v. Louisiana*). The Coastal Wetlands Environmental Levy of 1982 sought to tax all oil and gas passing through the coastal area of the state as a way of paying for the alleged environmental damage caused by the industry, but was never enacted. The current oil and gas processing tax was first proposed in 1992 during a state constitutional convention called for the purpose of revamping the state's tax structure.

87 *PON*, 15 December 2004: 5.

88 From 2006 to 2010, Congress would have to authorise that OCS revenues be appropriated for these states; from 2001 onwards, the revenues would be automatically distributed each year.

89 NPC 2003: 39.

90 Mommer 2002: 168–83; 228–35.

91 Chastko 2002: 261. On that occasion, Alberta threatened to cut the volume of crude that would be made available to consumers in Eastern Canada.

92 *FT*, 10 March 2003.
93 *Ibid.*
94 Jonathan Swift, *A Modest proposal for Preventing the Children of Ireland from Being a Burden to their Parents or Country* (1729).
95 Lal has now published a book-length treatment of the subject (Lal 2004).
96 Mommer 2002: 235.
97 *Ibid.*
98 This is a paraphrase of an ingenious aphorism coined by Baptista (1999: 43), regarding the economic development options open to his native Venezuela.

APPENDIX 1

CALCULATING FEDERAL TAX LIABILITIES ON INCOME FROM UPSTREAM ACTIVITIES IN THE GOM FEDERAL OCS

The only OCS payments that the Federal government receives as a fiscal authority come from the Federal income tax obligations that oil companies are liable for in connection with their profits from operations in the region (the current base rate for federal corporate income tax in the USA is 34 percent). However, since this tax is levied on corporations as opposed to ring-fenced fields, it is not possible to know *exactly* the share of an oil company's Federal income tax bill that is attributable solely to its GOM operations. The absence of a ring fence means that corporations can and do offset losses incurred in other activities against upstream OCS income. Published studies of effective tax rates for petroleum companies generally get around this problem by implicitly assuming that these companies' 'corporate-wide rates are essentially the same as those for their petroleum operations even though the companies are involved in wide range of other activities ... [and] this assumption is not always valid'.[1]

The tax data on the American petroleum industry compiled by the Internal Revenue Service (IRS) are corporate oriented, rather than line-of-business oriented. In contrast, the data collated in the DOE's Financial Reporting System (FRS; available from 1977 onwards) contain a wealth of detail that is not available elsewhere.[2] As the DOE itself stresses, FRS data 'are reported not only for corporate level operations but also by lines of business and segments within those lines. Thus, effective rates can be calculated for the aggregate of the companies for all their operations combined, as well as separately for their petroleum operations, petroleum production operations, petroleum refining/marketing operations, and for their other individual energy and nonenergy lines of business and business segments'.[3] Such calculations cannot be done with IRS data, which is in any case much harder to come by and use than the FRS data.

Although the FRS is based on an accounting system 'which promotes comparability, consistency and detail across companies by line of business', it has certain limitations. First of all, although FRS companies report their pretax incomes 'on a worldwide basis, [these] are not reported for aggregate foreign operations versus aggregate domestic productions even though taxes are so reported'.[4] There is publicly available data that can be used to fill these lacunae, but in a study such as this it is not necessary to do so, as the 'limitation does not apply to the individual petroleum line of business and business segments'.[5] A more serious – and unsolvable – problem is related to the fact that 'the tax and pretax income data at these disaggregated levels are limited to data that are allocated to those lines and segments. Substantial amounts are not allocated since they deal with corporate-wide activities and are essentially not traceable to particular activities'.[6] As some of these untraceable items are of critical importance (notably those related to finance), their absence is a cause for regret but, at least, the FRS data have the virtue of being 'treated consistently over time with respect to those which are allocated and those which are not'.[7] Finally, there is the issue of the inclusiveness of the data: only the largest oil companies have to file reports to the FRS, which means that a substantial proportion of US oil and gas output (and its associated income) is not covered by the data.

As explained more fully below, the methodology developed in this study to calculate Federal Income Tax obligations (inclusive of the oil depletion allowance) on OCS income relates the gross wellhead value of oil and gas with the gross nationwide E&P segment income of the FRS group of companies. Obviously, if FRS companies were inadequately represented in the GOM OCS, this would complicate matters. As it happens, though, there is a reasonable correspondence

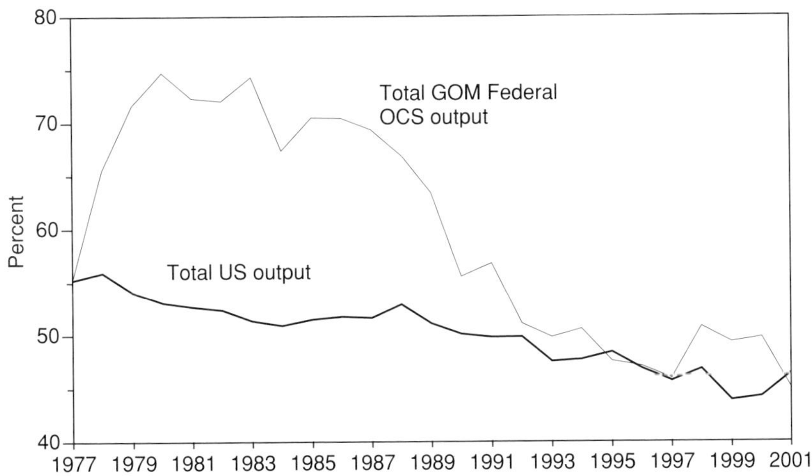

Sources: MMS, DOE

Figure A1.1: Share of FRS Companies in Total GOM Federal OCS Output and Total US Oil and Gas Output, 1977–2001

between both series because, throughout the period under consideration, the proportion of OCS oil and gas output accounted for by FRS companies was either similar or larger than these companies' share in total US oil and gas output (Figure A1.1). The Federal Income Tax obligations of non-FRS companies are calculated by the expedient of assuming that their production attracted the same effective tax rate as that of FRS companies.

Estimating the OCS Federal income tax liabilities for FRS firms requires the application of an apportionment rule similar to the ones that US states use to determine state corporate income on the basis of a given firm's Federal income tax form. States use a wide range of apportionment formulae granting different weights to in-state property, payroll and sales.[8] The apportionment formula chosen for OCS income is based on the one that the state of Iowa applies to determine the amount of total net income attributable to business conducted within its territory.[9] This is a simple formula that only considers the ratio of sales made in Iowa to gross sales elsewhere, thus:

(Iowa Gross Sales/Total Gross Sales)
× Federal Taxable Income = Iowa Income

Generally, sales figures on their own are not held to be good indicators for the income that is reasonably attributable to the trade or business conducted within a state (indeed, Iowa is the only state whose apportionment formula is based exclusively on sales).[10] Take, for instance, a company whose income in a given state derives exclusively from the sale of products manufactured outside that state, and carried out through a small number of modestly staffed retail outlets. Calculating this company's state income tax liabilities on sales alone would almost certainly result in a figure that grossly overstates the economic activity of the firm within that jurisdiction. In contrast, an apportionment formula that properly weighed payroll and property would lead to a more realistic figure.

Whatever the rights or wrongs of the way in which Iowa chooses to levy its state income tax on corporations, the fact is that disregarding payroll and property is entirely appropriate in the case of corporate income accrued in OCS upstream activities. This is because the totality of OCS gross income is attributable to production activities undertaken within the area under the jurisdiction of the Federal government, using assets entirely located within it (the whole of the OCS oil and gas output is exported to other jurisdictions, after all).

Taxable income for upstream activities carried out by FRS companies in the GOM Federal OCS is determined with the following apportionment formula:

OCS Taxable Income =
OCS Gross income/Segment total US income
× Segment US taxable income

Where:

OCS Gross Income: Gross value of GOM oil, natural gas and gas plant liquid sales (as reported by MRM)

Segment Total US Income: Operating revenues for domestic petroleum (i.e. upstream) activities of FRS companies

Segment US Taxable Income: Pre-tax income for domestic petroleum activities of FRS companies

Figure A1.2 shows GOM OCS taxable income for FRS companies as a proportion of their total US domestic petroleum activity income, and compares it against the share of total US oil and gas output accounted for by GOM output. As can be appreciated, given the weight of OCS

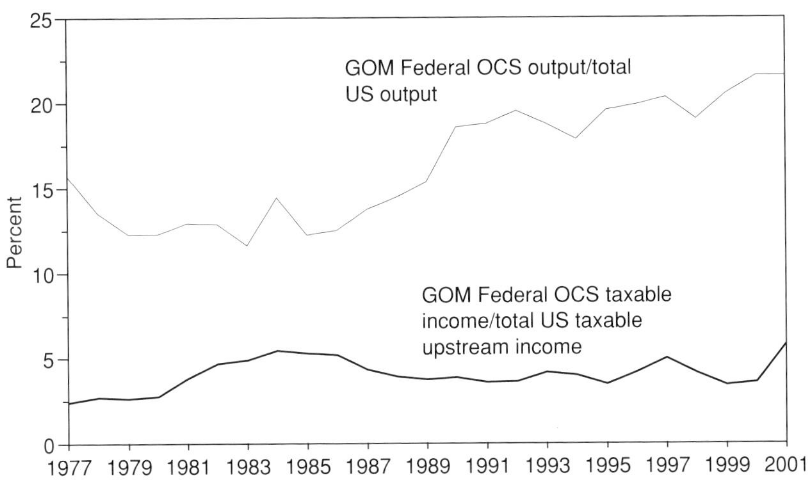

Sources: MMS, DOE

Figure A1.2: GOM Federal OCS Output and Taxable Income as a Proportion of Total US Oil and Gas Output and Upstream Taxable Income, 1977–2001

output in total US output, this ratio appears inordinately low. This is a reflection of the fact that OCS revenues are calculated on the basis of wellhead prices, which are significantly lower than landed prices due to relatively high processing and transportation costs (and also, as MMS has repeatedly alleged in the past, due to transfer price manipulation between affiliated parties).

Federal income tax liabilities for the GOM Federal OCS upstream income for FRS firms are calculated thus:

OCS Income Tax Liability =
OCS taxable income
× Federal Corporate Income Tax rate
– US Federal Investment Tax Credit applicable to OCS activities

Where:

US Federal Investment Tax Credit applicable to OCS activities = Total US Federal Investment Tax Credit × (OCS/Segment total US income)

In order to arrive at a figure that includes both FRS and non-FRS firms, the result obtained from the equation above then has to be divided by the share of FRS firms in GOM Federal OCS upstream output.

Between 1981 and 1988, upstream income derived from sales of crude oil was also liable for Windfall Profit Tax payments. Our simplistic apportionment formula for this tax (likely to overstate OCS Windfall Tax liabilities, as a significant proportion of OCS output after 1979 was Tier III oil, which attracted the lowest mark up) is as follows:

OCS Windfall Profit Tax Liability =
(OCS crude output/Total US crude output)
× Total US Windfall Profit liability

The behaviour through time of these various taxes and payments is shown in Figure A1.3, while Figure A1.4 plots their incidence as a percentage of OCS gross income. Figure A1.5 shows GOM OCS operating revenues by item, and also as a proportion of total US upstream revenue. Complete MRM data are available until the year 2002, which is the cut-off point for all figures in this appendix.

The apportionment method used to calculate total Federal income tax liabilities can also be used to estimate these liabilities for specific

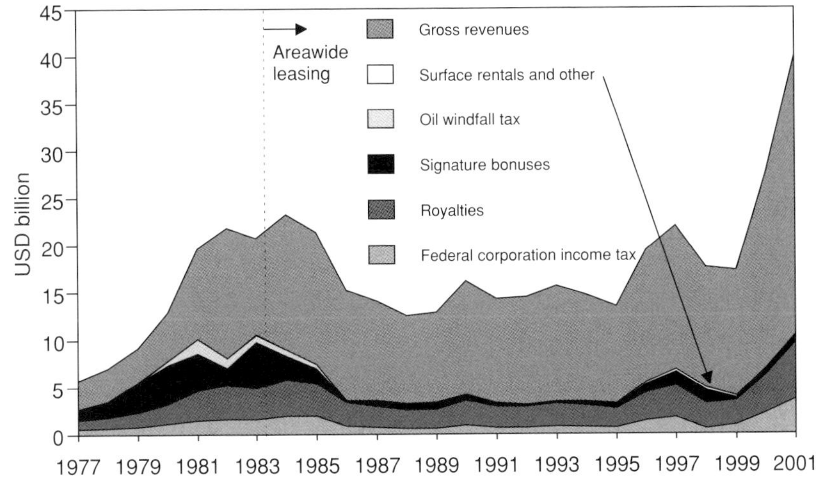

Sources: MMS, DOE

Figure A1.3: GOM Federal OCS. Gross Revenues and Fiscal Income, 1977–2001

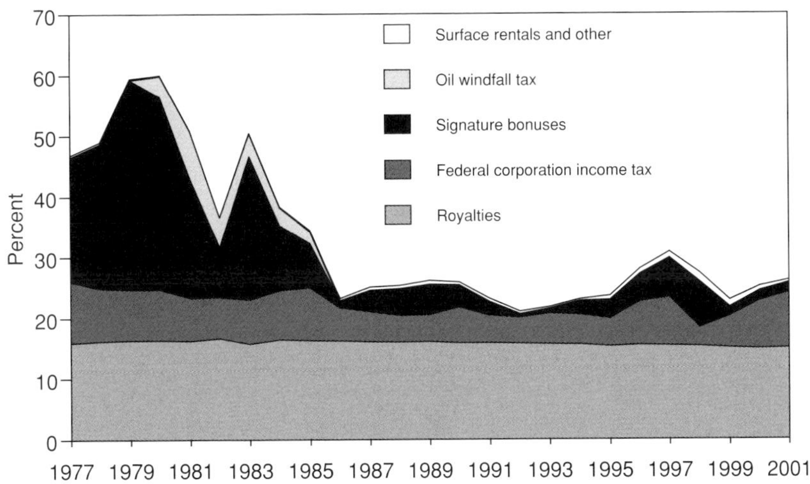

Sources: MMS, DOE

Figure A1.4: GOM Federal OCS. Lease Payments and Federal Income Taxes as a Percentage of Gross Income, 1977–2001

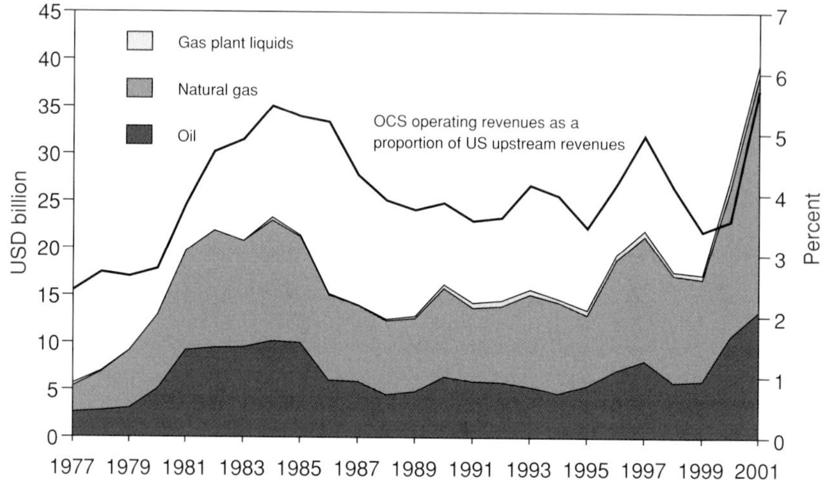

Sources: MMS, DOE

Figure A1.5: GOM Federal OCS. Operating Revenues by Item, and as a Proportion of Total US Upstream Operating Revenues, 1977–2001

development projects. That simply requires substituting OCS gross income by project gross income in the apportionment formula defined above. Project gross income, in turn, is obtained by multiplying the volumes of oil, condensates and natural gas liquids, and natural gas produced in specific leases by the wellhead price of these hydrocarbons, as reported by the MRM.

As an aside, it should be pointed out that in 1981–1982, DOI engaged in some blatant accounting sleight of hand to promote the case for the adoption of AWL. This involved the benefits that would supposedly be generated by the early receipt of corporate federal income tax on incremental oil and output upon implementation of the accelerated leasing programme. Strictly speaking, the magnitude of these benefits would have been given by the extent to which offshore production yielded tax revenues greater than what would have been received from taxing revenues generated by other investments. But in its calculations, DOI simply included all of the future income tax payments as a fiscal benefit even though, as GAO pointed out at the time, this was tantamount to assuming implicitly that 'capital not invested in offshore development [would] produce no taxable income'.[11] Naturally, by inflating the ratio between revenues from taxes and royalties, on the one hand and revenues from bonuses, on the other, this preposterous assumption significantly overstated not only the tax advantages of area-wide leasing but also the gains from the early receipt of revenues.

NOTES

1 DOE/EIA 1991: 1.
2 *Ibid.*: 7.
3 *Ibid.*
4 *Ibid.*
5 *Ibid.*: 8.
6 *Ibid.*
7 *Ibid.*
8 For details, consult Johnston and Reynolds 1985.
9 *Shell Oil Company v. Iowa Department of Revenue* 1987.
10 Johnston and Reynolds 1985.: 310.
11 GAO 1984: 17.

BIBLIOGRAPHY

Books, articles and government publications

Adelman, Morris A. (1972) *The World Petroleum Market.* Baltimore, The Johns Hopkins University Press.

Akerlof, George (1970) 'The Market for "Lemons": Qualitative Uncertainty and the Market Mechanism', *Quarterly Journal of Economics*, 89: 488–500.

Anderson, Robert O. (1984) *Fundamentals of the Petroleum Industry*. Norman (Okla.), University of Oklahoma Press.

Anderson, Roger N. (1993) 'Discovering Dynamic Gulf of Mexico Reserves and the US Energy Future', *O&GJ*, 91 (17): 85–92.

Anderson, Roger N. and Boulanger, Albert (2002) *Prospectivity of the Ultra–Deep-water Gulf of Mexico*. New York, Columbia University Lean Energy Initiative (mimeo).

Anderson, Roger N.; Flemings, Peter; Losh, Steven; Austin, John; and Woodhams, Richard (1994) 'Gulf of Mexico Growth Fault Drilled, Seen as Oil, Gas Migration Pathway', *O&GJ*, 92 (23): 97–104.

Antill, Nick and Arnott, Robert (2002) *Oil Company Crisis. Managing Structure, Profitability and Growth*. Oxford, Oxford Institute for Energy Studies.

Applegate, David (1997) 'Doughnut Holes in the Gulf of Mexico', *IBRU Boundary and Security Bulletin*, 5 (3): 70–2.

Applied Earth Sciences Institute (1994) *A National Deepwater Gulf of Mexico Initiative: Exploration and Technology Development in the Deep and Ultra–deepwater of the Gulf of Mexico. A White Paper submitted to the United States Department of Energy Office of Fossil Energy.* New York, Applied Earth Sciences Institute Lamont-Doherty Earth Observatory, Columbia University (mimeo).

Arnott, Robert (2003) 'Infrastructure Issues on the UKCS', in *Oxford Energy Forum*, Issue 53 (May): 16–18.

Austin, Diane; Carriker, Bob; McGuire, Tom; Pratt, Joseph; Priest, Tyler; and Pulsipher, Allan G. (2004.) *History of the Offshore Oil and Gas Industry in Southern Louisiana: Interim report; Volume I: Papers on) the Evolving Offshore Industry. Study MMS 2004-049.* New Orleans, MMS (Gulf of Mexico OCS Region).

Auty, Richard M. (1988) 'Oil Exporters' Disappointing Diversification into Resource Based Industry: the External Causes', *Energy Policy*, 16: 230–42.

—— (1989) 'The Internal Determinants of Eight Oil-exporting Countries' Resource Based Industry Performance', *Journal of Development Studies*, 34: 354–72.

—— (1990) *Resource Based Industrialization: Sowing the Oil in Eight Exporting Countries.* Oxford, Clarendon Press.

Ballam, John Bishop (1999) *The Oil and Gas Lease in Canada.* Toronto, University of Toronto Press.

Baptista, Asdrúbal (1999) 'Marcas en un calendario. En pos de la idea del capitalismo rentístico', en *Venezuela en Oxford. 25 años de la cátedra Andrés Bello en el St. Antony's College de la Universidad de Oxford.* Caracas, Banco Central de Venezuela.

Barents Group LLC (Barents) (2000) *Analysis of the Department of Interior, Minerals Management Service's Third Further Supplementary Proposed Rule Establishing Oil Value for Royalty Due on Federal Leases 64. Fed. Reg. 73820. December 30, 1999.* Washington D.C., Independent Petroleum Association of America.

Bartley, Ernest R. (1953) *The Tidelands Oil Controversy. A Legal and Historical Analysis.* Austin, University of Texas Press.

Baud, Richie D.; Doyle, Carey; Peterson, Robert H. and Richardson, G. Ed (2000) *Deepwater Gulf of Mexico: America's Emerging Frontier. OCS Report MMS 2000–22.* New Orleans, MMS (Gulf of Mexico OCS Region).

Baud, Richie D.; Doyle, Carey; Peterson, Robert H. and Richardson, G. Ed; French, Leanne S.; Regg, Jim; Montgomery, Tara: Williams, T. Scott; Doyle, Carye and Dorner, Mike (2002)) *Deepwater Gulf of Mexico 2002: America's Expanding Frontier. OCS Report MMS 2002–021.* New Orleans, MMS (Gulf of Mexico OCS Region).

Bascle, Barbara J.; Nixon, Lesley D.; and Ross, Catherine M. (2001) *Atlas of Gulf of Mexico Gas and Oil Sands. OCS Report MMS 2002–086.* New Orleans, MMS (Gulf of Mexico OCS Region).

Berman, Matthew D. (1997) 'Caveat Emptor: Purchasing Petroleum Industry Investment with Fiscal Incentives', *Journal of Energy Finance & Development,* 2: 25–44.

Boué, Juan Carlos (2002) *US Gulf Offshore Oil: Petroleum Leasing and Taxation and Their Impact on Industry Structure, Competition, Production and Fiscal Revenues.* Oxford, Oxford Institute for Energy Studies.

Bradbury, John (2002) 'When Will FPSOs Appear in the Gulf?', *Hart's E&P,* 75: 13–15.

Bradley Jr., Robert L. (1996) *Oil, Gas & Government - The US Experience.* Boston, Cato Institute/Rowman & Littlefield Publishers, 2 vols.

Bradley Jr., Robert L. and Mead, Walter J. (1998) 'Resolving the Federal Oil and Gas Royalty Valuation Dispute: a Free Market Approach', *The Journal of Energy and Development,* 23: 207–235.

Breeden, Richard (1976) 'Federalism and the Development of Outer Continental Shelf Mineral Resources', *Stanford Law Review,* 28: 1107–1159.

Brooks, James M; Kennicutt, Mahon C., II; and Thompson, Keith F. (1990) Classification of Offshore Gulf of Mexico Oils and Gas Condensates, *American Association of Petroleum Geologists Bulletin,* 74(2): 187–198.

Brown, Keith C. (1967) 'The Distribution of Louisiana Outer Continental Shelf Lease Bids', *Land Economics,* 43: 354–57.

Bryce, Robert (2004) *Cronies. Oil, the Bushes, and the Rise of Texas, America's Superstate.* New York, Public Affairs.

Byrne, Lloyd (2003) 'What is Normal?', *Energy Global Perspectives – March 2003* (Morgan Stanley): 31–9.

Camp, Wayne K and McGuire, Donn (1997) 'Mahogany Field, a Subsalt

Legend: A Tale of Technology, Timing and Tenacity, Offshore Gulf of Mexico', *Houston Geological Society Bulletin* 37 (2).

Campbell, David and Hulme, Ron (2001) 'The Winner-Takes-All Economy', *The McKinsey Quarterly*, 2001 (1): 82–93.

Capen, E.C.; Clapp, R.V. and Campbell, W. M. (1971) 'Competitive Bidding in High Risk Situations,' *Journal of Petroleum Technology*, 23: 641–53.

Chastko, Paul A. (2002) *Developing Alberta's Oil Sands, 1920–2002*. Athens (Oh.), Ohio University (PhD. Thesis).

Christopher, Warren M. (1953) 'The Outer Continental Shelf Lands Act: Key to a New Frontier', *Stanford Law Review*, 6: 23–68.

Coase, Ronald (1964) 'The Regulated Industries: Discussion', *American Economic Review*, 54(2): 194–7.

Congressional Budget Office (CBO) (2000) *Reforming the Federal Royalty Program for Oil and Gas*. Washington D.C., CBO.

Conn, Charles and White, David (1994) *Revolution in Upstream Oil and Gas*. Sydney and Melbourne, McKinsey & Company Asia-Pacific Energy Practice.

Conquest, Robert (1999)) *Reflections on a Ravaged Century*. London, John Murray.

Cox, James C.: Isaac, R. Mark and Smith, Vernon L. (1983) 'OCS Leasing and Auctions: Incentives and Performance of Alternative Bidding Institutions', *Supreme Court Economic Review*, 2: 43–87.

Crawford, T. Gerald; Burgess, Grant L.; Kinler, Clark J.; Prendergast, Michael T. (2003) *Outer Continental Shelf. Estimated Oil and Gas Reserves, Gulf of Mexico, December 31, 2000. OCS Report MMS 2003-050*. New Orleans, MMS (Gulf of Mexico OCS Region).

Danaczko, M; Pichini, P.Q. and Rowe, S.J. (1985) *Lena Guyed Tower Launch Studies*. Royal Institution of Naval Architects Symposium on Offshore Transportation and Installation, London, March 26–7.

Davis, John C, and Harbaugh, John W. (1980) 'Oil and Gas in Offshore Tracts: Inexactness of Resource Estimates Prior to Drilling', *Science*, 209: 1047–1048.

Davis, Randall E. and Neff, Shirley (1996) 'Deepwater Royalty Relief Product of $3^{1/2}$ Year U.S. Political Effort', in *Oil & Gas Journal*, 94, no. 14: 45–56.

Deacon, Robert; DeCanio, Stephen; Frech, H.E. and Johnson, M. Bruce (1990) *Taxing Energy. Oil Severance Taxation and the Economy*. Oakland, The Independent Institute.

de Chazeau, Melvin G. and Kahn, Alfred E. (1959) *Integration and Competition in the Petroleum Industry*. New Haven, Yale University Press.

Department of Trade and Industry (DTI) (1986) *Offshore Infrastructure Code of Practice. An Agreed Industry Document on the Rules and Procedures Governing Access to Offshore Infrastructure*. London, Her Majesty's Stationary Office.

Derman, Andrew and Johnston, Daniel (1998) *Bonuses Add Dimension to US OCS Fiscal System Analysis*. Dallas, Thompson and Knight LLP (mimeo).

DiMarco, S.F.; Howard, M.K.; Nowlin Jr.; W.D.; and Reid, R.O. (2004) *Subsurface, High-speed Current Jets in the Deepwater Region of the Gulf of Mexico. OCS Study MMS 2004-022*. New Orleans, MMS (Gulf of Mexico OCS Region).

Dougherty, Elmer L. and Lorenz, John (1976) 'Statistical Analyses of Bids for Federal Offshore Leases', *Journal of Petroleum Technology*, 28: 1377–90.

Dunn, Ward W. (1995) 'Ekofisk: a Giant Discovery of Oil and Gas in the North Sea', in *The Oil Finders. A Collection of Stories about Exploration*. Allen G. Hatley, Jr. (ed). Utopia (Tx.), Centex Press: 27–48.

Dyman, Thaddeus, and Cook, Troy A. (2001) 'Summary of Deep Oil and Gas Wells in the United States through 1998', in Dyman and Kuuskra (eds.), 1997: pp. B1–B7.

Dyman, Thaddeus S. and Kuuskraa, Vello A. (eds.) (1997) *Geologic Studies of Deep Natural Gas Resources. U.S. Geological Survey Digital Data Series 67*. Reston (Va.), USGS/ National Energy Technology Laboratory/Gas Technology Institute/Advanced Resources International.

Dyman, Thaddeus S.; Rice, D.D. and Westcott, P.A. (eds.) (1997) *Geologic Controls of Deep Natural Gas Resources in the United States. US Geological Survey Bulletin 2146-A*. Washington D.C., United States Government Printing Office.

Dyman, Thaddeus S.; Wilson, Michael S. and Beeman, William R. (1995) 'Deep Natural Gas Reservoirs and Conventional Plays in the United States', in Gautier, D.L., Dolton, G.L., Takahashi, K.I., and Varnes, K.L., (eds.), *1995 National Assessment of United States Oil and Gas Resources. Results, Methodology, and Supporting Data (U.S. Geological Survey Digital Data Series 30)*. Reston (Va.), USGS.

Energy Information Administration (EIA) (1991b) *Average Effective Corporate Income Tax Rates for Petroleum Operations 1977–1989*. Washington D.C., DOE/EIA.

—— (2000) *Accelerated Depletion: Assessing Its Impacts on Domestic Oil and Natural Gas Prices and Production*. Washington D.C., DOE/EIA.

Engelbrecht-Wiggans, Richard; Dougherty, Elmer L. and Lohrenz, John (1986) 'A Model for the Distribution of the Number of Bids on Federal Offshore Oil Leases', *Management Science*, 32: 1087–1094.

Ernst, David and Steinhubl, Andrew M.J. (1997) 'Alliances in Upstream Oil and Gas', *The McKinsey Quarterly*, 1997 (2): 144–155. (1999) 'Petroleum: After the Megamergers', *The McKinsey Quarterly*, 1999 (2): 49–57.

Eskew, Blake T. and Jones, Stephen L. (2001) 'Gulf Deepwater Crude Impacts Downstream Disposition', *Oil and Gas Journal*, 99 (18): 80–90.

Ezell, John S. (1979) *Innovations in Energy. The Story of Kerr-McGee*. Norman (Okla.), University of Oklahoma Press.

Farrow, F. Scott (with Boradus, James M.; Grigalunas, Thomas A.; Hoagland III, Porter and Opaluch, James) (1990) *Managing the Outer Continental Shelf Lands. Oceans of Controversy*. New York, Taylor and Francis.

Fiske, Leland E., Smith, Cecil L., Anderson, Guy W., and Jones, Morgan A. (1982) *Federal Taxation of Oil and Gas Transactions*. New York, Matthew Bender, 2 vols.

Fitzgerald, Edward A. (2001) *The Seaweed Rebellion. Federal-State Conflict over Offshore Energy Development*. Lanham (Ma.), Lexington Books.

—— (2002) 'The Seaweed Rebellion: Florida's Experience with Offshore Energy Development', *Journal of Land Use*, 18 (1): 1–72.

Flichman, Guillermo (1977) *La renta del suelo y el desarrollo agrario argentino*. México, siglo xxi.

French, Leanne S.; Richardson; G. Ed; Kazanis, Eric G.; Montgomery, Tara M.; Bohannon, Christy M. and Gravois, Michael P. (2006) *Deepwater Gulf of Mexico 2006: America's Expanding Frontier. OCS Report MMS 2006-022*. New Orleans, MMS (Gulf of Mexico OCS Region).

Galbraith, John Kenneth (1979) *The Great Crash 1929 (50th Anniversary Edition)*. London, André Deutsch.

George, Henry (1888) *Progress and Poverty. An Inquiry into the Cause of Industrial Depressions and of Increase of Want with Increase of Wealth: the Remedy*. London, C. Kegan Paul.

Gerking, Shelby; Kunce, Mitch; Kervliet, Joe; and Morgan, William (2000) *Mineral Tax Incentives, Mineral Production and the Wyoming Economy*. Casper (Wy.), Wyoming Legislature, Mineral Tax Incentives Subcommittee.

Gilley, Otis W. and Karels Gordon V. (1981) 'The Competitive Effects in Bonus Bidding, New Evidence'. *Bell Journal of Economics*, 12(2): 637–488.

Gilley, Otis W.; Karels, Gordon V. and Leone, Robert P. (1986) 'Uncertainty, Experience and the 'Winner's Curse' in OCS Lease Bidding', *Management Science*, 32: 673–682.

Gohrbrandt, Klaus H. (2001) 'West Florida Shelf and Slope: Prime Target for Gas, Oil in Eastern Gulf of Mexico', *O&GJ*, 99(24): 41–8.

Gold, Thomas) (1999)) *The Deep Hot Biosphere*. New York, Copernicus Press.

Gordon, Richard L. (1986) Review of *Offshore Lands Oil and Gas Leasing and Conservation in the Outer Continental Shelf*, in *The Energy Journal*, 7(4): 110–1.

Gramling, Robert (1995) *Oil on the Edge. Offshore Development, Conflict, Gridlock*. Albany, State University of New York Press.

Gurney, Judith (1997a) *The Gulf of Mexico. Revival and Opportunity in the Oil Industry*. London, Financial Times Energy Publishing.

—— (1997b) 'Migration or Replenishment in the Gulf', *Petroleum Review*, 51: 200–203.

Hedberg, H.D.; Moody, U.D.; and Hedberg, R.M. (1979) 'Petroleum Prospects of the Deep Offshore', *AAPG Bulletin*, 63: 286–300.

Hendricks, Kenneth and Porter, Robert (1992) 'Joint Bidding in Federal OCS Auctions', *The American Economic Review*, 82 (2), Papers and Proceedings of the Hundred and Fourth Annual Meeting of the American Economics Association (May 1992): 506–511.

Hendricks, Kenneth; Porter, Robert and Boudreau, Bryan (1987) 'Information, Returns and Bidding Behaviour in OCS Auctions: 1954–1969', *The Journal of Industrial Economics*. 35: 517–542.

Hendricks, Kenneth; Porter, Robert and Tan, Guofu (1993) 'Optimal Selling Strategies for Oil and Gas Leases with an Informed Buyer', *The American Economic Review*, 83 (2), Papers and Proceedings of the Hundred and Fifth Annual Meeting of the American Economics Association (May 1993): 234–239.

—— (2000) *Joint Bidding in Federal Offshore Oil and Gas Lease Auctions*. Mimeo.

—— (2003) *Bidding Rings and the Winner's Curse: the Case of Federal Offshore Oil and Gas Lease Auctions. Working Paper 0036*. Evanston (Ill.), The Center for the

Study of Industrial Organisation at Northwestern University.

Hendricks, Kenneth; Porter, Robert and Wilson, Charles A. (1994) 'Auctions for Oil and Gas Leases with an Informed Bidder and a Random Reservation Price', *Econometrica*, 62: 1415–1444.

Hewitt, Suzanne M. (1995) 'Seismic Stratigraphy and Salt Tectonics of the Alaminos Canyon Area, Gulf of Mexico', *Quarterdeck*, 3(1).

Holland, David S.; Leedy, John B.; and Lammlein, David R. (1999) 'Eugene Island Block 330 Field –U.S.A. Offshore Louisiana', in Beaumont, E.A. and Foster, N.H. (eds.), *Structural Traps III, AAPG Treatise of Petroleum Geology, Atlas of Oil and Gas Fields*. Tulsa, AAPG: 103–43.

Horsnell, Paul (1999) Review of *Offshore Pioneers: Brown & Root and the History of Offshore Oil and Gas*, in *The Journal of Energy Literature*, V, no.1: 61–64.

—— (2000) Review of *Oil in Deepwaters: Global Reserves and Production*, in *The Journal of Energy Literature*, VI, no.1: 77–79.

Horsnell, Paul, and Mabro, Robert (1993)) *Oil Markets and Prices. The Brent Market and the Formation of World Oil Prices*. Oxford, Oxford University Press.

Hunt, John M. (1996) *Petroleum Geochemistry and Geology*. New York, W.H. Freeman and Company.

International Maritime Organisation (2002) *Compilation of FPSOs and FSUs Submitted by the International Association of Oil and Gas Producers (OGP)*. London, IMO (Mimeo.).

Independent Petroleum Association of America (IPAA) (2002) *Testimony Submitted by the American Petroleum Institute, Domestic Petroleum Council, Independent Petroleum Association of America, International Association of Drilling Contractors, National Ocean Industries Association, Petroleum Equipment Suppliers Association, and the United States Oil and Gas Association to the U.S. Commission on Ocean Policy (New Orleans, Louisiana, March 8, 2002)*. Washington D.C., IPAA.

Johnston, Daniel (1994) *International Petroleum Fiscal Systems and Production Sharing Contracts*. Tulsa, PennWell Books.

—— (2002) *Petroleum Tax Design*. Paper Delivered at the World Bank Workshop on Petroleum Revenue Management, Washington D.C., October 23–24, 2002, pp. 28.

Johnston, James L. and Reynolds, Alan (1985) 'A Natural Resource Theory of Unitary Taxation', *The Energy Journal*, 6 (Special Tax Issue): 309–324.

Kaiser, Mark J. and Pulsipher, Allan G. (2004) *Fiscal System Analysis: Concessionary and Contractual Systems Used in Offshore Petroleum Arrangements. OCS Study MMS 2004-016*. New Orleans, MMS (Gulf of Mexico OCS Region).

Keefer, Donald L. (1991) 'Resource Allocation Models with Risk Aversion and Probabilistic Dependence: Offshore Oil and Gas Bidding', *Management Science*, 37: 377–395.

Kissinger, Henry (1999)) *Years of Renewal*. New York, Simon and Schuster.

Klara, Scott and Shuster, Erik (2004) *Tracking New Coal-Fired Power Plants. Coal's Resurgence in Electric Power Generation*. Pittsburgh, National Energy Technology Laboratory (mimeo).

Klemperer, Paul (2002) 'What Really Matters in Auction Design', *The Journal of Economic Perspectives*, 61 (1): 169–189.

—— (2003) *Using and Abusing Economic Theory. 2002 Alfred Marshall Lecture to the European Economic Association*. Mimeo.

Kobrin, Paul A. (1986)) 'The Cost of OCS Bid Rejection', *The Energy Journal*, 7 (1): 83–97.

Kosmo, Mark (1985) 'OCS Leasing Policy: Its Effects on the Structure of the Petroleum Industry', *The Energy Journal*, 6 (1): 79–96.

Kreidler, Tai (1997) *The Offshore Petroleum Industry: the Formative Years, 1945–1962*. Lubbock (Tx.), Texas Tech University (PhD. Thesis).

Kumins, Lawrence (2000) *Outer Continental Shelf: Oil and Gas Leasing and Revenue. CRS Issue Brief for Congress 10005*. Washington D.C., Congressional Research Service.

Kuntz, Eugene O.; Lowe, John S.; Anderson, Owen L; Smith, Ernest E. and Pierce, David E. (1998) *Cases and Materials on Oil and Gas Law*. St. Paul (Minn.) West Group.

Kvendseth, Stig S. (1988) *Giant Discovery. A History of Ekofisk through the First Twenty Years*. Stavanger, Phillips Petroleum Company Norway.

Laffont, Jean Jacques and Tirole, Jean (1993) *A Theory of Incentives in Procurement and Regulation*. Cambridge (Mass.), The MIT Press.

Laherrère, Jean (2003) 'Future of Oil Supplies', *Energy Exploration & Exploitation*, 21: 227–267.

Lajous, Adrián (2004) 'Las reservas de gas natural de México', in *Este país. Tendencias y opiniones*, no. 158: 22–30.

Lal, Deepak (2004) *In Praise of Empires: Globalization and Order*. New York, Palgrave Macmillan.

Leland, Hayne E. (1978) 'Optimal Risk Sharing and the Leasing of Natural Resources, with Application to Oil and Gas Leasing on the OCS', *The Quarterly Journal of Economics*, 92 (3): 413–438.

Logue, Dennis E.; Sweeny, Richard James and Willett, Tomas D. (1975) 'Optimal Leasing Policy for the Development of Outer Continental Shelf Hydrocarbon Resources', *Land Economics*, 51: 191–207.

Lore, Gary L. (1992) 'Conventional Gas Resources of the Gulf of Mexico Outer Continental Shelf – Past Experience, Current Activities, Future Potential' in *The Future of Energy Gases, US Geological Survey Professional Paper 1570*: 157–170.

Lore, Gary L.; Marin, David A.; Batchelder, Eric C.; Courtwright, William C.; Desselles Jr., Richard P.; and Klazynski, Ralph J. (2001) *2000 Assessment of Conventionally Recoverable Hydrocarbon Resources of the Gulf of Mexico and Atlantic Outer Continental Shelf as of January 1, 1999. OCS Report MMS 2001-087*. New Orleans, MMS.

Mabro, Robert; Bacon, Robert; Chadwick, Margaret; Halliwell, Mark and Long, David (1986) *The Market for North Sea Crude Oil*. Oxford, Oxford University Press.

Macho-Stadler, Inés and Pérez Castrillo, David (1997) *An Introduction to the Economics of Information. Incentives and Contracts*. Oxford, Oxford University Press.

Martin, Ray J. and Foote, Richard Q. (1981a) 'Petroleum Geology in the Gulf of Mexico', in Powers (1981): 156–183.

—— (1981b)) 'Geology and Geophysics of the Maritime Boundary assessment areas', in Powers (1981): 30–67.)

Martin, Steve (1997) *Tax or Technology? The Revival of UK North Sea Oil Production*. Oxford, Oxford Institute for Energy Studies.

Maskin, E. and Riley, J. (1984) 'Optimal Auctions with Risk Averse Buyers', *Econometrica*, 52: 1473–1518.

Matharu, Andrew and Lucas, Fred (2002) *To Shovel or to Drill?* London, Cazenove Global Oil and Gas Research.

McBride, Barry C.; Weimer, Paul and Rowan, Mark G (1998) 'The Effect of Allochthonous Salt on the Petroleum Systems of Northern Green Canyon and Ewing Bank (Offshore Louisiana), Northern Gulf of Mexico', *AAPG Bulletin*, 82: 1083–1112.

McDonald, Stephen L. (1979) *The Leasing of Federal Lands for Fossil Fuels Production*. Baltimore, Johns Hopkins University Press.

McKay, M. and Nides, J. (eds.) (2005) *Proceedings: Twenty-second Annual Gulf of Mexico Information Transfer Meeting, January 2003. OCS Study MMS 2003-073*. New Orleans, MMS (Gulf of Mexico OCS Region).

McKellar, Barbara J. (1988) *The Role of Giant Fields in Future U.S. Oil Production*. Madison, The University of Wisconsin (Ph.D Thesis).

McLean, John H. and Haigh, Robert W. (1954) *The Growth of Integrated Oil Companies*. Boston, Harvard University Press.

McMahon, Neil (2004) *Global Integrated Oils: Declining Deepwater Discoveries, Another Datapoint for Continued High Oil Prices?* London, Sanford C. Bernstein & Co. LLC.

Mead, Walter J. (1993)) 'Oil and Gas Leasing Policy Alternatives', in *The Environment of Oil*, Richard J. Gilbert, (ed.). Dordrecht, Kluwert Academic Publishers: 215–60.

—— (1994) 'Toward an Optimal Oil and Gas Leasing System', *The Energy Journal*, 15 (4): 1–18.

Mead, Walter J. and Sorensen, Philip E. (1980a) *Competition and Performance in OCS Oil and Gas Lease and Lease Development, 1954–1969. Final Report to the U.S. Geological Survey, Contract #14-08-0001-16552*. Reston (Va), USGS.

—— (1980b) *Additional Studies of Competition and Performance in OCS Oil and Gas Lease and Lease Development, 1954–1969. Final Report to the U.S. Geological Survey, Contract #14-08-0001-18678*. Reston (Va), USGS.

Mead, W.J., Moseidjord, Asbjorn; and Sorensen, Philip E. (1983) 'The Rate of Return Earned by Lessees under Cash Bonus Bidding for OCS Oil and Gas Leases', *The Energy Journal*, 4: 37–52.

Mead, W.J., Moseidjord, Asbjorn; Muraoka, Dennis; and Sorensen, Philip E. (1985) *Offshore Lands: Oil and Gas Leasing and Conservation on the Outer Continental Shelf*. San Francisco, Pacific Institute for Public Policy Research.

Melancon, J. Michael and Durr, Alvin G. (2004) *Design and Rationale of the Final Rule on Royalty Relief for Deep Gas Production from Shallowwater Leases in the Gulf of Mexico*. Paper presented at the *Institute on Federal and Indian Oil and Gas Royalty Valuation and Management*, Houston (Tx.), 11–13 February 2004: 11 pp.

Melancon, J. Michael; Bongiovanni, Roy; and Baud, Richie D.; (2003) *Gulf of Mexico Outer Continental Shelf Daily Oil and Gas Production Rate Projections From 2003 Through 2007. OCS Report MMS 2003-028.* New Orleans, MMS (Gulf of Mexico OCS Region).

Melancon, J. Michael; and Baud, Richie D.; G. Boice, Angela: Bongiovanni, Roy; DeCort, Thierry; Desselles, Richard; Kazanis, Eric G. (2004) *Gulf of Mexico Oil and Gas Production Forecast: 2004 – 2013. OCS Report MMS 2004-065.* New Orleans, MMS (Gulf of Mexico OCS Region).

Merrill Lynch (2002) *Global Octane – 2002. Global Oils – A Beacon of Light in Today's Turbulent Market.* London, Merrill Lynch Global Securities Research & Economics Group, Global Fundamental Equity Research Department.

Milsapps, Steven W. and Ott, Mack (1985) 'Risk Aversion, Risk Sharing and Joint Bidding: A Study of Outer Continental Shelf Petroleum Auctions', *Land Economics*, 61: 372–386.

Minerals Management Service (MMS) (1983) *From Policy to Production. Offshore Leasing and Operations.* Washington D.C., MMS/DOI.

—— (1999) *Boundary Development on the Outer Continental Shelf (OCS Report MMS 99–0006).* Lakewoood (Co.), US Department of the Interior (Minerals Management Service, Mapping and Boundary Branch).

—— (2001a) *The Promise of Deep Gas in the Gulf of Mexico. OCS Report MMS 2001-037.* New Orleans, MMS (Gulf of Mexico OCS Region).

—— (2001b) *FPSO Environmental Impact Statement Frequently Asked Questions.* New Orleans, MMS.

—— (2001c) *Proposed Use of Floating Production, Storage, and Offloading Systems on the Gulf of Mexico Outer Continental Shelf, Final Environmental Impact Statement. OCS Report MMS 2000–090*, New Orleans, MMS.)

—— (2003) *Gulf of Mexico OCS Deep Shelf Gas Update: 2001–2002. OCS Report MMS 2003-026*, New Orleans, MMS (Gulf of Mexico OCS Region).

—— (2004a) *Five Year Royalty In Kind Business Plan.* Washington D.C., MMS.

—— (2004b) *History of the Offshore Oil and Gas Industry in Southern Louisiana: Interim Report. MMS Publication 2004-049-051. Volume I: Papers on the Evolving Offshore Industry. Volume II: Bayou Lafourche – An Oral History of the Development of the Oil and Gas Industry. Volume III: Samples of Interviews and Ethnographic Prefaces.* New Orleans, MMS (Gulf of Mexico OCS Region).

—— (2005) *Royalty in Kind Fiscal Revenue Performance: Fiscal Year 2004.* Washington D.C., Minerals Revenue Management.

Mitchell, John (1996) *The New Geopolitics of Energy.* London, Royal Institute for International Affairs.

Mitchell, John with Morita, Koji; Selley, Norman and Stern, Jonathan (2001) *The New Economy of Oil. Impacts on Business, Geopolitics and Society.* London, Earthscan Publications (Royal Institute of International Affairs, Energy and Environment Programme).

Mommer, Bernard (1999) *Oil Prices and Fiscal Regimes.* Oxford, Oxford Institute for Energy Studies.

—— (2001) *Fiscal Regimes and Oil Revenues in the UK, Alaska and* Venezuela. Oxford, Oxford Institute for Energy Studies.

—— (2002) *Global Oil and the Nation State.* Oxford, Oxford University Press.

Montgomery, Scott L. and Moore, Dwight (1997) 'Subsalt Play, Gulf of Mexico: A Review', *AAPG Bulletin*, 81 (6): 871–896.

Moody, Carlisle E. (1994) 'Alternative Bidding Systems for Leasing Offshore Oil: Experimental Evidence', *Economica*, 61: 345–353.

Moody, Carlisle E. and Kruvant, William J. (1988) 'Joint bidding, entry and the price of OCS leases', *RAND Journal of Economics*, 19 (2): 276–84.

—— (1990) 'OCS Leasing Policy and Lease Prices', *Land Economics*, 66 (1): 30–9.

Moore, Charles A. (2000) *Upstream Legal Structure from the US Perspective*. Presentation to the seminar *Energía y Derecho 2000*, Mexico City, Instituto Tecnológico Autónomo de México/Asociación Mexicana de Derecho Energético.

National Energy Board (2003) *Short–term Natural Gas Deliverability from the Western Canada Sedimentary Basin 2003–2005*. Calgary, National Energy Board.

—— (2004) *Looking Ahead to 2010. Natural Gas Markets in Transition*. Calgary, National Energy Board.

National Ocean Industries (NOI) (1984) *Area-Wide Leasing: National Boon or Industry Boondoggle?* Washington D.C., NOI.

National Petroleum Council (NPC) (1999) *Meeting the Challenges of the Nation's Growing Natural Gas Demand*. Washington D.C., NPC, 3 volumes.

—— (2003) *Balancing Natural Gas Policy – Fueling the Demands of a Growing Economy*. Washington D.C., NPC, 5 volumes.

National Research Council of the National Academies (NRU) (1978) *OCS Oil & Gas. An Assessment of the Department of the Interior Environmental Studies Program*. Washington D.C., National Academy of Sciences.

—— (2003) *Summary on a Workshop on US Natural Gas Demand, Supply and Technology. Looking Toward the Future*. Washington D.C., The National Academies Press.

Nehring, Richard (2001a) 'Gulf of Mexico: Reservoir Temperatures, Low Thermal Gradient Limit US Gulf's Deepwater Gas Potential', *Offshore*, 61 (1).

—— (2001b) *The Gulf of Mexico: Rising Star or Over the Hill?* Presentation at the Annual Energy Outlook 2001 Conference, Washington D.C. (March 27, 2001).

Odell, Peter (2001) 'The Global Energy Market in the Long-Term: the Continuing Dominance of Affordable Non-Renewable Resources', in *Oil and Gas: Crisis and Controversies 1961–2000. Studies and Commentaries by Peter R. Odell. Volume 1: Global Issues*. Brentwood, Multi–Science Publishing Company: 183–201.

Office of Technology Assessment (OTA), US Congress (1975a) *Analysis of the Feasibility of Separating Exploration from Production of Oil and Gas on the Outer Continental Shelf*. Washington D.C., US Government Printing Office.

—— (1975b) *An Analysis of the Department of the Interior's Proposed Acceleration of Development of Oil and Gas on the Outer Continental Shelf*. Washington D.C., US Government Printing Office.

—— (1985) *Oil and Gas Technologies for the Arctic and Deepwater*. Washington D.C.,

US Government Printing Office.
—— (1990) *The Royalty Management Program's Auditing and Financial System: Technical Issues*. Washington D.C., OTA.
Olien, Roger M. and Olien, Diana Davids (1984)) *Wildcatters. Texas Independent Oilmen.* Texas Monthly Press.
Osmundsen, Petter (1995) 'Taxation of Petroleum Companies Possessing Private Information', *Resource and Energy Economics*, 17: 357–77.
Paddock, James L.; Siegel, Daniel R.; and Smith, James L. (1988) 'Option Valuation of Claims on Real Assets: The Case of Offshore Petroleum Leases', *The Quarterly Journal of Economics*, 103:479–508.
Pechman, Joseph A. (1987)) *Federal Tax Policy*. Washington D.C., The Brookings Institution.
Pérez Alfonzo, Juan Pablo (1960) *Venezuela y su petróleo. Lineamientos de una política.* Caracas, Imprenta Nacional.
Petroleum Economist (2001) *Fundamentals of the Global Offshore Industry*. London, Petroleum Economist Publications.
Porfir'yev, V.B. (1974)) 'Inorganic Origin of Petroleum', *AAPG Bulletin*, 58 (1): 3–33.
Porter, Robert H. (1995) 'The Role of Information in US Offshore Oil and Gas Lease Auctions', *Econometrica*, 63 (1): 1–27.
Powers, R.B. (ed.) (1981) *Geologic Framework, Petroleum Potential, Petroleum Resource Estimates, Mineral and Geothermal Resources, Geologic Hazards, and Deep-water Drilling Technology of the Maritime Boundary Region in the Gulf of Mexico: U.S. Geological Survey Open File Report 81-265*. Reston (Va.), US Geological Survey.
Prato, Anthony A. and Miller, Ronald R. (1981) 'Evaluating the Energy Production Potential of the United States Outer Continental Shelf', *Land Economics*, 57: 77–90.
Pratt, Joseph A. (2004) 'The Brave and the Foolhardy: Hurricanes and the Early Offshore Oil Industry', in Austin *et al.* (2004): 7–28.
Pratt, Joseph A.; Priest, Tyler and Castaneda, Christoper J. (1997) *Offshore Pioneers. Brown & Root and the History of Offshore Oil and Gas*. Houston, Gulf Publishing Company.
Priest, Tyler (2004) 'History of U.S. Oil and Gas Leasing on the Outer Continental Shelf', in Austin *et al.* (2004): 29–52.
Prindle, David F. (1981) *Petroleum Politics and the Texas Railroad Commission*. Austin, University of Texas.
Pulsipher, Allan G.; Iledare, Omowumi O. and Baumann, Robert H. (1996) *Modeling the Structure and Performance of Integrated and Independent Producers in the Gulf of Mexico OCS Region. OCS Study MMS 95-0056*. New Orleans, Centre for Energy Studies (Louisiana State University)/MMS.
Pulsipher, Allan G.; Iledare, Omowumi O. and Mesyanzhinov, Dmitry V., (2003) *Changing Patterns of Ownership and Control in the Petroleum Industry: Implications for the Market for Oil and Gas Leases in the Gulf of Mexico OCS Region, 1983–1999. OCS Study MMS 2003-041*. New Orleans, Centre for Energy Studies (Louisiana State University)/MMS.
Pursell, David (1998) *Depletion. The Forgotten Factor in the Supply and Demand*

Equation: Gulf of Mexico Analysis. Houston, Simmons and Company International.

Rainey, David (2002) *Changing the Game in the Deepwater Gulf of Mexico: The Road to Crazy Horse and Beyond.* Michel T. Halbouty Lecture, Annual Meeting of the American Association of Petroleum Geologists.

Reece, Douglas K. (1978) 'Competitive Bidding for Offshore Petroleum Leases', *The Bell Journal of Economics*, 9(2): 369–384.

Reeves, Scott R., Kuuskraa, Jason and Kuuskraa, Vello (1998) 'Deep Gas Poses Opportunities, Challenges to U.S. Operators', *O&GJ*, 96 (18): 133–146.

Rice, Dudley D.; Schenk, Christopher J.; Schmoker, James W.; Fox, James E.; Clayton, Jerry L.; Dyman, Thaddeus S.; Higley, Debra K.; Keighin, C. William; Law, Ben E. and Pollastro, Richard M. (1997) 'Deep Natural Gas Resources in the Eastern Gulf of Mexico', in Dyman and Westcott (1997): 219–229.

Richardson, G. Ed; French, Leanne S.; Baud, Richie D.; Peterson, Robert H.; Roark, Carla D.; Montgomery, Tara M.; Kazanis, Eric G.; Conner, G. Michael and Gravois, Michael P. (2004) *Deepwater Gulf of Mexico 2004: America's Expanding Frontier. OCS Report MMS 2004-021.* New Orleans, MMS (Gulf of Mexico OCS Region).

Richardson, Jeremy J. (1982) 'Problems of Controlling Public Sector Agencies: the Case of Norwegian Oil Policy', *Political Studies*, 29 (1): 35–50.

Riddle, Joel; Snyder, Matthew and George, Polly (2001) *The Gulf of Mexico: an Overview.* Arthur Andersen Petroleum Services.

Robertson, Steve and McFarlane, Georgie (2004) *Strong Growth Forecast in Floating Production Sector.* London, Douglas Westwood Ltd.

Robinson, Silvan (1989)) 'Real Cost Base of Oil Isn't What You Think', *PIW*, April 3: 6–7.

Rockwood, Alan (1983)) 'The Impact of Joint Ventures on the Market for OCS Oil and Gas Leases'. *Journal of Industrial Economics* 31(4):453–68.

Roesset, José M. (1998) 'The Challenges of Deepwater Oil Production', in *Nuevas tecnologías en exploración y producción de hidrocarburos y su impacto en el nivel mundial de reservas.* Madrid, Fundación Repsol.

Roll, Richard (1986) 'The Hubris Hypothesis of Corporate Takeovers', *Journal of Business*, 59: 197–216.

Rose, M.B., Farndon, T.W., and Fraser, S.J. (1998) *Design and Rationale of the Final Rule on the Deep Water Royalty Relief Act.* Paper presented at the 1998 Offshore Technology Conference, Houston, Texas (4–7 May).

Rothkopf, Michael H. (1969) 'A Model of Rational Competitive Bidding', *Management Science*, 15: 362–73.

—— (2000) *Tales from a Nonstandard Career in Operations Research. RUTCOR Research Report RRR 48-2000.* Piscataway (NJ), Rutgers University.

Rutledge, Ian (2003) 'Profitability and Supply Price in the US Domestic Oil Industry: Implications for the Political Economy of Oil in the 21st Century', *Cambridge Journal of Economics*, 27: 1–23.

—— (2005) *Addicted to Oil. America's Relentless Quest for Energy Security.* London and New York , I.B. Tauris.

Saidibaghgandomi, Gholamreza (1987) *Testing the Assumptions of the Competitive Bidding Models in Outer Continental Shelf Auction Lease Markets*. Lexington (Ky.), University of Kentucky (Ph.D Thesis).

Saidi, Reza and Marsden, James R. (1992) 'Number of Bids, Number of Bidders and Bidding Behaviour in Outer Continental Shelf Oil Lease Auctions'. *European Journal of Operations Research*, 58: 335–343.

Sandrea, Iván (2004) 'Deepwater Oil Discovery Rate May Have Peaked: Production Peak May Follow in 10 Years', *O&GJ*, 102 (28): 18–23.

Scott, Richard (1994–5) *IEA The First 20 Years*. Paris, OECD/IEA, 3 vols.

—— (1994) Vol. 1: *Origins and Structure*.

—— (1995) Vol. 2: *Major Policies and Actions*.

—— (1995) Vol. 3: *Principal Documents*.

Selley, Richard C. (1985)) *Elements of Petroleum Geology*. New York, W.H. Freeman and Company.

Seydlitz, Ruth; Sutherlin, John and Smith, Samantha (1995) *Characteristics and Possible Impacts of a Restructured OCS Oil and Gas Industry in the Gulf of Mexico. OCS Study MMS 95-0055*. New Orleans, Environmental Social Science Research Institute (Department of Sociology, University of New Orleans)/MMS (Gulf of Mexico OCS Region).

Sherrill, Robert (1983) *The Oil Follies of 1970–1980. How the Petroleum Industry Stole the Show (and Much More Besides)*. New York, Anchor Press/Doubleday.

Siegel, Daniel R. and Smith, James L. (1984)) 'Does Profit-sharing Leasing for Outer Continental Shelf Leasing Need Finer Tuning?', *O&GJ*, 82: 144–52.

Simmons, Matthew R. (2003) *The Gathering Storm: Our Natural Gas Crisis*. Presentation at North American Natural Gas Summit, Houston (Tx.), 19 November 2003.

Smith, James L. (1982) 'Risk Aversion and Bidding for Offshore Petroleum Leases', *The Journal of Industrial Economics*, 30: 251–269.

Smith, James L.; Siegel, Daniel R. and Cheng, C.S. Agnes (1988) 'Failure of the Net Profit Share Leasing Experiment for Offshore Petroleum Resources', *The Review of Economics and Statistics*, 70: 199–206.

Smith, Michael A. (2002) 'Geological Controls and Variability in Pore Pressure in the Deep-Water Gulf of Mexico', in A.R. Huffman and G.L Bowers (eds.), *Pressure Regimes in Sedimentary Basins and Their Prediction: AAPG Memoir 76*. Tulsa (Ok.), AAPG: 107–113.

Sollen, Robert (1998) *An Ocean of Oil. A Century of Political Struggle over Petroleum off the California Coast*. Juneau, The Denali Press.

Stiff, John J. and Singelmann, Joachim (2004)) *Economic Impact in the U.S. of Deepwater Projects: A Survey of Five Projects. OCS Study MMS 2004-041*. New Orleans, MMS (Gulf of Mexico OCS Region).

Stiglitz, Joseph E. (2002) 'Information Economics and Paradigm Change', *The American Economic Review*, 92: 460–501.

Stouffer, Rick and Knight, Patrick (2002)) 'Oil from Deep Waters', *Energy Economist Briefings*, May 2002.

Strange, Susan (1998)) *States and Markets*. London, Pinter.

Sullivan, Robert E. (1955)) *Handbook of Oil and Gas Law*. New York, Prentice-Hall.

Teisberg, Thomas J. (1980)) 'Federal Management of Energy and Mineral Resources on the Public Land', *Bell Journal of Economics*, 11: 448–65.

Thaler, Richard H. (1988) 'Anomalies: The Winner's Curse', *The Journal of Economic Perspectives*, 2 (1): 191–202.

Thiel, Stuart E. (1988) 'Some Evidences on the Winner's Curse', *The American Economic Review*, 78: 884–95.

Traynor, JJ; Slorer, Ole (1998) *Deep Water West Africa. Monsters of the Deep.* London, NatWest Markets.

Traynor, JJ; Aldridge, Nicholas and Cook, Caroline (2002) *Deep Water Oil and Gas. Taming the Monster.* London, Deutsche Bank Global Equity Research.

Uman, Myron F.; James, William R. and Tomlinson, Holly R. (1979) 'Oil and Gas in Offshore Tracts: Estimates Before and After Drilling', *Science*, 205: 489–91.

United States Government Accountability Office (GAO) (1975) *Outlook for Federal Goals to Accelerate Leasing of Oil and Gas Resources on the Outer Continental Shelf. Report to Congress.* Washington D.C., GAO.

—— (1982) *Statement of Douglas L. McCullough, Deputy Director, Energy and Minerals Division, before the Subcommittee on the Panama Canal and the Outer Continental Shelf, House Committee on Merchant Marine and Fisheries.* Washington D.C., GAO.

—— (1984) *Evaluation of Department of the Interior's Comments on GAO's Report on Alternative Bidding Systems for Leasing Offshore Lands.* Washington D.C., GAO.

—— (1985a) *Early Assessment of Interior's Area-wide Program for Leasing Offshore Lands. Report to the Chairman, Subcommittee on Oversight and Investigations, Committee on Energy and Commerce, House of Representatives.* Washington D.C., GAO.

—— (1985b) *Selectively Reducing Offshore Royalty Rates in the Gulf of Mexico Could Increase Production and Federal Government Revenue. Report to the Chairman, Committee on Energy and Natural Resources, United States Senate.* Washington D.C., GAO.

—— (1986) *Offshore Oil and Gas. Views on Interior's Comments to GAO Reports on Leasing Offshore Lands. Briefing to the Chairman, Subcommittee on Oversight and Investigations, Committee on Energy and Commerce, House of Representatives.* Washington D.C., GAO.

—— (1990) *Additional Petroleum Production Tax Incentives Are of Questionable Merit. Report to the Chairman, Subcommittee on Energy and Power, Committee on Energy and Commerce, House of Representatives.* Washington D.C., GAO.

—— (1998) *Federal Oil Valuation. Efforts to Revise Regulations and an Analysis of Royalties in Kind.* Washington D.C., GAO.

—— (1999) *Federal Oil Valuation. Efforts to Revise Regulations and an Analysis of Royalties in Kind. Statement of Susan D. Kladiva, Associate Director, Energy, Resources and Science Issues, Resources Community and Economic Development Division.* Washington D.C., GAO. (2000) *Briefing on Natural Gas Royalty Payments. Prepared for Rep. Carolyn B. Maloney.* Washington D.C., GAO.

United States Congress, Joint Committee on Taxation (1981) *General Explanation of the Crude Oil Windfall Profit Tax Act of 1980 (H.R. 96th Congress; Public Law 96–223).* Washington D.C., Government Printing Office.

—— (1998) Oversight Hearing on Outer Continental Shelf Oil and Gas Leasing before the Subcommittee on Energy and Mineral Resources of the Committee on Resources, House of Representatives, One Hundred and Fifth Congress, Second Session, Mat 14, 1998 (mimeo).

United States Senate (2002) *Gas Prices: How Are They Really Set? Report Prepared by the Majority Staff of the Permanent Subcommittee on Investigations*. Washington D.C., Government Printing Office.

Veldman, Hans and Lagers, George (1997) *50 Years Offshore*. Delft, Foundation for Offshore Studies.

Verleger, Philip K. (1987) 'The Evolution of Oil as a Commodity', in *Energy: Markets and Regulation. Essays in Honor of M.A. Adelman*. R. Gordon, H. Jacoby and M. Zimmerman (eds.). Cambridge, Mass., MIT Press: 161–86.

Vickers, John and Yarrow, George (1988) *Privatization. An Economic Analysis*. Cambridge (Mass.), The MIT Press.

Wade, William W.; Plater, Jason R. and Kelley, Jacqueline Q. (1999) *History of Coastal Alabama) Natural Gas Exploration and Development, Final Report. OCS Study MMS 99-0031*. New Orleans, Foster Associates Inc./MMS (Gulf of Mexico OCS Region).

Wallace, Barbara; Duberg, John and Kirkley, James (2003) *Dynamics of the Oil and Gas Industry in the Gulf of Mexico: 1980–2000. Final Report Prepared under MMS Contract 1435–01–00–CT–31058*. New Orleans, MMS (Gulf of Mexico OCS Region)/Tech Law Inc.

Weimer, Paul; Rowan, Mark G; McBride, Barry C.; and Kligfield, Roy (1998) 'Evaluating the Petroleum Systems of the Northern Deep Gulf of Mexico through Integrated Basin Analysis: an Overview', *AAPG Bulletin*, 82: 865–877.

Weissman, Andrew (2003) 'Puncturing Natural Gas Myths', online in *Energy Pulse* (*www.energypulse.net*); Part I (November 21); Part II (November 24); Part III (December 3).

—— (2005) 'The Critical Need to Examine More Carefully the Role of Liquefied Natural Gas (LNG) in Meeting Future U.S. Energy Needs', online in *Energy Pulse* (*www.energypulse.net*); Part I (May 17); Part II (May 18); Part III (May 19).

Wu, Shengyu (1993) *Salt and Slope Tectonics Offshore Louisiana*. Houston, Rice University (Ph.D Thesis).

Young, K. T., Notfors, C.D, Meng, X.G., and Montecchi, P. (2000) *Subsalt Imaging in Walker Ridge, Gulf of Mexico*, Veritas Marine Surveys DGC (Mimeo)

US Government Statistical Publications

Annual Energy Outlook (EIA/DOE)
Federal Offshore Statistics (MMS)
Petroleum Marketing Monthly (EIA/DOE)
Revenue Collection Activity on Public Lands (Minerals Revenue Management)

Security Offering Documents

Prospectus for Medium Note Term Program 1999 (Companhía Petrolífera Marlim)

Legal Cases

Conoco Inc. v. United States, 35 Fed. Cl. 309 (1996)
Marathon Oil Co. & Mobil Oil Exploration and Producing Southeast Inc. v. United States, (United States Court of Appeals for the Federal Circuit, 97–5146)
OXY USA, Inc. v. FERC, 64 F.3d 679, 691 (D.C. Cir. 1995)
Santa Fe Snyder Corporation et al. v. Norton et al. (U.S.D.C., W.D. La, Docket No.2:00-CV-1641)
Shell Offshore v. Babbitt, No. 99-30532 (5th Cir.2001)
Shell Oil Company v. FERC, No. 92–1634 (D.C. Cir.1995)
Shell Oil Company v. Iowa Department of Revenue (No. 87-984 in the Supreme Court of the United States, October Term 1987, on appeal from the Supreme Court of Iowa)
Tesoro Alaska Petroleum Company v. FERC, No. 99–1223 (D.C. Cir. 2000)
Williams Gas Processing-Gulf Coast Company, L.P v. FERC, No. 97–1182 (D.C. Cir. 1998)
Winstar Corp. v. United States, (64 F3.d 1531, Fed. Circuit 1995; affirmed 518 US Supreme Court 839, 1996)

Annual Reports

Amerada Hess
BHP
BP
Chevron
ChevronTexaco
Conoco
ConocoPhillips
ENI
Exxon

ExxonMobil
Kerr-McGee
Marathon
Mobil
Petrobrás
Royal Dutch/Shell
Texaco

DOE Freedom of Information Act Decisions

Case No. VFA–0503, 27 DOE ¶ 80,216 (July 1999)
Case No. VFA–0522, 27 DOE ¶ 80,236 (October 1999)

FERC Hearings and Orders

52 FERC 61,268 (1990) for Amerada Hess Corp.
61 FERC 61,050 (1992) for Bonito Pipe Line Company.
61 FERC 61,051 (1992) for Oxy Pipeline, Inc.
74 FERC 61,896 (1996) for Shell Gas Pipeline Co.
91 FERC 61,182 (2000) for ExxonMobil Pipeline Co.

Periodicals and Trade Journals

AAPG Explorer
Federal Register
Financial Times (FT)
FT Energy World
Hart's E&P
Hart's Oil and Gas Investor
IndustryWeek
Inside F.E.R.C.'s Gas Market Report
The Journal of Energy Literature (JEnL)
JP Morgan Weekly Oil Data Report
The Leading Edge
Offshore
Oil and Gas Journal (O&GJ)
Platt's Oilgram News (PON)
Tulsa Geospectrum

Internet Resources

Alabama State Oil and Gas Board (*www.ogb.state.al.us*)
Energy Information Agency, Department of Energy (*www.eia.doe.gov*)
Federal Energy Regulatory Commission (*www.ferc.fed.us*)
Louisiana Department of Natural Resources (*www.dnr.state.la.us*)
Louisiana Midcontinent Oil and Gas Association (*www.lmoga.com*)
Minerals Management Service (*www.mms.gov*)
Minerals Management Service, Gulf of Mexico Region (*www.gomr.mms.gov*)
Minerals Revenue Management (*www.mrm.mms.gov*)
Railroad Commission of Texas (*www.rrc.state.tx.us*)
Shell Exploration & Production Company (*www.shellus.com/sepco*)

INDEX

Office of Technology Assessment (OTA) 5